Acoustics in the Built Environment

Acoustics in the Built Environment

Advice for the design team

Duncan Templeton (Editor)
MSc(Acoustics), BArch(Hons), MIOA, RIBA

Peter Sacre
BSc(Hons), MSc, MIOA, CEng, CPhys, MInstP

Peter Mapp
BSc(Hons), MSc(Acoustics), MIOA, MInstP, FInstSCE, AMIEE

David Saunders
BSc(Hons), PhD

Butterworth Architecture

Butterworth Architecture
An imprint of Butterworth-Heinemann Ltd
Linacre House, Jordan Hill, Oxford OX2 8DP

A member of the Reed Elsevier group

OXFORD LONDON BOSTON
MUNICH NEW DELHI SINGAPORE SYDNEY
TOKYO TORONTO WELLINGTON

First published 1993

British Library Cataloguing in Publication Data
Templeton, Duncan
Acoustics in the Built Environment:
Advice for the Design Team
I. Title
729.29

ISBN 0 7506 0538 3

Library of Congress Cataloguing in Publication Data
Acoustics in the Built environment: advice for the design team/
Duncan Templeton, editor . . . [et al.]
p. cm.
Includes bibliographical references and index.
ISBN 0 7506 0538 3
1. Architectural acoustics. I. Templeton, Duncan.
NA2800.A29 1993
729'.29–dc20 92–39206
CIP

Composition by Genesis Typesetting, Laser Quay, Rochester, Kent
Printed and bound in Great Britain

Contents

Acknowledgements vi

Contributors vii

Introduction 1

Chapter 1: Environmental acoustics 7
Peter Sacre
Environmental appraisals; Site analysis; Transportation noise; Construction noise; Industrial noise; Leisure noise; Groundborne vibration; New developments as a noise source; References

Chapter 2: Design acoustics 34
Duncan Templeton
Sound insulation; Sound absorption; Criteria for different building types; References

Chapter 3: Services noise and vibration 83
Peter Sacre, Duncan Templeton
Background; Setting design objectives; Design considerations; Vibration; Installation; References

Chapter 4: Sound systems 107
Peter Mapp
Introduction; System planning; Design principles; System design and components; Speech intelligibility; References

Chapter 5: Technical information 128
David Saunders
Definitions; Equivalent Standards; International Standards; German National Standards; American National Standards Institute; American Society for Testing and Materials Standards; French Standards; British Standards

Index 156

Acknowledgements

We are indebted to colleagues who have given us assistance and advice during the compilation of this book including as follows: Calvin Beck of United Cinemas International; David Belton of BDP; Jeff Charles of Bickerdike Allen Partners; Richard Cowell of Arup Acoustics; Niels Jordan of Jordan Akustik; Professor Peter Lord of the University of Salford Department of Applied Acoustics and BDP Acoustics; Eve Templeton; Jo Webb of Vibronoise.

Contributors

Duncan Templeton

As a specialist practising architect, Duncan Templeton, MSc(Acoustics), BArch(Hons), MIOA, RIBA, is Director of BDP Acoustics Ltd (a subsidiary of Building Design Partnership, the largest architectural practice in the UK) based in BDP's London and Manchester offices. Consultancies include work at the Royal Festival and the Royal Albert Halls, London, new theatres at High Wycombe and Llandudno, and conference halls at Warrington and Limerick, Eire. He is the co-author of three books on architectural acoustics: *Detailing for Acoustics*, *The Architecture of Sound*, and *Acoustic Design*.

Peter Sacre

Peter Sacre, BSc(Hons), MSc, MIOA, CEng, CPhys, MInstP, has been employed in the field of acoustic consultancy for 20 years. Prior to becoming a Senior Consultant with BDP Acoustics Ltd, Peter was head of the Acoustics Department at Wimpey Laboratories Ltd. He is involved primarily with environmental and planning projects. Tasks he has undertaken include environmental assessments for the Channel Tunnel fixed link and the subsequent acoustic design of the Folkestone Terminal, extensive noise monitoring around RAF airfields, design and supervision of acoustics for the Queen Elizabeth Conference Centre, predictions for the South Warwickshire Prospect coal mine, and a coal-loading facility and open-cast mine in NSW Australia when he was resident there.

Peter Mapp

Peter Mapp is an independent acoustics and sound systems design consultant. He has a BSc in Applied Physics and a Masters in Acoustics.

Peter set up his own practice in 1984, to specialize in sound system design and room acoustics. Before this he worked for two of the UK's largest general acoustic consultancies, where he was involved with all types of noise control and architectural projects.

He has a particular interest in the loudspeaker/room interface, and in speech-intelligibility measurement and prediction. He has presented papers and seminars on these topics both in the UK and abroad.

Peter regularly carries out technical reviews for a number of publications, and is the author of more than 40 articles and papers. He has contributed to three international references on acoustics and electroacoustics. Sound systems with which Peter has been involved include: the Queen Elizabeth Conference Centre; the Royal Hong Kong Jockey Club Stadium; plus the Broadgate Arena and the British Museum.

David Saunders

David Saunders graduated from the University of Nottingham in 1964 with a first class honours degree in Physics. He was awarded a PhD from the University of St Andrews in 1967 for a research project in solid state physics.

He then joined the Physics Department at the University of Salford to work with a small group doing research in the field of building acoustics. This group developed and in 1975 the Department of Applied Acoustics was formed. It is now the second largest acoustics research and teaching department in the UK, and David is now a senior lecturer there.

His original research was concerned with subjective reaction to noise and vibration and general building acoustic problems. However, for the last eight years his interest has been in studying the propagation and effects of high level impulsive noise.

His consultancy experience covers a wide range of environmental and noise control problems, in particular the assessment of the impact of transportation and industrial development. He has carried out work for industry, local government, building and architectural firms and legal organizations, and has represented clients at planning applications and appeals.

Introduction

Figure I.1 *RAC Walsall – external. Challenge of keeping out motorway noise*

Books on acoustics fall into several stereotypes: the primers, the mathematical/theoretical, the glossy, and the practical. The glossy may centre on auditoria, the practical on services noise, but it is difficult to find a really useful day-to-day reference covering a range of acoustics issues in the building technology in its widest sense. Sound is different to each discipline: to a sociologist it is a stimulus eliciting a range of subjective responses, to a physicist it is a measurable phenomenon with varying propagatory character, to the structural engineer vibration is the issue, to the mechanical engineer it is noise control. Environmental noise matters – transportation, industry – may impinge on the planner's considerations. The architect may come across sound as a characteristic of key spaces (e.g. studios, auditoria) and in providing adequate isolation and privacy to areas within a building. What seems to be wanted, and does not exist, is a technical thesaurus covering practical reference needs without flannel and undue mathematics, offering concise guidance and assisting in design methodology. A designer does not want to calculate from basic theory how a partition assembly will achieve certain sound insulation values; he will want to check his design intent against performance tables, have some idea from other tables about the internal needs and external noise exposure, and either adjust his design or use the data as a performance specification for suppliers to implement. The need for value for money in noise control and general acoustic design, i.e. to justify a level of performance to a client and avoid overkill, should be recognized, with the emphasis on consistent standards, evenly applied, for maximum effect.

The scope for advice is more and more apparent: the

Figure I.2 *RAC Walsall – internal. Challenge of separating engine test beds from office areas*

environment is getting noisier, the standards demanded higher, ventilation and sound systems more sophisticated, computer-aided instrumentation and prediction techniques more reliable and accurate (Figures I.1 and I.2). We want this book to appeal to a wide audience – clients, project managers, and students, as well as architects, mechanical, electrical and sound system engineers.

Each of the headings generates a separate approach and different disciplines can refer to the relevant chapter, although there is a great deal of overlap (Diagram I.1). The clean ideals of theory are inevitably compromised on the rack of fast track site progress; judgements and advice have often to be given based on half-truths and inadequate information. Good acoustic study techniques are sometimes too cumbersome: on a recent auditorium physical modelling exercise, the project was tendered before research results could be applied. If advice in this book nudges designers and engineers in the right direction once in a while, that is as much as we could expect. Timely advice during the design, construction, and early use of buildings is the aim (Diagrams I.2–I.4).

As practicing acousticians we come across 'runs' of design issues in design sectors. In offices it may be 'how noisy can it possibly be before the building has to be sealed rather than naturally ventilated to the perimeter?' The cost constraints are such that developers are very reluctant to have sealed buildings in speculative offices. Similarly, hotels sprout on busy interchanges to catch the passing trade, and commercial business parks crowd the airports. Leisure centres group innovative combinations of noisy activity.

Many developments are of such a scale now that new infrastructure – transport, landscape and topography – is entailed long before the building work starts. Many building complexes are of such a scale that the initial contract is a mere shell for fit-out contracts by numerous tenants, so there is a shift in the approach to looking after the client's interests. Novel building types, like trading rooms, microchip production facilities, multiplex cinemas and theme parks, demand assessment in the absence of published data. Relationships to other professions can get complex (Diagram I.5).

Legislation is a key issue, not only because of the closer compatibility to European standards, but also because of recent far-reaching statutes, for example the Noise at Work Regulations 1989 and the 1988 Town and Country Planning (Assessment of Environment Effects) Regulations. The former tightens the legal duties of employers, designers, manufacturers and suppliers, to minimize hearing damage; the latter defines environmental assessment for any major projects of more than local importance or projects in sensitive areas. New or amended legislation is available relating to key aids, for example part E of the Building Regulations (June 1992), BS 4142 and BS 6472.

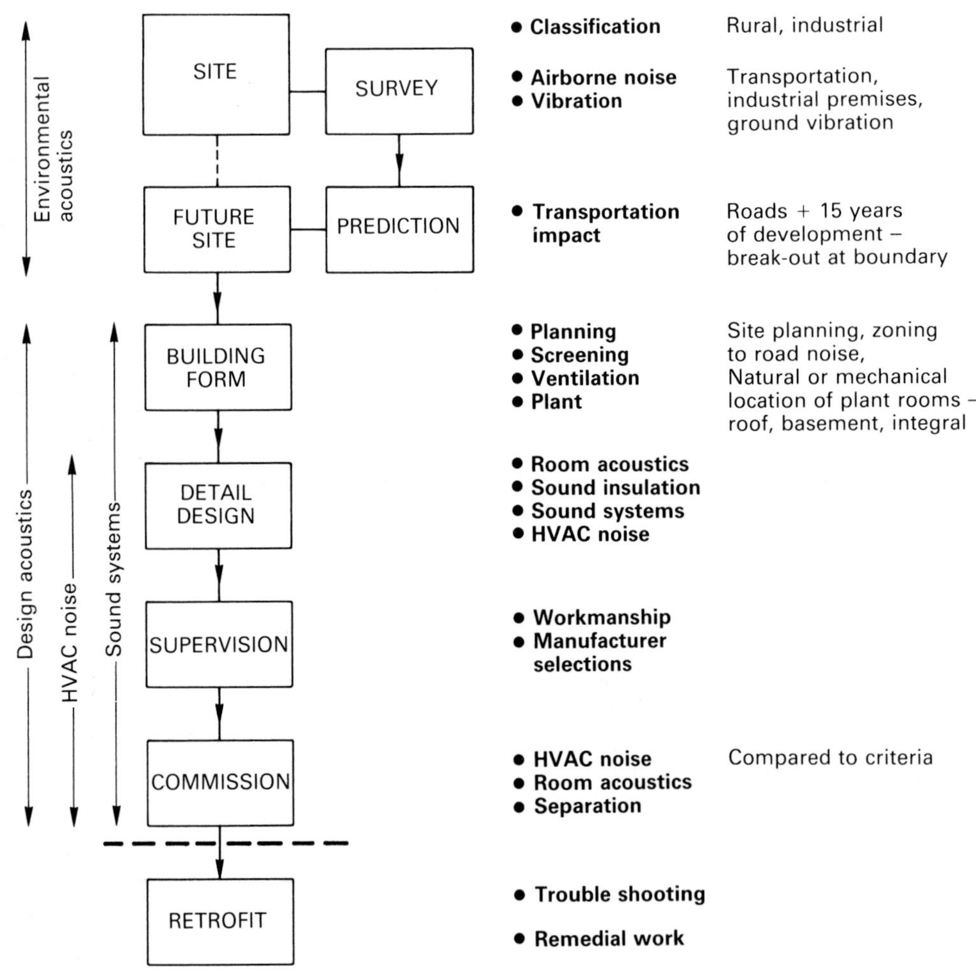

Diagram I.1 *Checklist: stages of design*

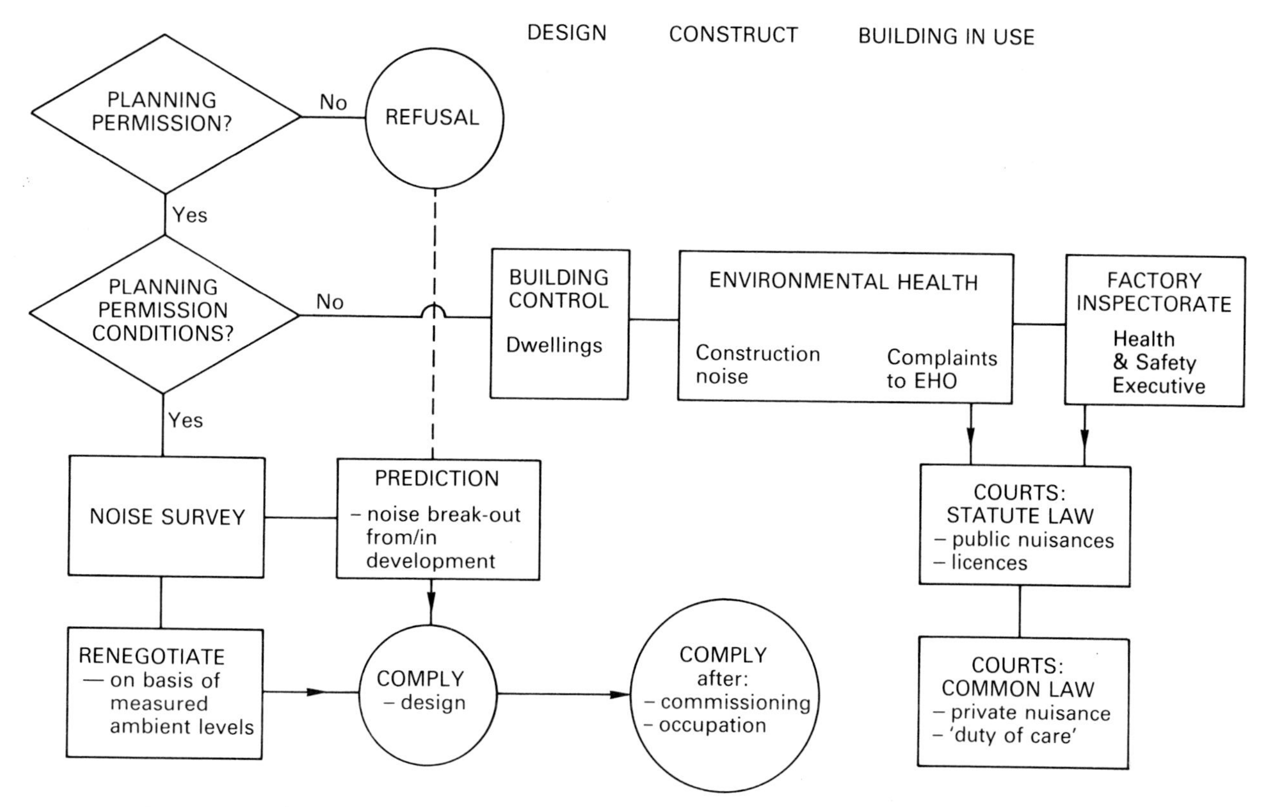

Diagram I.2 *Regulatory authorities and examples of legislation*

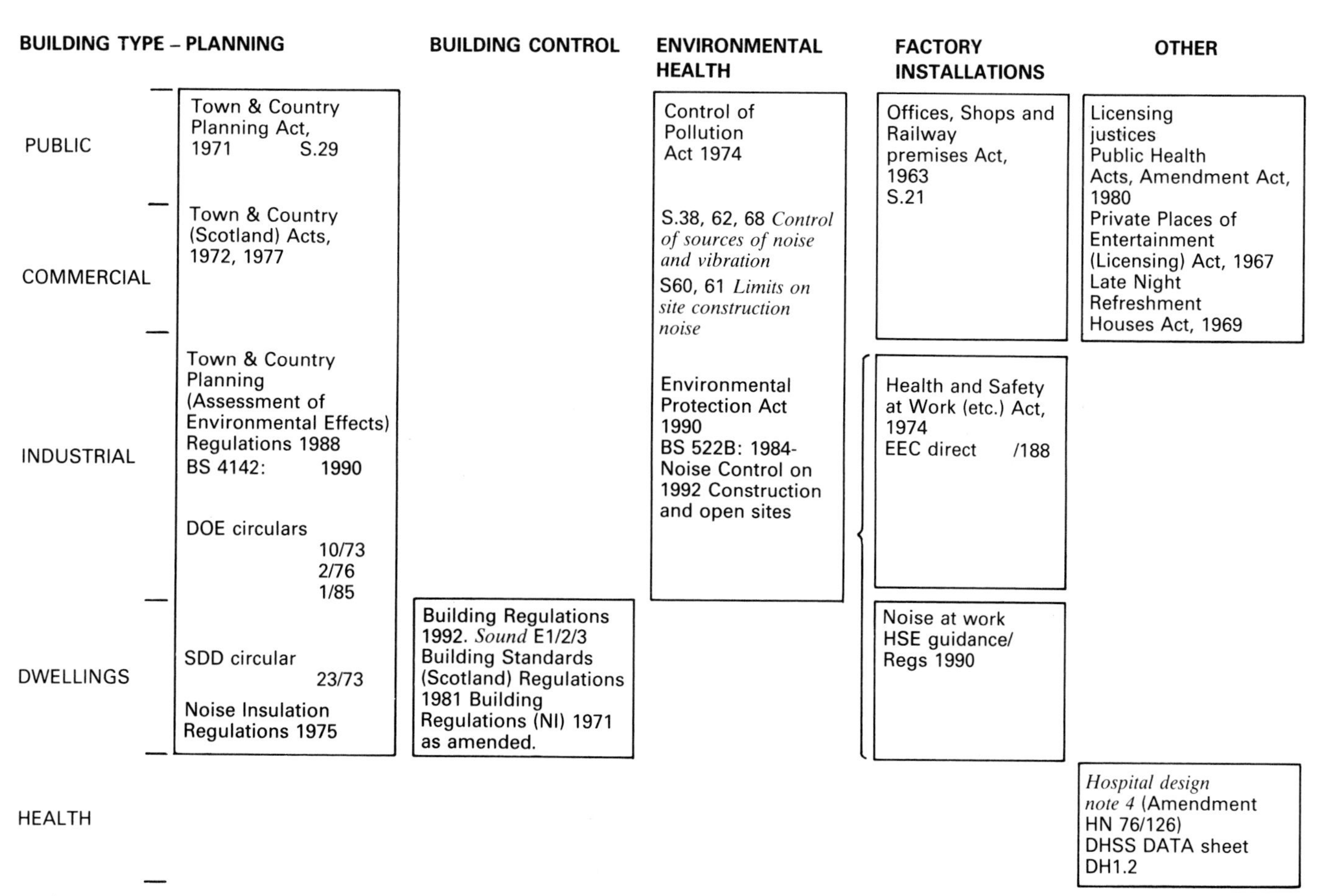

Diagram I.3 *Statutes: sample reference publications*

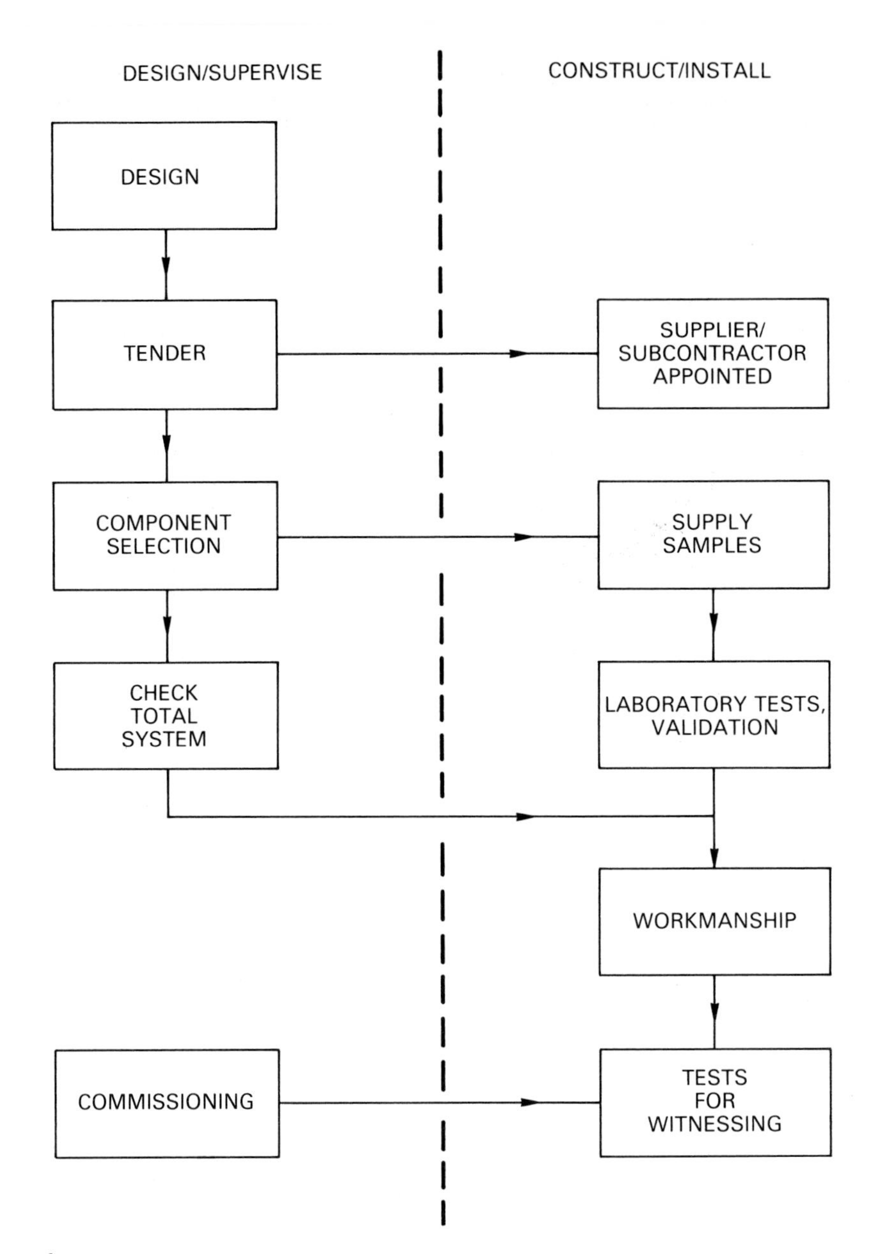

Diagram I.4 *Design and construction stages*

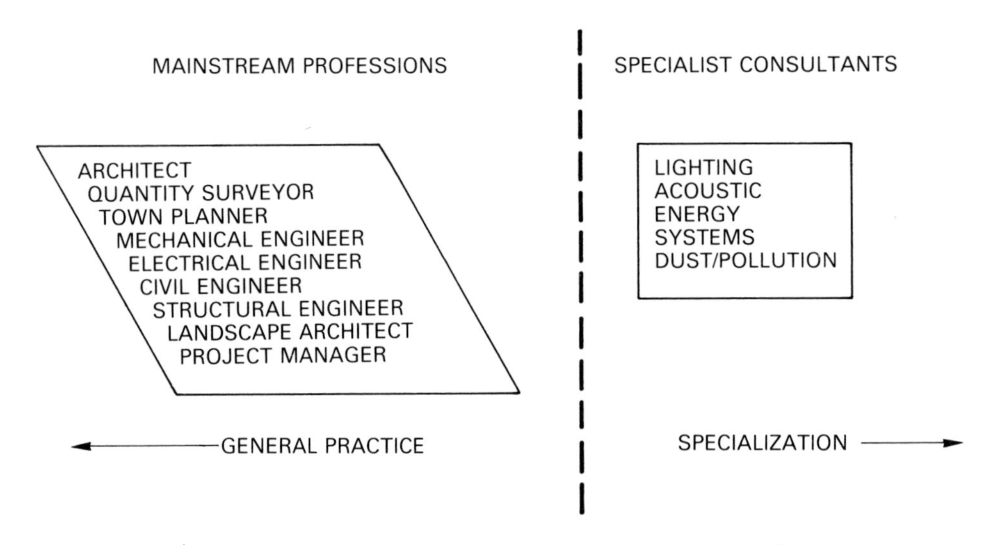

Diagram I.5 *Design disciplines: context of acoustic consultancy to other skills in the building industry*

'Environmental acoustics' covers a topic largely overlooked to date but is a growth area in consultancy because of the real concern that Green issues raise. Increasingly, road noise is universal and there are few truly quiet spots left on the mainland UK. Protection techniques to properties alongside roads vary from UK rustic timber to Swiss curved glazing. Environmental noise has become an everyday issue: in 1989 environmental health officers received over 50 000 complaints about noise. An Environmental Protection Act in Parliament increases penalties for excessive noise from industrial premises ten-fold.

'Design acoustics' too has been well served in publications but the authors feel there is increasing need for more detailed advice specific to buildings' uses; one can no longer generalize and suggest a single set of criteria for say studios or practice rooms. More and more, the designer is setting performance criteria only, for specialist suppliers and installers to implement.

'Services noise and vibration' have been reasonably served by a number of publications to date, for example the Sound Research Laboratories' *Noise in Building Services*. Chapter three complements existing advice rather than competes. An increasing proportion of buildings are mechanically ventilated, and economic and space pressures lead to a tendency to higher velocity duct systems where good control is critical.

'Sound systems' have been considered in systems manuals and electronics guides but the applications here of interest relate to speech intelligibility (PA), audibility (fire alarms), sound quality in particular spaces (sound reinforcement) and electroacoustics (modifying the way auditoria sound), applications directly related to building projects and to acoustics, rather than attempting to cover the rapidly-changing equipment field. There are a number of professions with a useful half-knowledge – electrical engineers, systems specialists, theatre consultants – but the area falls dangerously between professions as regards full and reliable documentation; it becomes all too tempting to 'leave it to the trade'. Sound systems are an intrinsic part of any modern performance space.

Professor Stephen Hawking, who popularized cosmology and astrophysics with a bestseller, *A Brief History of Time*, was advised by his publisher that 'each equation included would halve the sales'. Fortunately in Butterworth-Heinemann we have an enlightened publisher whose objective has been to produce reference material of the greatest use in an attractive format – with minimal essential formulae supporting the methodology.

In Chapter 5 we address the working knowledge of acoustic terminology, relevant standards in the UK and worldwide, and up-to-date information sources. Some of the topics arise from the course notes for a university degree in acoustics. Such a course is a general grounding for acoustics, as opposed to being specifically related to the built environment. The technology and analysis techniques are advancing quickly, so there will be in the near future more data available to analyse, define and accurately commission criteria set on projects. Chapter 5 is a summary of definitions overlapping with the topics covered in the chapters, intended as a quick reference source to ensure that terms quoted in performance specification documents are correctly ascribed, or alternatively to interpret, in a dictionary style, terms come across in contract documents or technical reports.

The most relevant standards have been selected and it has been a difficult decision to decide how much to include on this; database keywords generate many hundreds of standards but the 'first port of call' reference should be given otherwise any oracle referred to will be too broad and meaningless. A problem with quoting large numbers of standards is the constant updating; any standard quoted herein should therefore be checked for any amendments subsequent to publication.

The humorist Max Frisch defined technology as 'the knack of so arranging the world that we don't have to experience it'. This book tries to make design acoustics less of a black art or science, by giving concise and economically reasonable advice, topic by topic.

Chapter 1 Environmental acoustics

Peter Sacre

Environmental appraisals

Introduction

The initial assessment that needs to be made for any development takes account of its location in the environment. Thus environmental acoustics needs to be considered at the outset, whether it is the consideration of planning issues which take account of the possible effect of a development on its surroundings or whether it is the effect of an external noisy climate on that proposed development.

The need to consider environmental acoustics has been given more emphasis now that environmental issues generally are of universal concern. The recent publication of the Government's White Paper on the Environment [1] adds weight to this consideration by requiring an environmental assessment for significant schemes.

This chapter looks at the need for an acoustic appraisal, what needs to be considered for a site inspection or survey, the types of environmental noise sources that could be encountered, and identifies those factors that need to be considered when investigating the impact of a development on its surroundings, including environmental impact assessments.

Need for an acoustic appraisal

An appraisal may be required for one of the following reasons:

- to assess possible site constraints
- as part of an Environmental Statement to accompany a Planning Application.

In the latter case, an acoustics assessment could be one of several issues to be covered or it may be required as an independent study.

The route taken to identify the need for an appraisal is shown on Diagram 1.1.

Specialist help

Once the need for an acoustic appraisal has been agreed, specialist advice is available from a number of acoustics consultants. These can be found via professional bodies such as the Institute of Acoustics or the Association of Noise Consultants. Often the local authority will keep a register of acoustics consultants able to undertake appraisals within their area.

The appointment of an acoustic consultant can be direct to the client or on a subcontract basis to the lead consultant (often the architect), as part of a design team.

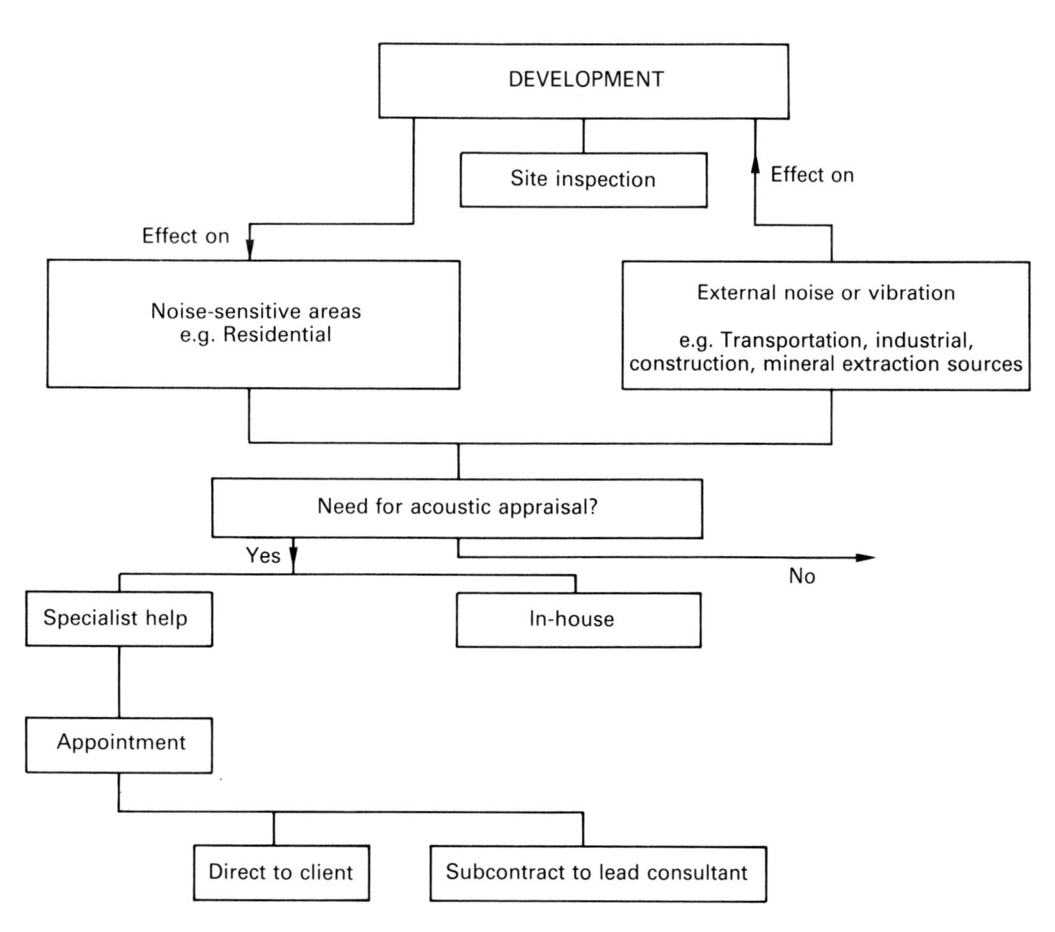

Diagram 1.1 *Acoustic appraisal: need*

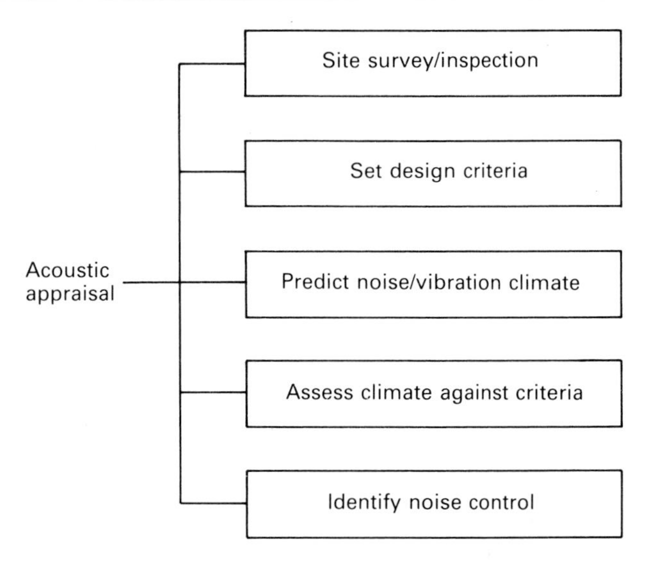

Diagram 1.2 *Acoustic appraisal: requirements*

Method of acoustic appraisal

The basic requirements of an acoustic appraisal are shown in Diagram 1.2.

The site survey or inspection will enable important site-specific information to be obtained, such as whether there are any local noise and/or vibration sources which may affect any new development, e.g. transportation routes or industry, or whether there are any nearby noise-sensitive areas for example housing.

In setting design criteria for a development, reference will need to be made to such documentation as British Standards, to establish acceptable intrusive noise or vibration levels in a development, or possibly planning conditions, which ensure that a development will not affect a nearby noise-sensitive area. In some cases, research studies may need to be referred to in addition to, or in the absence of, relevant standards.

A prediction exercise would, in the majority of cases, be based on measured data taking account of site-specific details. Although it would be possible to undertake prediction without measured site data in situations where there is sufficiently reliable published information, e.g. noise due to road traffic, a site inspection would provide site-specific factors which would assist in the exercise. Prediction of vibration on a site is extremely complicated and measurement must always be the preferred method.

The differences between the predicted noise or vibration climate and the design criteria will identify the scale of any potential problem. An assessment will need to determine whether any additional control measures are necessary and practicable. Obviously, small differences may not warrant huge expenditure and agreement must be sought with all interested parties to determine the best course of action. Where control methods are required, the appraisal should identify the best methods.

Site analysis

Site inspection

The site survey is probably the most important part of an acoustic appraisal, whether it is only a site inspection or a full measurement survey, since it will determine the location of noise-sensitive areas and noise sources and other local factors needed to make an accurate assessment, e.g. local shielding. If a full survey is being undertaken, the initial site inspection or pilot survey will identify the preferred measurement locations.

The items that need to be considered in undertaking an inspection are identified in Diagram 1.3. Diagram 1.3 also gives a checklist of the likely aims of an inspection. This includes reference to local topography, particularly embankments or cuttings, which would provide significant acoustic shielding but the details of which would not easily be determined from maps or plans. In determining noise from transportation routes, data such as type and gradient of road or type of railway track proximity to airports – civil or military – need to be obtained.

Measurement locations need to be selected to be representative of the local noise or vibration climate and take account of site practicalities. This would include whether noise measurements need to be made at heights greater than 1.5 m to obtain appropriate data. Short-period indicative measurements taken during a site inspection are helpful and can establish the preferred monitoring locations.

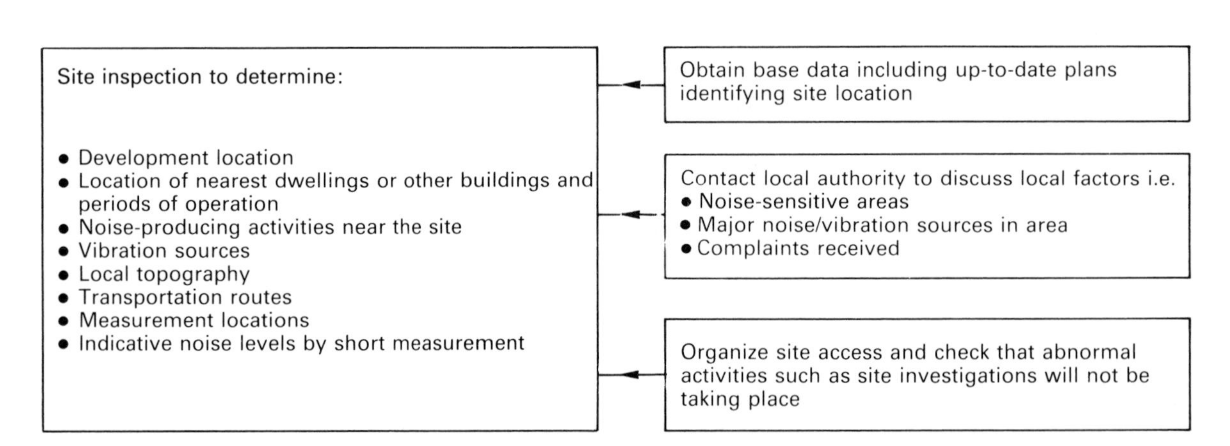

Diagram 1.3 *Site inspection*

The site inspection also serves to identify whether there are any local activities which could affect a full survey, e.g. transportation maintenance or industrial down-time period.

Contact with local authority

Contact with the local authority, normally the Environmental Health Department (EHD), to discuss acoustic appraisal is necessary at some time during the contract. It is desirable therefore to agree any local factors that could affect the survey, including planning conditions. The EHD can identify the nearest noise-sensitive areas and/or any major noise/vibration sources in the area. It is also beneficial to know the pattern of complaints arising from noise nuisance.

Activities affecting the site

It is necessary to obtain, preferably before a full survey, the likely operating hours of a nearby industrial development or the likely movements on a transportation route, e.g. for railways the number of passenger and freight trains during certain periods should be obtained from British Rail or the local railway, since the information collected on a particular day may not represent the total picture.

Prior to any site visit, access to the site must be ensured possible. This will normally be by contact with the landowners or estate/letting agents. This contact will also need to be made before a site survey is undertaken to ensure that there are no other site activities taking place to invalidate the measurements, for example a clash with any site investigation must be avoided.

Survey procedure

Once site access has been arranged and the presence of any activities either on or close to the site has been checked, the basic survey requirements are as illustrated in Diagram 1.4 and as discussed below.

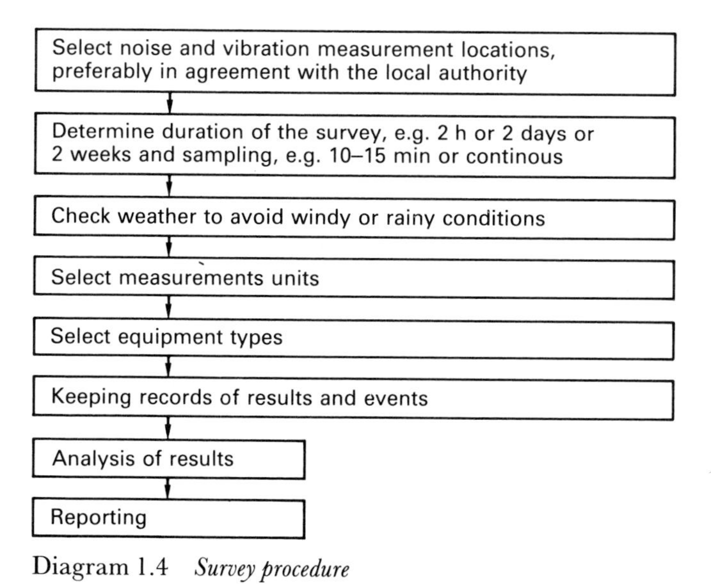

Diagram 1.4 *Survey procedure*

Measurement locations

Measurement locations should be agreed between all parties. Their selection will be based on the site inspection and take account of site practicalities. For example, it is not always best to set up equipment close to houses where dogs are present (although it could be argued that they could be considered as part of the environment, it is likely that barking is caused by the presence of the surveyor) but to select a representative equivalent location. Temporary shielding of a potential noise source may also affect the measurements, and locations should be avoided if they screen a noise source. An example of temporary shielding is a builder's stockpile of materials.

A noise measurement location should always be selected with an unobstructed view of the proposed development and preferably at least 3.5 m from a reflecting surface. In the case of nearby housing, it is often the first floor windows that are the most sensitive, i.e. bedrooms.

Typically the heights of microphones will be set at 1.2 m or 1.5 m above ground which correspond to a reception point at the ground floor level of a building. For a reception height at first floor level, a microphone height of 4.0 m or 4.5 m above ground using a stand extension may be more appropriate.

If greater heights need to be considered for a reception point, e.g. to represent the third floor height of a building affected by road or railway shielded by a barrier at ground floor height (see Figure 1.1), then a hydraulic mast, which can typically go up to 10 m to 12 m in height, may be required.

In order to reduce the amount of measurement equipment needed to measure at several locations, a primary location could be selected where continuous monitoring is carried out, together with satellite locations where regular but not continuous measurements are obtained. Typically, the primary location would be unmanned and the satellite locations manned.

It may be possible, if a refurbished development is proposed, to use the existing building and position a microphone out of a window, at a distance of about 1 m from the facade. In this case an allowance for facade reflection will need to be made of approximately 3 dBA. Once the distance is greater than 3.5 m from a reflecting surface, the measurements obtained at these locations will be relatively free-field. Sometimes a distance of 10 m from a reflecting surface is adopted for free-field measurements.

Determination of groundborne vibration would typically be in order to assess its impact on a proposed development. Monitoring would need to take place at the part of the proposed development nearest the source and then at regular distances away from the source.

Vibration measurements should be made ensuring that the transducer is coupled effectively with the ground, or the surface to be measured. A heavy block placed in the soil is typically used to fix a transducer for ground measurements with the ability to allow measurements in the vertical, radial and transverse directions.

Duration of survey

To determine the measurement periods required, it is necessary to identify the periods of operation of the proposed development and any noise/vibration sources nearby. For example, the likely operation of different

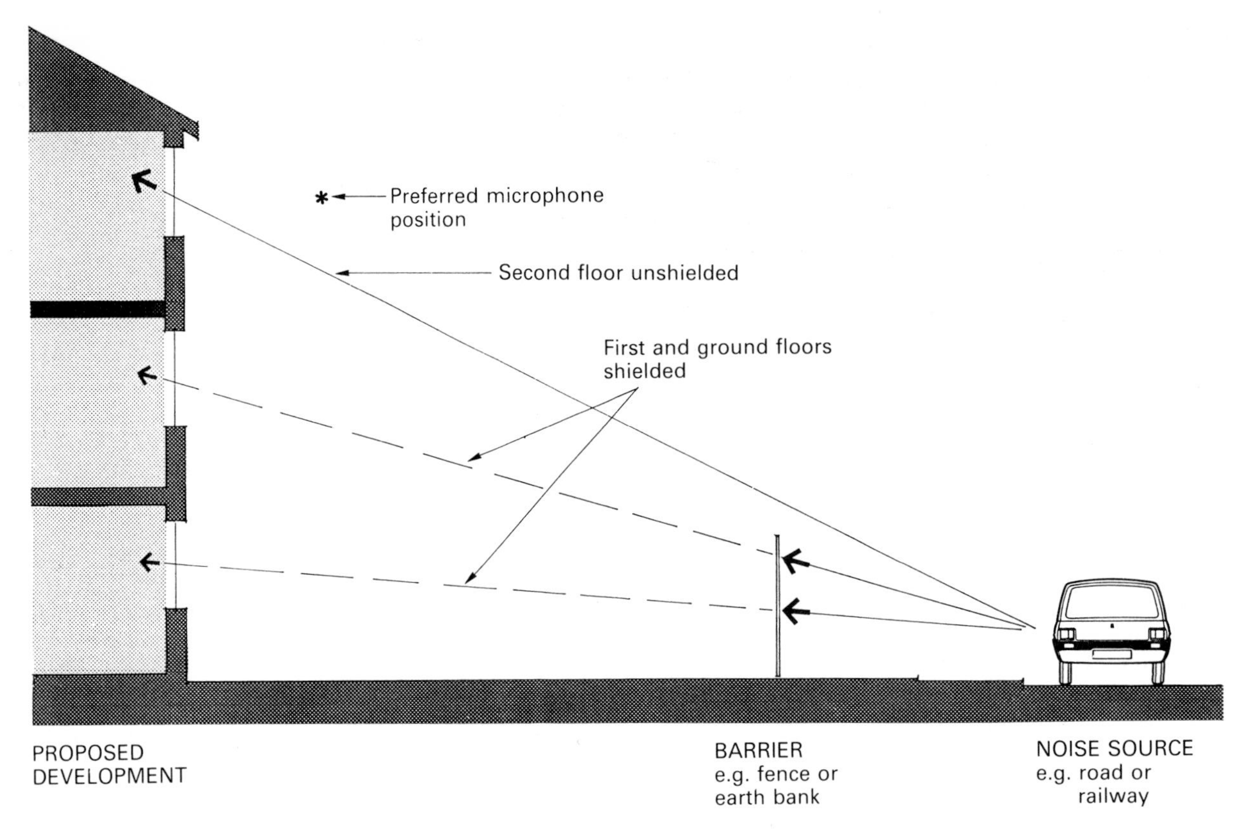

Figure 1.1 *Elevated monitoring need*

mechanical services plant for an office development will affect the noise climate at different times of the day. The use of a development 24 h/day, 7 days/week, would also identify the need to survey at weekends.

Typically, consideration needs to be given to assessment at night and the lowest ambient/background noise levels normally occur between 02:00 and 04:00 hours. However, if the background noise levels are very low between 02:00 and 04:00 hours, it may be acceptable from a sleep disturbance point of view to take 22:30 to 00:00 hours and 05:30 to 07:00 hours as the most sensitive periods.

In describing noise/vibration climates, the period 07:00 to 19:00 should invariably be used as the 'daytime', 19:00 to 23:00 hours as the 'evening' and 23:00 to 07:00 hours as the 'night-time'. These accord with periods defined in the Department of the Environment's *Report of the Noise Working Party 1990* [2].

The duration of the survey will be dependent not only on the hours of operation but also on the site of a development and/or the noise-sensitive areas and a survey carried out over a number of days would average out any differences occurring due to weather conditions. For example, to establish the existing noise climate for a development on the scale of the Channel Tunnel project [3] required monitoring twice per year over a 2-week period including weekends at approximately 15–20 locations surrounding the proposed Terminal development area. A small housing development affected only by a single industrial noise source known to maintain a continuous noise level may only take 2–3 h at one location.

The duration of samples will be dependent on the noise climate; 10–20 min/h at different locations is normally sufficient.

Vibration monitoring would typically be of short duration since it is normally only the effect of vibration on the proposed development that is of interest. Thus measurements only need to take account of the maximum levels that would occur during, for example, train pass-bys or quarry blasting and the number of occurrences in a given period.

Weather

The preferred monitoring conditions are on a dry and clear day or night with a light wind blowing from the source towards the measurement location, or when it is calm. If the monitoring periods are over a long duration then the effect of weather should not be important, provided reasonably accurate information relating to the weather can be obtained, and it will only be necessary to avoid long spells of windy and rainy weather.

High winds and heavy precipitation must be avoided during surveys. High or even moderate winds result in increased background noise levels due to leaves rustling in trees or hedges and wind noise in fences. Even with a windshield, there can be wind 'roar' effects at the microphone itself. Therefore, conditions where wind speeds are greater than 5 m/s should be avoided. Rain could affect the measurement equipment and would create higher noise levels due to its impact on roofs or trees or causing the surface of a road to become wet (in wet conditions, tyre noise increases). Temperature inversions could also affect

Table 1.1 *Measurement units*

Noise source	Parameters to be determined	
	Noise unit	Other data
Rail	SEL (to determine L_{Aeq}) $L_{Amax,T}$	Number and type of trains
Road	$L_{A10,T}$ $L_{Aeq,T}$	Traffic counts, light and heavy vehicles
Aircraft	SEL (to determine L_{Aeq}) $L_{Amax,T}$	Number and types
Industrial	$L_{Aeq,T}$ $L_{A90,T}$ $L_{Amax,T}$ L_{Apeak} (if impulsive)	Occurrences of different activities and periods of operation
Construction	$L_{Aeq,T}$ $L_{A1,T}$	Occurrences of different activities and periods and likely duration of events

UK:
$L_{Aeq,T}$ A-Weighted equivalent continuous sound level over a stated time period, *T*; the preferred measure of environmental noise varying with time.
$L_{A90,T}$ A-Weighted sound level exceeded for 90% of a measurement period, *T*; widely used as the descriptor of background, or ambient, noise.
$L_{A10,T}$ A-Weighted sound level exceeded for 10% of a measurement period, *T*; used for road traffic noise measurement.
$L_{A1,T}$ A-Weighted sound level exceeded for 1% of a measurement period, *T*; used to describe the maximum noise climate.
$L_{Amax,T}$ A-Weighted maximum sound level which describes the maximum level measured during a measurement period, *T*.
USA:
L_{DN} Used widely to assess community noise. To determine L_{DN}, $L_{Aeq,T}$ must be monitored during both daytime (07:00–22:00 hours) and night time (22:00–07:00 hours).

monitoring where long distances are involved but it is likely that variations due to wind would have more effect.

Reasonably reliable and up-to-date information can always be obtained from regional weather centres. Weather information should always be recorded during any environmental survey and include wind speed and direction, temperature, humidity and cloud cover.

Measurement units

The various units and parameters for measuring environmental noise are defined in Chapter 5. In undertaking an environmental noise survey the values identified in Table 1.1 should be determined for different noise sources. The table also suggests additional parameters that should be obtained.

Consideration may need to be given to obtaining frequency spectra of distinct noise sources, e.g. industrial plant, for subsequent design development purposes.

When the impact of a particular noise source on a development is being assessed in isolation it may be possible to limit the range of parameters measured (see Table 1.1) but since most equipment records the full range of units, it may be preferred to discount unwanted parameters at a later date.

If vibration levels that are to be measured are steady then r.m.s. acceleration and/or r.m.s. velocity should be determined. Where the vibration levels are caused by intermittent or impulsive sources then the peak acceleration and/or peak velocity should be measured. For subsequent analysis, frequency spectra should also be obtained.

Equipment

The basic instrumentation for noise or vibration measurements, together with a checklist of requirements for instrumentation, is given in Table 1.2. Measurement equipment must be regularly calibrated, at least once every 2 years, and this calibration must be traceable via a laboratory accredited for testing by the National Measurement Accreditation Service (NAMAS).

To determine environmental noise levels, a calibrated sound level meter complying with the requirements of preferably type 1 but at least type 2 as given in BS 6698 [4], or BS 5969 [5] should be used. The microphone selected should always be protected by a windshield and shielded from heavy rain.

In addition to using equipment calibrated to a National Standard, the equipment should always be calibrated on site before and after any survey and at the beginning and end of any tape recording, using a reference sound source – typically either an electronic calibrator or pistonphone.

Table 1.2 *Equipment selection*

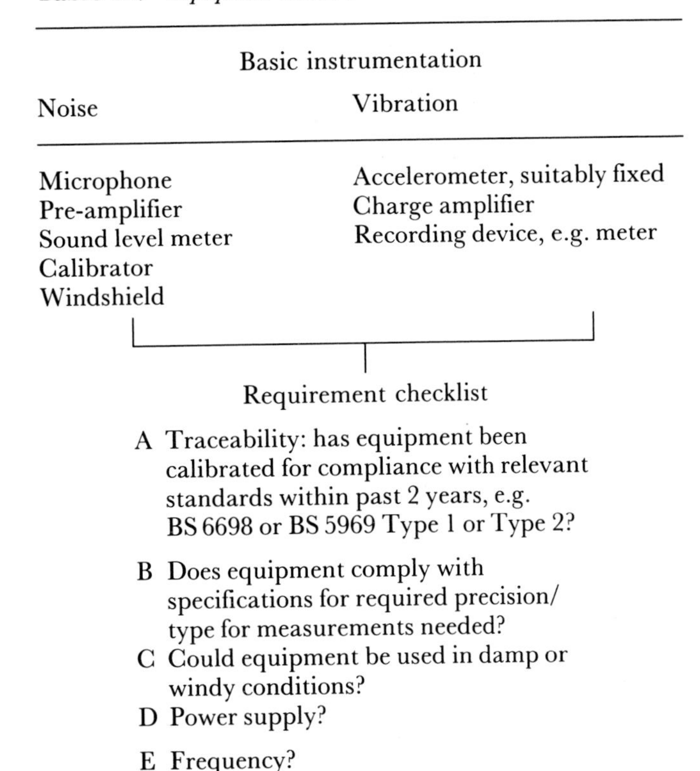

Basic instrumentation	
Noise	Vibration
Microphone Pre-amplifier Sound level meter Calibrator Windshield	Accelerometer, suitably fixed Charge amplifier Recording device, e.g. meter

Requirement checklist

A Traceability: has equipment been calibrated for compliance with relevant standards within past 2 years, e.g. BS 6698 or BS 5969 Type 1 or Type 2?

B Does equipment comply with specifications for required precision/type for measurements needed?

C Could equipment be used in damp or windy conditions?

D Power supply?

E Frequency?

F Time history?

Where noise levels are being monitored over a long period of time and are therefore unmanned some of the time, a reliable data logger coupled to a calibrated sound level meter or equivalent is required.

Detailed consideration will need to be given to the power supply of long term monitoring equipment, and batteries may need to be regularly changed in colder weather.

Where frequency spectra need to be obtained from a steady sound source this can readily be achieved on site using a filter set coupled to the sound level meter. If the source is intermittent and/or impulsive, it may be necessary to tape record the occurrence for subsequent analysis or to use a Real Time Analyser for on-the-spot analysis. The tape recorder or other recording device should be selected so as not to affect the accuracy of the measurements. Tape recordings should ideally be made linearly, i.e. not A-weighted, in order to improve the signal at low frequencies.

Vibration levels can be recorded directly onto meters and time history records kept using chart recorders. Any frequency analysis of intermittent and/or impulsive sources of vibration should be undertaken using either a tape recorder, or a real time analyser which ideally has a lower limiting frequency of 1 Hz or below. On-site calibration is normally achieved using an electronic signal, but it is preferable to use an accelerometer calibrator.

Record keeping and reporting

An important part of the survey procedure is to keep records of all necessary data. This could include a time log of events, weather information, measurement locations and sample periods, in addition to the measured data.

Data need to be summarized and sound level histograms are a good visual method of achieving this. Charts showing sound/vibration level versus time can also be useful. The results of frequency analyses can be described more effectively on a sound pressure or vibration level versus frequency (octave band or 1/3 octave band) graph.

A report should clearly identify the main results of any survey. If it is necessary to show all results, they can be provided in appendices.

During a site survey, the types of environmental noise and vibration that are likely to be encountered are due to transportation, construction or industrial sources. A brief description of each is given below with relevant legislation criteria, suggested methods of prediction and noise control.

A flowchart summarizing the process of acoustic appraisal is included in Diagram 1.5.

Transportation noise

Road traffic

Noise sources

Fighting Noise in the 1990s [6], an Organisation for Economic Corporation and Development publication, observes that road traffic noise is a major source of disamenity as between 32% and 80% of OECD populations were exposed to 18-h levels above 55 dBA.

Noise sources of individual road vehicles can be basically broken down to power train noise, which include the engine and transmission, and rolling noise, which is due to aerodynamics and tyre/road surface interaction.

The effect of speed on the contribution of the power train noise and rolling noise to the overall noise level from a single vehicle is shown in Figure 1.2. For light vehicles, engine noise in low gears at low road speeds dominates up to about 30 km/h where at higher revolutions/min rolling noise starts to become dominant. For heavy vehicles, noise from the diesel engine, exhaust and cooling fans dominates up to about 50 km/h, before rolling noise becomes a significant factor. Above 50 km/h, rolling noise increases at a rate of about 9 dBA per doubling of speed for all vehicle categories. Thus the noise level due to a single vehicle can be determined if its speed is known.

The noise level due to road traffic with a mixed flow of light and heavy vehicles can be determined from *Calculation of Road Traffic Noise* [7]. It is basically dependent on the flow of vehicles during a period of either 1 or 18 h (06:00–24:00 hours), their speed and the proportion of heavy vehicles. Additional factors are the texture of the road surface which affects rolling noise and the road gradient which affects engine noise. In wet conditions tyre noise increases; however, road traffic noise assessments assume dry road conditions.

In addition to the engine/rolling noise there may be occasions when noise from refrigeration equipment or reversing signals need to be considered. Data for modern refrigerated vehicles, i.e. not diesel engine powered, indicates that noise levels from equipment would typically be 65 dBA at 10 m.

Measurement unit

The measurement unit that has historically been used to described road traffic noise is L_{A10}. L_{A10} is the A-weighted sound level which is exceeded for 10% of the time period.

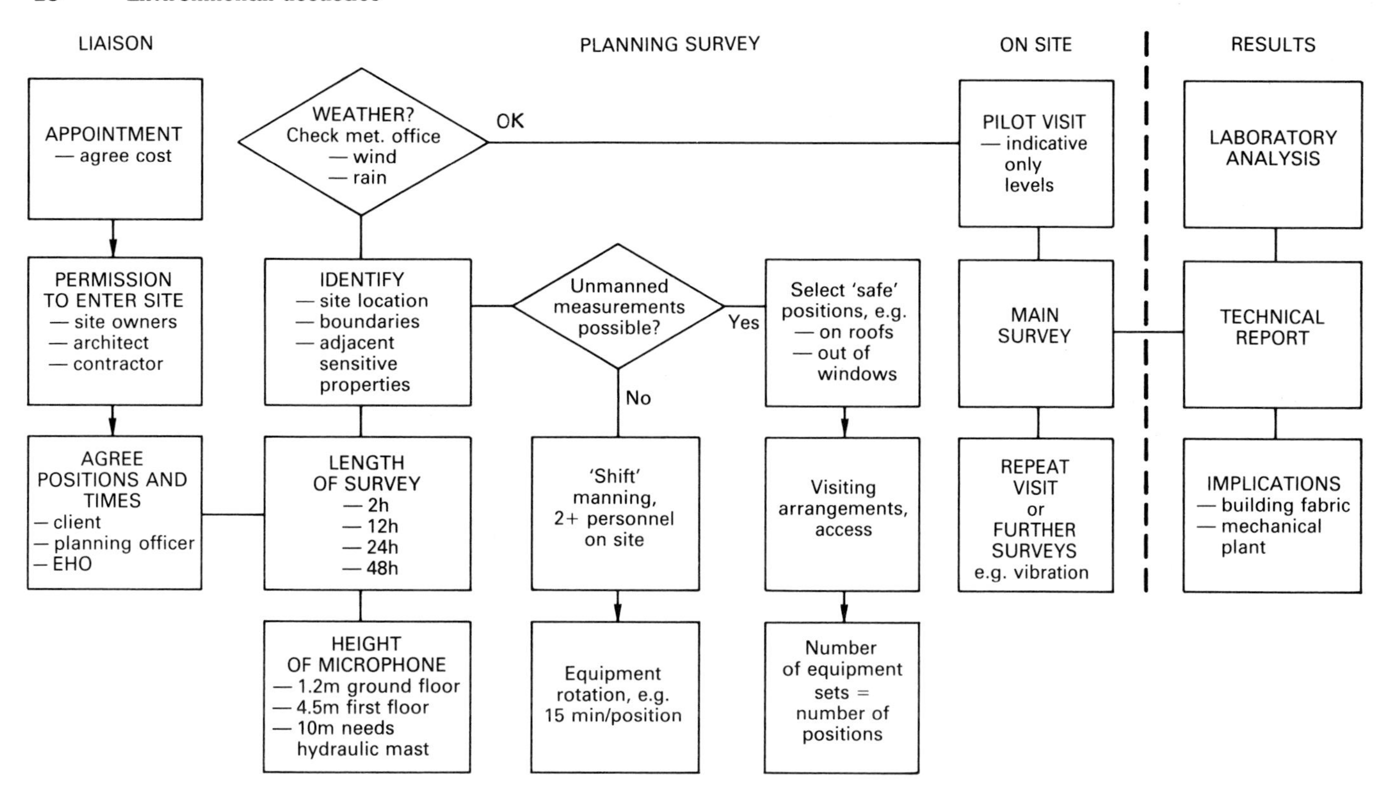

Diagram 1.5 *Summary of acoustic appraisal process*

The period normally used is 18 h (06:00–24:00 hours). The $L_{A10,18h}$ noise level is the basis for determining eligibility under the Noise Insulation Regulations 1975 (see Legislation criteria, below) [8].

$L_{Aeq,T}$ is the preferred unit for measuring environmental noise generally and is the A-weighted equivalent continuous sound level. However, it will probably be several years before L_{Aeq} is adopted as the descriptor for road traffic noise. Periods of 1 h are often used.

For most situations: $L_{Aeq,T} \simeq L_{A10,T} - 3$ dB. In 95% of such conversions the estimated $L_{Aeq,T}$ is likely to be within ±2 dB of the 'true' value.

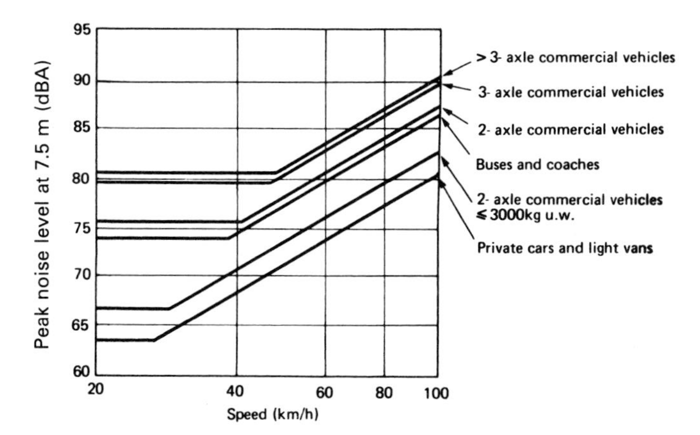

Figure 1.2 *Generalized sound level/speed characteristics for different vehicle categories*

Legislation and criteria

In the UK, the main legislation dealing with road traffic noise is the Noise Insulation Regulations 1975 [8]. This is issued under the Land Compensation Act 1973 [9]. These regulations were brought into force to compensate residents subjected to additional noise due to the use of new roads. Road construction noise is also included. If additional noise is at or above a specified level the affected residents receive a grant for acoustic double windows, supplementary ventilation, and, where appropriate, venetian blinds to control solar gain in south-facing windows, and double or insulated doors. The specified level is 68 dB $L_{A10,18h}$.

These regulations do not apply to new housing. New housing or development should be appraised by DOE Circular 10/73 [10]. This Circular proposes the following based on a 15-years-ahead predicted traffic flow:

- there should be a strong presumption against permitting residential development where external noise levels are in excess of 70 dB $L_{A10,18h}$
- internally, noise levels should be maintained below 50 dB $L_{A10,18h}$ with windows closed and every effort should be made to achieve 40 dB $L_{A10,18h}$ as a 'good' standard with windows closed.

The latter requirement necessitates the introduction of an alternative means of ventilation other than via the windows or indicates that external levels should be no greater than 60 dB L_{A10} and preferably no greater than 50 dB L_{A10} if windows are to be opened. However, it will often be better to achieve a higher standard.

There are no regulations governing acceptable noise levels in offices. However, BS 8233:1987 suggests that for private offices 40–45 dB $L_{Aeq,T}$ and in open-plan areas

45–50 dB $L_{Aeq,T}$ should be the aim. This indicates that where external noise levels are in excess of 60 dB L_{Aeq} or 63 dB L_{A10}, then a sealed office building with some form of mechanical ventilation will be required.

Prediction

An accurate procedure for the prediction of noise due to freely-flowing road traffic is given by *Calculation of Road Traffic Noise* (CORTN) [7]. To determine noise levels in accordance with CORTN, it is necessary to know detailed information about the road geometry and surface, topography and likely future traffic parameters. The traffic flow 15 years after the date of interest should be considered. Depending on the road geometry and topography, the road is broken down into segments and the resultant noise level at a reception point is calculated for each segment and then combined to give an overall level. A flowchart showing the process is shown in Diagram 1.6. This calculation method is available as a computer program such as RoadNoise by W. S. Atkins & Partners (Epsom, Surrey), and a large number of acoustic consultants have their own in-house programs.

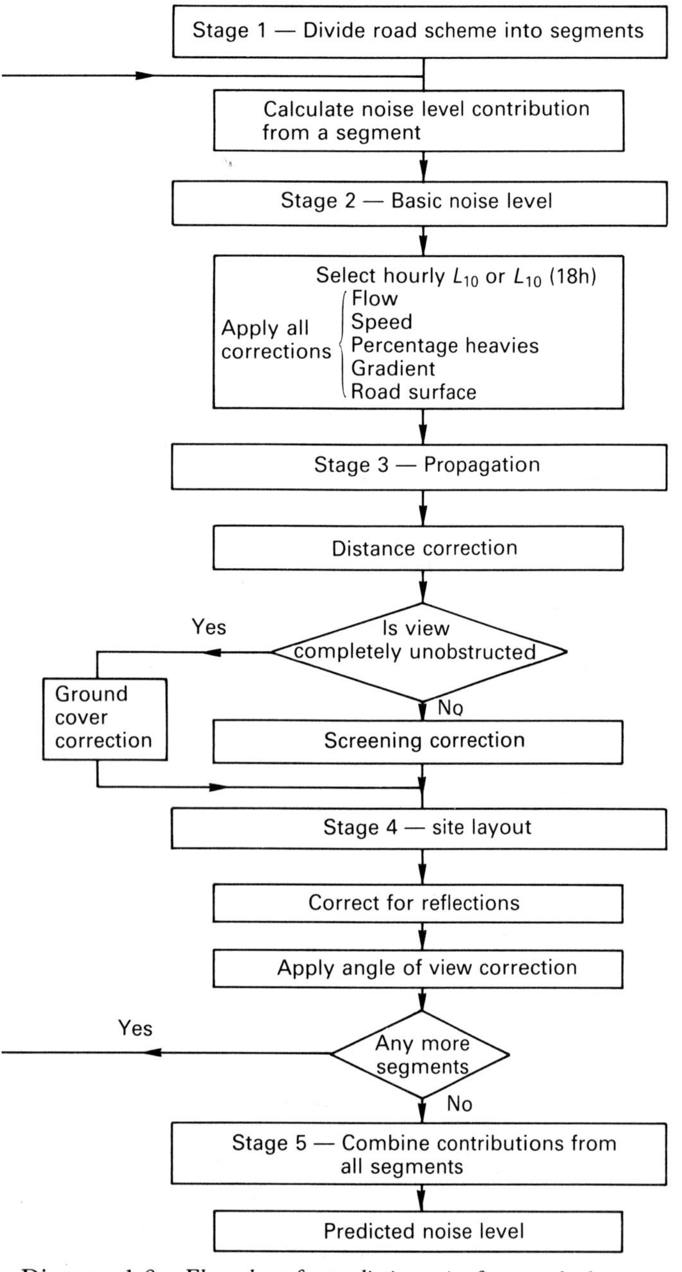

Diagram 1.6 *Flow chart for predicting noise from road schemes*

Table 1.3 *Typical road traffic noise levels based on BS 8233: 1987*

Situation	$L_{A10,\,18\,h}$	$L_{Aeq,\,18\,h}$
At 20 m from the edge of a busy motorway carrying many heavy vehicles, average traffic speed 100 km/h, intervening ground grassed	80	77
At 20 m from the edge of a busy main road through a residential area, average traffic speed 50 km/h, intervening ground paved	70	67
On a residential road parallel to a busy main road and screened by the houses from the main road traffic	60	57

Table 1.3 taken from BS 8233 gives an indication of traffic noise levels for different road types. Figure 1.3 from CORTN gives the basic noise level at 10 m from the nearside carriageway edge for traffic containing about 10% heavy goods vehicles (those over 1.5 tonnes) at up to 60 km/h. This also assumes that the road surface is bitumen and relatively level (gradient less than 3%). If the traffic speed exceeds 60 km/h then the noise level will increase at a rate of approximately 6 dBA/doubling of speed. If the percentage of heavy vehicles is greater than 10% than an approximate factor of 2 dBA/doubling of heavy vehicle content could be used.

The propagation of traffic noise with distance is predominantly based on distance to the source, angle of view of the road, intervening ground cover, and whether any barriers exist between source and receiver. Typically, over ground covered with vegetation and a reception point not more than 4 m above ground, the reduction in noise level could be as much as 7 dBA/doubling of distance. Over hard ground or an acoustically reflective surface such as concrete or water, the reduction in noise level will be 3 dBA/doubling of distance.

The effect of barriers depends on the path difference and it is important to check that the line of sight between source, 0.5 m above road surface, and receiver, typically 1.5 m or 4.0 m above ground (ground and first floor reception heights respectively) cuts the barrier (Figure 1.4). The source for light and mixed traffic is taken to be 0.5 m above the road surface, and for heavy vehicles only can be taken to be typically 1.5 m above the road surface. The barrier needs to extend a significant way alongside the road to provide effective screening but if this is achieved the reduction can be determined from Figure 1.4.

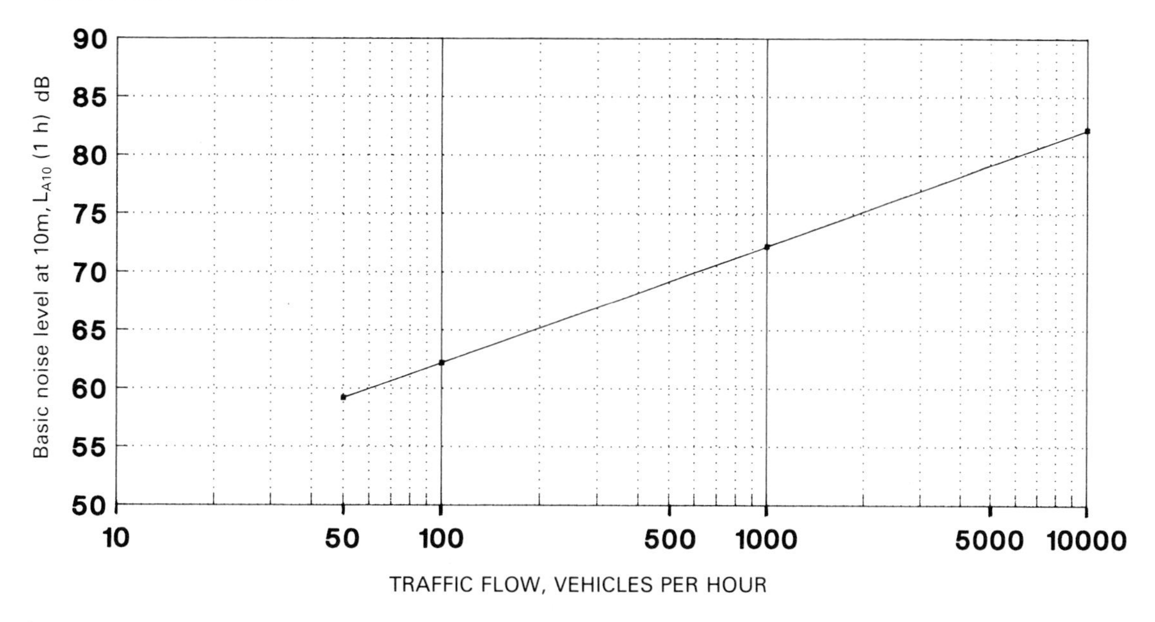

Figure 1.3 *Mixed flow road traffic noise at 10 m (based on Reference 7)*

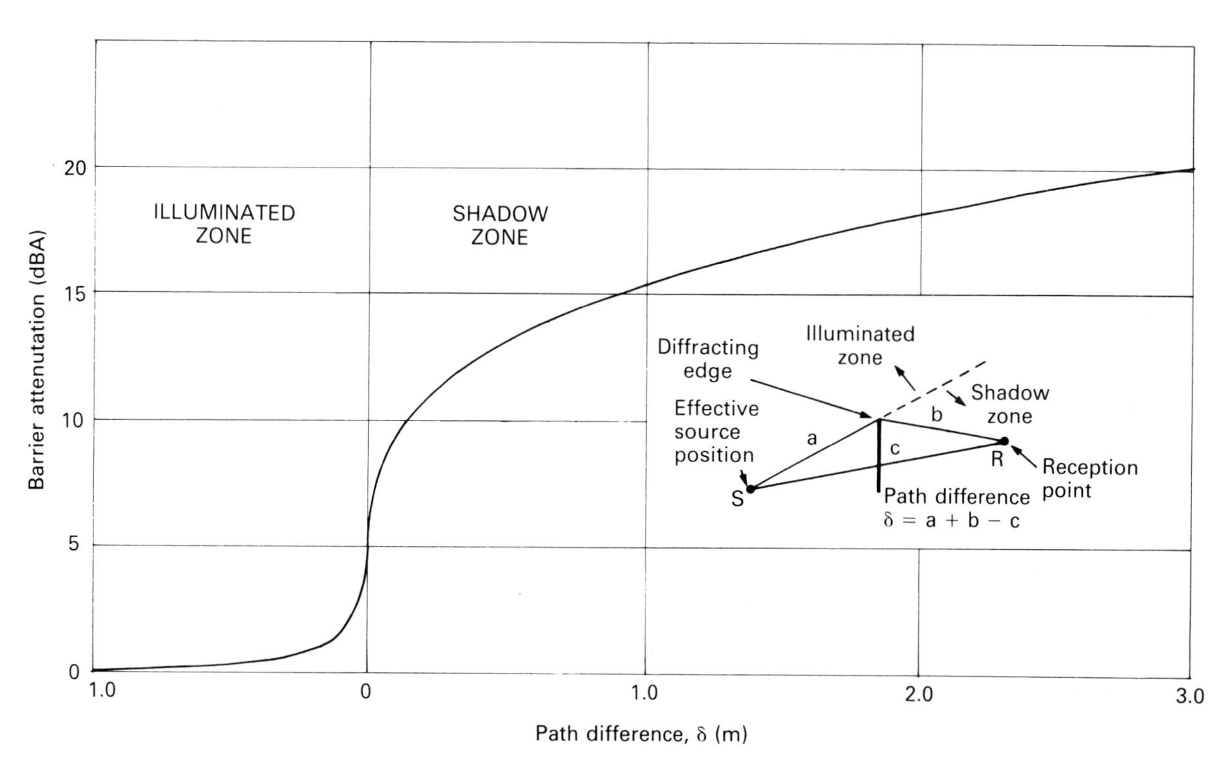

Figure 1.4 *Barrier attenuation for road traffic noise* [7]

Alternatively, if vehicle flows are low and measurements difficult to make, it may be preferable to use the following method based on the Noise Advisory Council guidance [11].

Figure 1.5 provides an estimate of the single-event exposure level at a distance of 10 m for light and heavy vehicles for different speeds. This can be used to determine the overall noise level in terms of L_{Aeq} at 10 m for a number of vehicles. The overall level at a distance greater than 10 m can then be estimated based on attenuation of 3–7 dBA/doubling of distance depending on the ground cover. There are also occasions when a predictive exercise may become complicated and a measurement is the only available course of action, such as at a traffic-lit junction or a roundabout.

The resultant noise levels are normally given in terms of a level at a particular point. However, provision of contours, particularly on a site where the best location for a building is being determined, can be helpful. The predicted levels for road traffic noise will be overall A-weighted single figures which can be converted to typical octave band levels using the graphs shown in Figure 1.6.

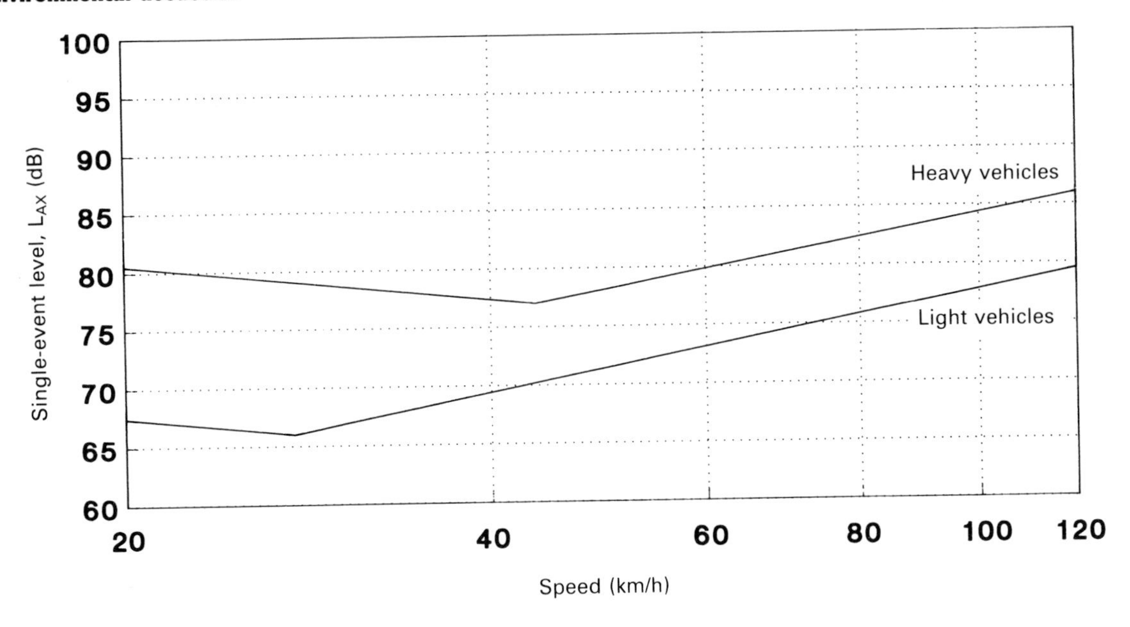

Figure 1.5 *Single-event road vehicle noise level at different speeds at a distance of 10 m [11]*

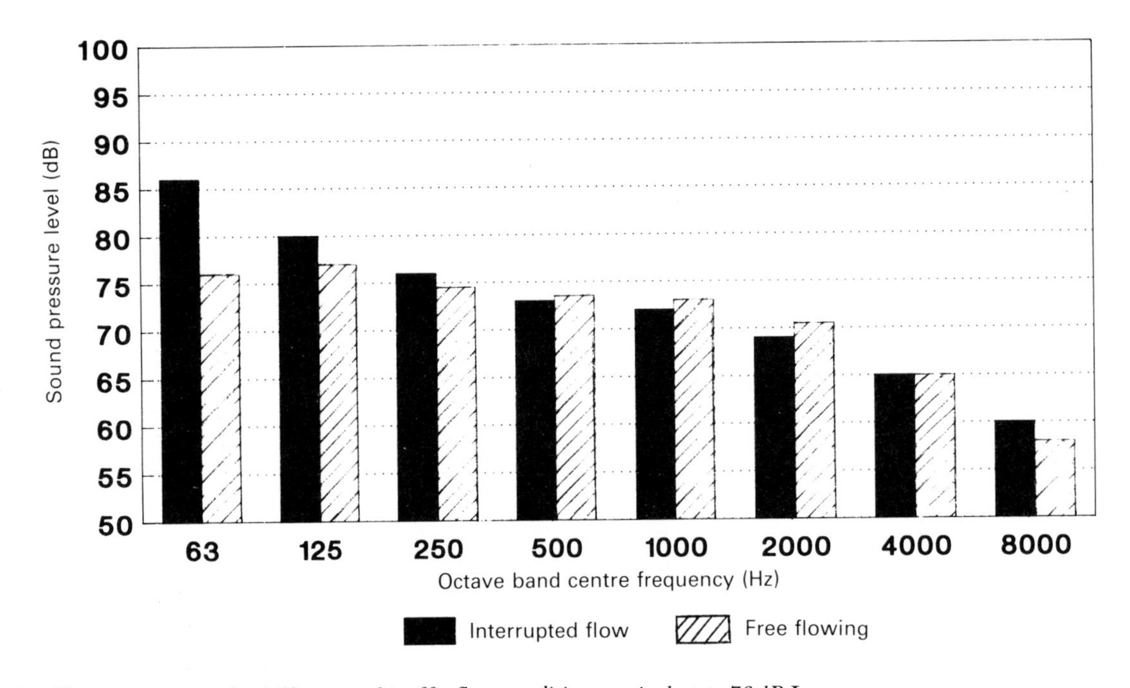

Figure 1.6 *Frequency spectra for different road traffic flow conditions equivalent to 76 dB* $L_{A10,T}$

Railways

Noise sources

The predominant sources of noise due to train movement are propulsion equipment and wheel/rail interaction. The propulsion equipment includes diesel locomotives and diesel multiple units; noise from electric locomotives and electric multiple units is significantly lower than from diesel equivalents. In addition, auxiliary equipment, such as ventilation systems and other carriage-mounted components, can be sources of noise, and elevated structures, such as bridges, tend to increase noise levels but both are typically insignificant in the UK compared to diesel locomotive and wheel/rail noise. In the US there are many steel elevated structures causing high noise levels. A similar situation also occurs with Docklands Light Railway but it is predominantly at low frequencies[12]. The maximum noise level at 25 m from diesel locomotives is typically 85–95 dBA [13].

Wheel/rail noise is due to the vibration of both caused by the action of one rolling over the other. The parameters that can affect this noise are the type of track, i.e. continuously welded rail (CWR) or jointed (+5 dBA), the type of braking system, i.e. disc- or tread-braked, and maintenance of track/wheels, i.e. removal of corrugations. Noise due to tread-braked rolling stock can be 10 dBA higher than disc-braked, and badly corrugated track could cause

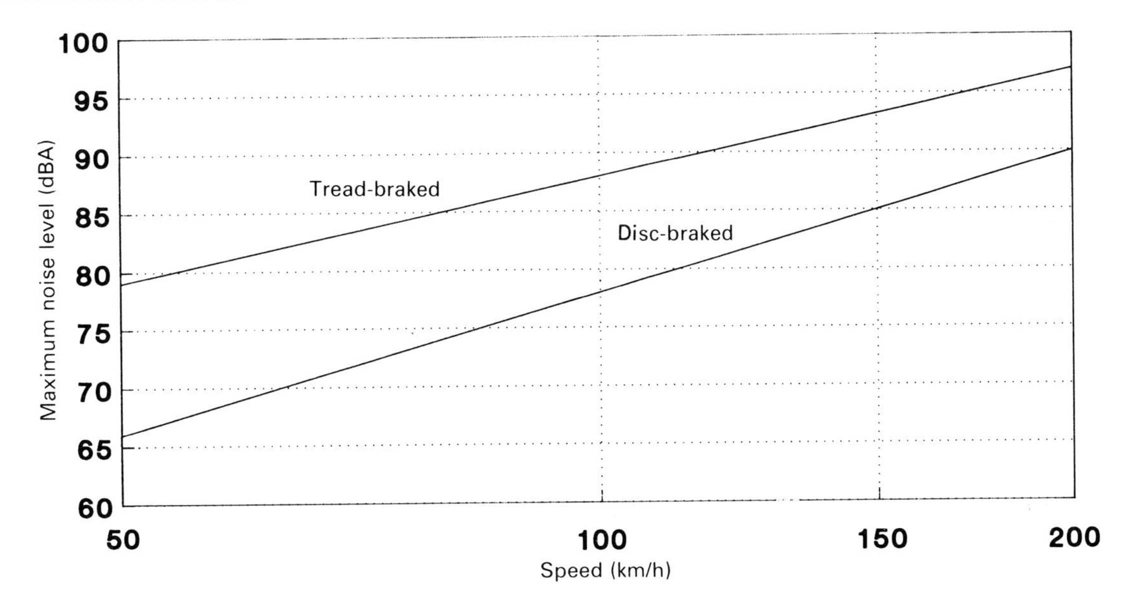

Figure 1.7 *Noise levels of passenger trains at different speeds at 25 m [13]*

increases of 10 dBA. Therefore rolling stock with disc brakes on CWR that is regularly maintained will result in the lowest noise levels. Typical noise levels of the different train types hauled by electric locomotives are shown in Figure 1.7. Noise control measures to railways are being brought in, in the form of 'Hush' rails, beneficial through their smaller cross section, and wheels which are damped to reduce 'ringing'. In future, as train speeds get higher, aerodynamic noise may become significant, but that stage has not yet been reached in the UK.

Measurement unit

Train noise is measured in terms of the A-weighted equivalent continuous noise level, $L_{Aeq,T}$. The only parameter which seems to vary is the period T selected for consideration. It is often taken over the full 24-h daily period but day (07:00–23:00 hours) and night (23:00–07:00 hours) periods are often used separately.

In order to determine the L_{Aeq} over a given time period, it is often preferred to undertake a calculation using individual train pass-by levels. Thus the event noise exposure level, L_{Ax}, is measured for different train types; typically this is at a distance of 25 m. In addition, the maximum noise level is often measured in order to assess the effect of train pass-bys on conversation and telephone use, for example.

Legislation and criteria

There is no direct reference to railway noise in DOE Circular 10/73 *Planning and Noise* [10] and local authorities tend to set their own noise standards. Based on information obtained from a number of local authorities, the Midland Joint Advisory Council for Environmental Protection have set the following criteria for housing: the situation is considered satisfactory where external free-field noise levels are below 55 dB L_{Aeq} (07:00–23:00 hours) by day and 50 dB L_{Aeq} (23:00–07:00 hours) by night. If these acceptable noise levels are exceeded by 10 dB L_{Aeq}, then action should be taken to control noise. No development is recommended where the external free-field noise levels exceed 65 dB L_{Aeq} (07:00–23:00 hours) day or 60 dB L_{Aeq} (23:00–07:00 hours) night.

In some instances an overall 24-h time period has been used as a criterion, e.g. the GLC recommended an external level of 65 dB $L_{Aeq,24h}$ [14].

Maximum noise levels to avoid sleep disturbance should be limited to 50–55 dBA internally. Thus with housing and its ubiquitous opening windows this suggests an external free-field maximum noise level of 60–65 dBA at night. For developments with general offices where the impact on communication, either verbal or by telephone, must be considered, a maximum internal level of 55–60 dBA is suggested in guidance given by the US Environmental Agency [15]. This relates to an external free-field noise level of 65–70 dBA unless it is a sealed building.

Prediction

The overall L_{Aeq} noise level can be determined from the single event exposure levels L_{Ax}, as described in Chapter 5. In order to assess noise levels at other distances over grassland, the chart shown in Figure 1.8 can be used. Typically this is 5 dBA/doubling of distance. Intervening properties such as semi-detached and terraced housing could provide the following noise reduction:

- single row of semi-detached houses 8 dBA
- subsequent rows, each 4 dBA
- terraced housing 13 dBA

Cuttings or intervening ground barriers could be assessed in a similar way to that described for roads. It is possible to determine single-event levels, L_{Ax}, for the rail/wheel source given the maximum noise level, L_{Amax}.

$$L_{Ax} = L_{Amax} + 10\log(Lt/V) + C$$

where Lt = train length (in m),
V = train speed (in km/h) and
C is determined from Figure 1.9.

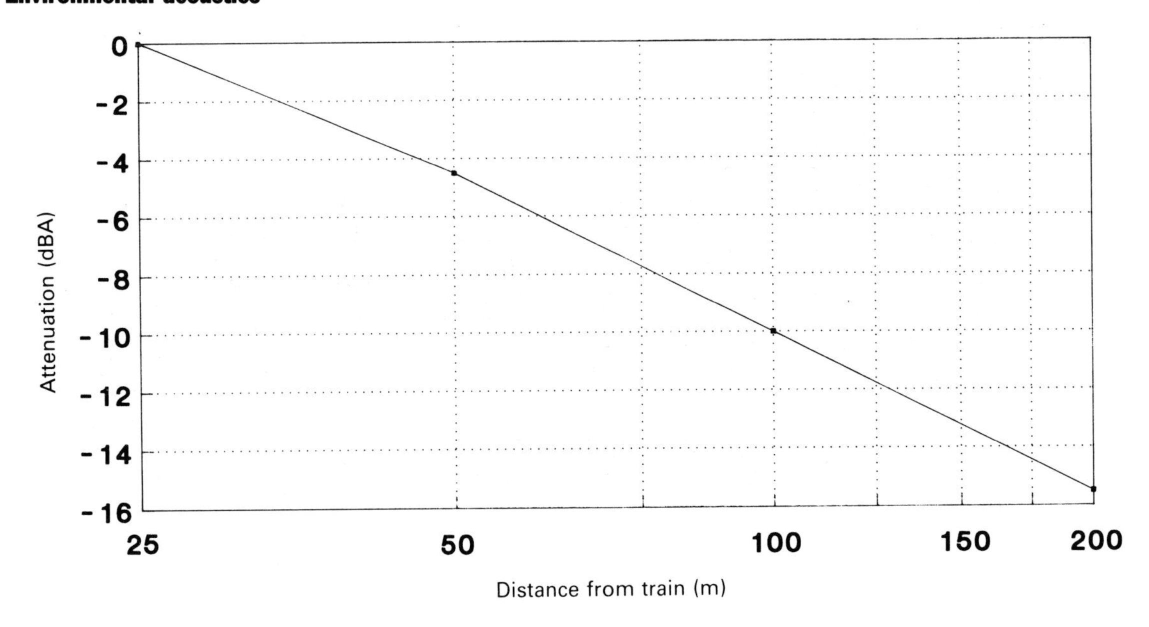

Figure 1.8 *Attenuation of train noise with distance over grassland [13]*

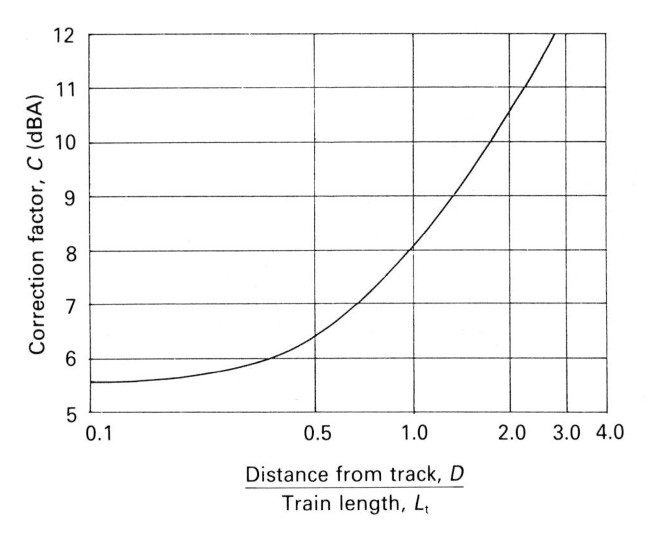

Figure 1.9 *Factor to determine single event train noise levels from maximum noise levels [11]*

Table 1.4 *Noise levels for a typical railway*[a]

Distance from track over open grassland (m)	$L_{Aeq,24h}$ (dB)
25	67
50	64
100	59
200	54

[a]Based on BS 8233. Typical railway traffic is assumed to consist of a mixture of a total of 120 high-speed diesel-hauled passenger and freight trains, per day.

Computer programs can be derived to produce either spot levels or noise contours based on charts and information given by the Noise Advisory Council. BS 8233[16] contains estimated noise levels for a track carrying diesel-hauled passengers and freight trains at different distances over open grassland and is reproduced here as Table 1.4.

The typical octave frequency band levels can be determined either from on-site measurements or the typical noise spectra given in Figures 1.10a and 1.10b for both diesel electric locomotives and tread- or disc-braked rolling stock hauled by electrically-powered locomotives.

Aircraft

Source noise

The main concern relating to aircraft noise is associated with take-offs and landings near an airport. In addition, ground operation noise may also need to be considered. In terms of noise due to flying operations, the main factor is the type of aircraft. The maximum noise levels for various types of aircraft under different operating conditions are given in Table 1.5. Data are given for the UK reference distance of 152 m used to determine the Reference Noise Level. Military airfields may also need to be considered not only for noise near the airfield but sometimes due to other operations such as low flying. Data in this case needs to be obtained from the Ministry of Defence.

Measurement units

Historically, the Noise and Number Index (NNI) has been used as the noise unit for measuring aircraft noise. It takes into account the maximum perceived noise level of each aircraft for the number of aircraft movements during a 12-h period (06:00–18:00 hours). However, in September 1990 the Department of Transport changed to the use of L_{Aeq} to describe aircraft noise over a 16-h period (07:00–23:00 hours).

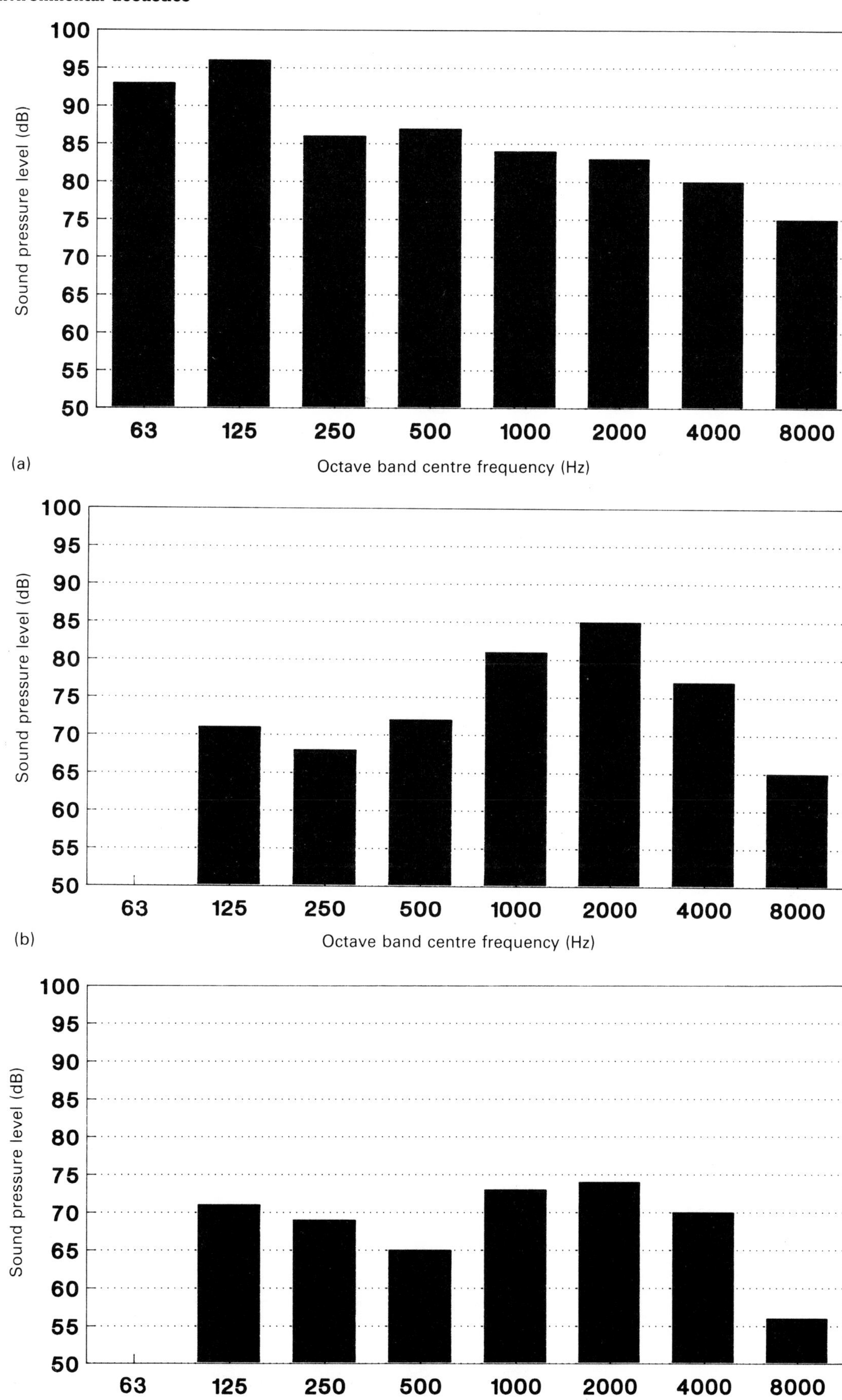

Figure 1.10 *Frequency spectra for trains: (a) a diesel locomotive at 25 m; (b) tread-braked passenger train at 25 m with electric locomotive at 100 km/h; (c) disc-braked passenger train at 25 m with electric locomotive at 100 km/h*

Table 1.5 *Typical noise levels of aircraft at a distance of 152 m*

Aircraft type	Example	Operation	L_{Amax} (dB)	L_{Ax} (dB)
Supersonic long range	Concorde	Take-off	126	
		Departure	116	
		Approach	107	
Old technology long range	B707	Take-off	112–120	120
	DC-8	Departure	105–112	
	VC-10	Approach	99–102	107
Old technology jet	Trident	Take-off	111–115	113–115
	B727	Departure	107–110	
	B737	Approach	94– 99	97–100
	BAe1-11			
	DC-9			
New technology long range	B747	Take-off	103–107	110
	DC 10-30	Departure	99–104	108
		Approach	91– 96	97
New technology medium range	Tristar	Take-off	96–104	
	B737-300	Departure	93–100	
	B757	Approach	85– 92	
	DC10-10			
	B767-200			
New technology feeder/commuter	BAe146-100/200	Take-off	92	
		Departure	87	
		Approach	85	
Hushed jet	BAe1-11	Take-off	108–110	
	400/500	Departure	102–106	
		Approach	91– 93	
STOL medium/large	Dash 7	Take-off	82	
		Departure	79	
		Approach	73– 78	
STOL small	Twin Otter	Take-off	91	
		Departure	80	
		Approach	76	

Table 1.6 *Frequency spectra for typical jet aircraft movements at approximately 5 km from airport*

	Octave band centre frequency								
	63 Hz	125 Hz	250 Hz	500 Hz	1 kHz	2 kHz	4 kHz	8 kHz	
Take-off: 96 dBA	92	94	96	95	92	84	68	56	dB
Landing: 89 dBA	79	85	86	84	82	83	80	71	dB

Legislation and criteria

Department of the Environment Circular 10/73, *Planning and Noise*[10] states that minimum noise routes should be established by busy airports and land subject to significant aircraft noise should not be used for noise-sensitive development. To achieve this, NNI criteria were recommended for the control of development (Table 1.7). The NNI levels on a site should be determined for a period 15 years ahead. For planning purposes, an areas such as a residential estate should be considered as lying wholly within the highest NNI contour that cuts the site. Circular 10/73 is to be updated and is likely to delete reference to NNI.

There are a number of airport grant schemes under which grants are paid towards the cost of sound insulation in existing dwellings within defined areas around major airports. The sound insulation package is similar to that offered under the Noise Insulation Regulations for road traffic noise, plus increased roof insulation.

Prediction

Although a simple calculation can be carried out using the sound source data L_{Ax} from Table 1.5, the number of aircraft types and a correction for distance, the error in accuracy is likely to be great. In reducing the error, reliance will have to be placed on published contour maps which are based on accurate flight profile data. However, consideration may need to be given to the maximum noise level on a site from a design point of view. Measurement is obviously the easiest method of determining maximum noise levels, but if this is not possible, then the maximum noise level at a particular location can be obtained by calculating the slant distance as shown in Figure 1.11 and applying the correction 8 dB/doubling of distance. An indication of the frequency content for typical jet aircraft types is shown in Table 1.6.

NNI contour maps are available from the Civil Aviation Authority or via the local authority or airport authorities; examples are Gatwick, Heathrow and Manchester. It is possible to estimate the approximate equivalent $L_{Aeq,16h}$ value within ±2 dB from the following table:

NNI	35	40	45	50	55	60
$L_{Aeq,16h}$	57	60	63	66	69	72

Ground operation noise can be taken to be approximately 85 dBA at 300 m and to reduce at a rate of 12 dBA/doubling of distance.

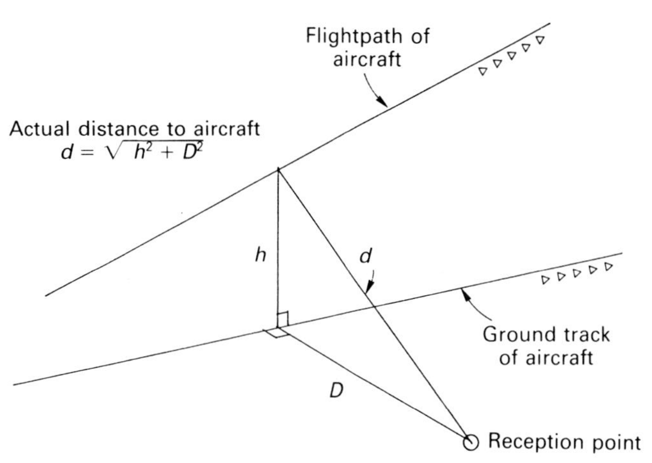

Figure 1.11 *Estimation of slant distance from an aircraft flight path*

Table 1.7 *Recommended criteria for control of development in areas affected by aircraft noise*[a]

Level of aircraft noise to which site is, or is expected to be, exposed	60 NNI+ >71 dB $L_{Aeq,16h}$	50 NNI–59 NNI 66–71 dB $L_{Aeq,16h}$	40 NNI–49 NNI 60–65 dB $L_{Aeq,16h}$	35 NNI–39 NNI 57–59 dB $L_{Aeq,16h}$
Dwellings	Refuse	No major new developments. Infilling only with appropriate sound insulation		Permission not to be refused on noise grounds alone
Schools	Refuse	Most undesirable. For exceptions, sound insulation measures to DES Guidelines[b]	Undesirable	Permission not to be refused on noise grounds alone
			Sound insulation to a standard consistent with DES Guidelines[b]	
Hospitals	Refuse	Undesirable	Each case to be considered on its merits	Permission not be refused on noise grounds alone
		Appropriate sound insulation required		
Offices	Undesirable	Permit	Permit but advise insulation of conference rooms, depending on position, aspect	
	Full insulation to be required			
Factories, warehouses	Permit (Up to occupier to take mitigation measures)			

[a]Table derived from DOE Circular 10/73 [10].
[b]Department of Education and Science.

Control of transportation noise sources

Unless a new transportation route is under discussion, the design of the route cannot be influenced and noise control can only be achieved by increasing the sound insulation of the building under consideration or by the introduction of a noise barrier.

The introduction of noise barriers in the case of aircraft noise where ground running is a potential problem will only be of limited benefit if the development is near an airfield. Appropriate mufflers or noise testing pens/hush houses will be needed to control noise from engine testing.

A noise barrier should ideally be located as close to the noise source as possible. In some cases this may lead to maintenance problems since it may need to be sited on someone else's land. For the determination of the preferred location for a noise barrier, if it cannot be positioned close to the source, sections across the site will be invaluable. Using a sight line between the source height, for example for roads it will be 0.5 m above the road surface, and the reception point, typically at a window on the highest floor of a building, the most effective position of a barrier can be decided upon. The performance of a noise barrier is given by the path length difference as illustrated in Figure 1.4 for road traffic noise. In order to take account of the different frequency spectra of train noise compared to traffic noise, Figure 1.12 can be used.

The noise reduction achieved by a noise barrier along a road or railway is typically between 5 and 10 dBA and to achieve greater reductions is often quite difficult. The effect of excess attenuation due to soft ground, which was probably included in determining the noise level on site from transport, is negated in determining the overall performance provided by a barrier, unless it is on earth mounding with shallow-sloped sides. The design of the barrier should ensure that the length of the barrier is sufficient to protect the whole development.

The barrier need only be relatively lightweight and normally a close-boarded timber fence is quite adequate. Other barrier types include metal sandwich construction or precast concrete unit assemblies. The performance of barriers alongside railways can be reduced by as much as 5 dBA where the side closest to the track is acoustically reflective. Consideration should be given to a barrier type with an acoustically-absorbent surface facing the track. There are barriers, of metal sandwich construction and precast concrete faced with woodwool slabs, which will achieve this requirement.

Construction noise

Sources

Major noise sources involved with construction activities include piling rigs, earthmoving equipment such as dozers and excavators, and concrete pouring plant such as concrete pumps and truck mixers. A range of construction equipment is given on Table 1.8 which includes the

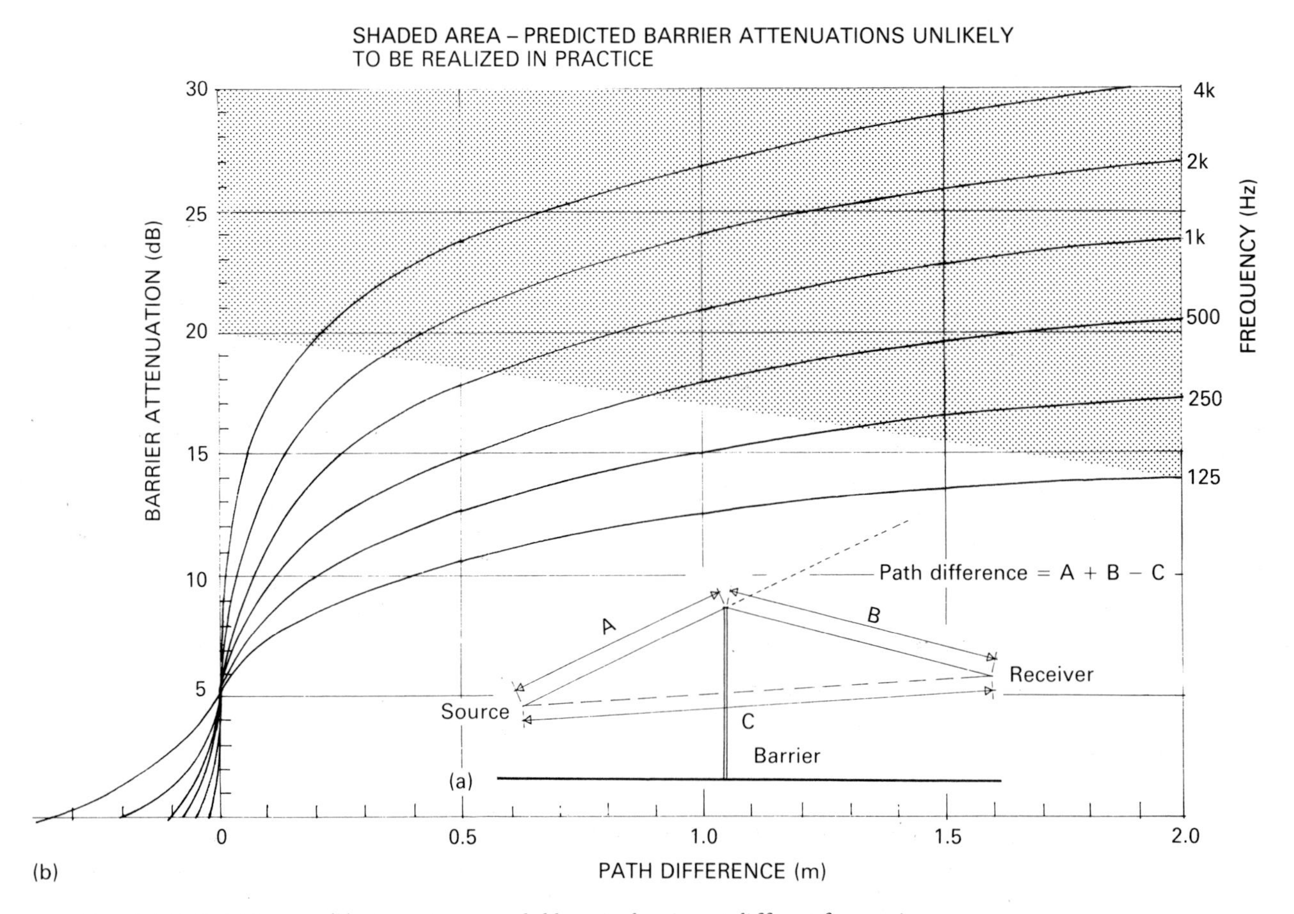

Figure 1.12 *(a) Path difference; (b) attenuation provided by noise barriers at different frequencies*

Table 1.8 *Typical construction plant noise*

Equipment	Approx. L_{WAeq} (dB)
Tracked loader	109
Tracked excavator	109
Dozer	111
Piling:	
Diesel hammer	130
Drop hammer + wooden dolly	115
Auger bored	112
Pneumatic breaker	116
Concrete pump	110
Truck mixer	110
Concrete mixer	95
Batching plant	105
Poker vibrators plus compressors	102
Compressors:	
4 m^3/s	98
7 m^3/s	101
17 m^3/s	111
Generator	104
Pump	103
Crane	103

approximate average sound power level during the activity of each item. In most cases, diesel engine noise predominates but consideration needs to be given to piling and material handling noise.

In addition to the above equipment which is operating normally out in the open, structureborne noise due to hand-operated drills or breakers may cause potential noise problems in buildings coupled to the construction under consideration. The noise level due to structureborne noise varies significantly depending on local site conditions and an estimate of the noise level likely to occur cannot easily be provided, although noise levels of 55–60 dBA in nearby areas during percussive drilling could be anticipated.

Measurement unit

The A-weighted equivalent continuous sound level, $L_{Aeq,T}$, is the preferred unit for describing construction noise. However, in addition, to take account of isolated events and impulsive sources such as piling, it is recommended that the maximum noise level, L_{pAmax}, is also considered. In describing site noise, the particular period of the day should always be stated.

Legislation and criteria

Noise from construction sites is specifically referred to in Sections 60 and 61 of Part III of the Control of Pollution Act 1974 [17].

Section 60

Under Section 60 of the Act, a local authority may serve a Notice on the contractor specifying one or more of the following:

- plant or machinery which is, or is not, to be used
- hours during which works may be carried out
- noise limits

However, in specifying any of the above, a local authority should have regard to:

- relevant Codes of Practice issued under this part of the Act, viz. BS 5228: Parts 1–4: 1984/92 [18];
- the need to ensure that the best practicable means ('practicable' meaning reasonably practicable having regard amongst other things to local conditions, the current state of technical knowledge, and the financial implications; 'means' includes design, maintenance and manner and periods of operation of plant and machinery and the design, construction and maintenance of buildings and acoustic structures; this is provided safety and safe working conditions are met and regard paid to any provision of BS 5228) are employed to minimize noise;
- the interest of the recipient before specifying any particular methods or plant or machinery, i.e. where alternative methods or plant more acceptable to the construction operator would be substantially as effective in minimizing noise as those proposed by the local authority;
- the need to protect any persons from the effect of noise.

Any person served with such a notice may appeal to a magistrate's court within 21 days from receipt. The grounds for appeal and form of notices are outlined in Department of the Environment's Circular 2/76 [19].

Section 61

The other approach, outlined in Section 61 of the Act, places the onus on the contractor or other responsible persons. In this section, the contractor can notify the local authority of his methods of working and noise control procedures, and apply for a consent. The local authority may grant such a consent or have the power to apply conditions to the consent. Thus the contractors can have some certainty about their position and the risk of interruption to works that have started is removed as far as possible. The local authority should reply to applications for consent within 28 days. If this reply is not forthcoming, or the conditions atached are not acceptable, the contractor can appeal to a magistrate's court within 21 days from the end of that period. Applications for consent should be made at the same time as or, where it is necessary, after application for Building Regulations approval. However, this could have certain implications on normal tendering procedures, and it is being suggested that local authorities should be prepared to give advice as early as possible in respect of their proposed noise limitations. It is essential to the working of this legislation for both contractors and local authorities to have consultation prior to any formal procedures occurring.

Contravention

If the contractor or other responsible person knowingly allows work to be carried out, in contravention of either any conditions attached to a consent or any requirements of a notice, they will be guilty of an offence against Part III of the Control of Pollution Act. It should be noted that the contractor would also be responsible for a subcontractor operating on the project and their attention must be drawn to any requirements of a consent or notice on that project.

Code of practice

The relevant Code of Practice relating to construction noise is BS 5228. The aim of the Code is to recommend methods of noise control in respect of construction and other open sites and to enable developers, architects, engineers, planners, designers, site operators and local authorities to control noise. One of the factors which complicates any assessment is the relative sensitivity of different individuals in the same neighbourhood to the same noise.

Site noise is normally described in terms of the equivalent continuous A-weighted sound level L_{Aeq} over a stated time, for example 1 h or 12 h. In addition to L_{Aeq}, the site noise may be described in terms of the maximum sound level, L_{pAmax} or the one-percentile level L_{A1}. Whichever measure is selected to describe the site noise, the period of the day to which the particular value of the measure applies must also be stated. In assessing whether noise from a site is likely to constitute a problem, in addition to site location and the existing ambient noise levels, consideration should be given to the following:

- Duration of site operations: higher noise levels may be accepted by local residents provided they are aware that the work is only of short duration; good public relations are important in this.
- Hours of work: certain periods of the day are more sensitive than others; as an indication of the difference between daytime and evening time working, the noise level may have to be up to 10 dBA quieter in the evening. Even during the daytime period, certain times are likely to be more sensitive than others in offices and other workplaces. The code states that at noise sensitive premises the L_{Aeq} may need to be as low as 40–45 dBA during night-time. It is possible, however, that the level may need to be even lower to avoid sleep disturbance.
- Attitude of site operator: noise from the site may be more readily accepted if the site operator is doing all he can to avoid unnecessary noise. The acceptability of the project itself may also be a significant factor.
- Noise characteristics: impulse or tonal characteristics may make the site noise less acceptable. There is no detailed information available on assessing the acceptability of site noise.

Construction site noise prediction

Once a practical noise limit has been specified by the local authority, it will be necessary for that local authority and also the developer, architect, engineer and contractor to know whether the intended site operations will cause problems. The noise levels for different operations will have to be predicted at tender stage so that appropriate allowances can be made in the tender for noise control. Once those site operations that exceed the noise limit are known, the contractor will be required to include for the necessary noise control to achieve the limit. The noise limit would normally be quoted as a site boundary level in terms of L_{Aeq} over a given time period, typically a 12-h working day, and/or in terms of either an overriding short period, e.g. 5 min, or maximum levels measured with a sound level meter set at slow or fast response.

In order to estimate the noise level at a given location, the procedure basically consists of the following factors:

- sound power outputs of processes and plant
- distances from source to receiver
- presence of screening by barriers and the reflection of sound
- periods of operation of processes and plant.

An example of the data needed is shown in Table 1.9. In this case the $L_{Aeq,\,12\,h}$ level would be:

$$L_{Aeq,\,12\,h}$$
$$= 10\log \tfrac{1}{12}[4 \times 10^{58/10} + 1 \times 10^{69/10} + 8 \times 10^{71/10} + 6 \times 10^{73/10}]$$
$$= 73\,\text{dB (to the nearest dB)}.$$

The method of prediction can be represented by a diagram taken from BS 5228, reproduced here as Diagram 1.7.

Setting suitable criteria

Information from local authorities in the UK indicates that noise from road-works, and construction and demolition sites causes relatively few cases of noise nuisance, i.e. between 5 and 10% of the total nuisances confirmed. However, the effect of construction site noise as a possible nuisance must not be overlooked. There are three different

Table 1.9 *Example of construction site noise prediction based on BS 5228: 1984/92*

Plant and size	Operating sound power level, L_{WAeq} (dB)	Distance (m)	Distance attenuation (dB)	Screening attenuation (dB)	Mobile plant correction (dB)	Resultant sound power level (dB)	On-time in 12 h (h)
Compressor 7 m^3/min	100	50	42	0	0	58	4
Pneumatic breaker	116	50	42	5	0	69	1
Tracked excavator	109	30	38	0	0	71	8
Tracked loader	109	15	32	0	4	73	6

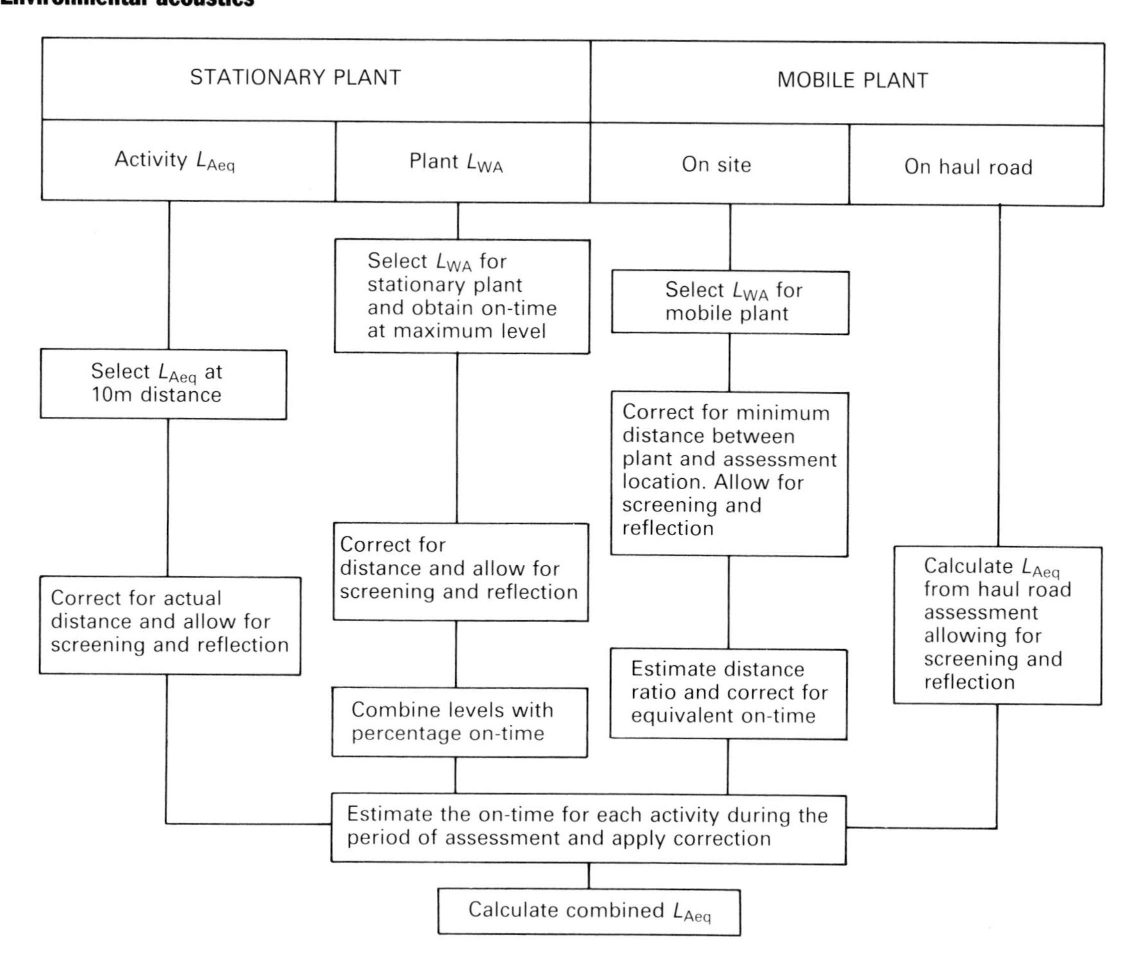

Diagram 1.7 *Prediction of site noise (after BS 5228: Part 1)* [18]

approaches available to local authorities in dealing with construction noise:

- encourage application for prior consent
- serve a Notice
- allow work to be carried out and if any cases of noise nuisance arise dealing with them either under Section 58 or Section 60 of the Control of Pollution Act. Although construction noise nuisance should be dealt with under Sections 60 or 61, individuals may complain to local authorities, who could then serve a Notice under the Environmental Protection Act to abate the nuisance.

Although there is no requirement to set a noise limit for any of the three approaches, noise limits can be set based on those aspects described earlier in the 'Code of Practice' section above.

Noise nuisance caused by construction noise is normally of short duration. Often, there are no means available to apply control, e.g. structureborne noise due to hand-held tools, and so it is very important that the contractor has a good public relations policy. If a noisy activity has to take place which could affect neighbouring properties, the neighbours should be warned, particularly with regard to the duration of the activity. It may even be possible to agree to periods of operation that are acceptable to both parties.

Noise control

Noise from construction equipment can be controlled at source or by controlling the way it propagates or spreads. Control at source is by:

- *Substitution.* Noisy plant and operations including piling should be replaced by less noisy alternatives where reasonably practical.
- *Modification.* Machines can be made quieter by modification, but this should only be carried out after consultation with the manufacturers.
- *Enclosures and screening.* In designing an enclosure to control noise from a machine, consideration must be given to the ventilation requirements in order to prevent overheating. Suitable materials and some examples of enclosure design are included in the British Standards section (Chapter 5). Alternatively, screens around the noisy area can be used or, if screening cannot be provided by site buildings or by earth mounds, temporary screening can be constructed with materials such as external-quality plywood or prescreeded woodwool slabs.
- *Use and siting of equipment.* Plant should be used in accordance with the manufacturer's instructions. In situations where operation is intermittent, plant should be shut down or throttled down to a minimum

when not in operation. Care should be taken to position noisy equipment away from noise-sensitive areas, and in cases where an item of plant is known to emit noise strongly in one direction, it should be orientated so that the noise is directed away from noise-sensitive areas. Engine covers should be kept closed during use.

- *Maintenance.* Regular and effective maintenance of plant is essential and will assist in keeping noise levels to a level similar to that from a new item of plant. It is particularly important to effectively maintain the silencing systems, for example engine exhaust silencers.
- *Periods of use.* It may be possible to operate certain items of noisy equipment to avoid sensitive periods. In some city areas, agreements have been reached for piling not to take place during the periods 10:00–12:00 hours and 14:00–16:00 hours, thus enabling normal office activities to take place during these hours for a limited period without any likelihood of disturbance.

Project supervision

Noise control should be considered at each project stage in order to meet the necessary requirements. Early consultation between developer, architect or engineer, and the local authority, should be held to ascertain the likely noise limitations. Processes and equipment involved with the site operations should be considered in order to keep those particular noisy operations to a minimum.

Issues include planning the hours of working, ensuring the use of the most suitable plant, economy and speed of operations, on-site monitoring, and the provision of prominent warning notices where high noise levels exist.

Local authorities may wish to lay down requirements relating to the work programmes, plant to be used, siting of plant, and working hours, rather than (or in addition to) specifying site noise limits. This approach will often be preferred by both authorities and site operators because specified requirements can be easily monitored.

Industrial noise

Noise sources

Industrial noise is caused by a wide variety of sources. Some general noise-producing activities are quarrying or other mineral extraction, material handling, metal fabrication, and building services plant operation.

The situations and modes of operation of the sources can also vary widely, and it can be a single machine or an array of machines, operating either internally within a building or externally, which are of concern. The sources may even emit noise levels which fluctuate with time, for example mobile plant, machines on- and off-load. Examples of the various types of industrial noise sources are:

- mineral extraction: blasting and mobile earthmoving type equipment with reversing signals, truck loading;
- materials handling: fork lifts or cranes, loading/unloading trucks with associated impact noise which can take place internally or externally;
- metal fabrication: cutting of steel sheet by guillotine, press operation and associated material handling including waste disposal into bins possibly via cyclone units;
- building services plant operation: this could serve an office building or industrial premises with the plant sometimes mounted on the roof.

Due to the variety of sources, typical noise levels cannot be given reliably.

Measurement unit

BS 4142: 1990[20] identifies the A-weighted equivalent continuous noise level, L_{Aeq}, as the preferred measurement unit, although if the industrial noise is reasonably steady, an average A-weighted noise level measured with a sound level meter set to 'slow' time weighting is acceptable. It is necessary to obtain the background noise level when considering the impact of industrial noise and this is defined as the A-weighted noise level exceeded for 90% of a time period, $L_{A90,T}$. This should be measured by a noise analyser operating with a fast time weighting. Although detailed consideration is not given in BS 4142 to the likely impact of impulsive-type noise sources, except in applying a fixed correction, the maximum noise levels of a process should always be obtained for subsequent assessment.

Legislation and criteria

For industrial development, Section 80 of the Environmental Protection Act 1990[21] and the Department of the Environment's Circular 10/73[10] are relevant (Circular 10/73 is to be amended).

Circular 10/73, paragraphs 24–34 inclusive, provides guidance for the local authorities to determine the suitability of a site for either a proposed industrial development or the effect of existing industrial premises on a proposed development. The fact that an industrial development is proposed to be sited on land specifically allocated for industry should not automatically result in permission being granted if it may give rise to high levels of noise. Careful siting of the noisier activities away from noise-sensitive areas should be a major consideration. There are, however, occasions when some form of industrial or similar development has to be allowed near noise-sensitive areas. Every precaution should be taken in these instances to control the noise. Guidance on the assessment of the noise impact of proposed industrial or other fixed installations is given in BS 4142[19]. Basically the noise level likely to be generated by the development is compared to the existing background noise level. The resulting difference gives an indication of the likelihood of complaints. If by this method, noise from the proposed development 'is likely to give rise to complaints' then permission is unlikely to be granted. In determining the predicted noise level from a proposed development, it will be necessary to take account of the plant operating at its maximum capability.

Circular 10/73 proposes that local authorities should prevent increases in ambient noise levels affecting residential and other noise-sensitive areas, wherever possible. Even where existing noise levels are already high, it will scarcely ever be justifiable to allow new development which is likely to cause an equivalent rating level of 75 dBA by day or 65 dBA by night in a noise-sensitive area.

If a local authority gives permission for the development, they will need to ensure that:

- noisier processes than those proposed by the developer are not allowed, and
- all physical features of the submitted plans which control noise levels are incorporated in the finished development.

Appropriate conditions will need to be imposed to meet these requirements and model conditions are given in Appendix 1 of Circular 10/73. Appropriate conditions have subsequently been published in Appendix A of the Department of the Environment's Circular 1/85 [22] and are reproduced here in Table 1.10. Although conditions relating to the physical characteristics of the development, the type and intensity of activity to be carried out, and hours of operation, are preferable, in some instances a condition laying down a maximum noise level at a particular location or possibly different levels for different periods of the day may be appropriate.

If a proposed development were shown by a noise assessment to be acceptable during normal working hours but not at other times, it would be reasonable to apply a condition restricting operation to certain specified hours rather than reject the application altogether.

Using this guidance, permission will be given for developments against which the local authority is unlikely to find it necessary to serve a noise abatement notice under the Environmental Protection Act. However, it will not necessarily provide protection against legal action by private citizens.

Under Section 80 of the Environmental Protection Act, the local authority is empowered to deal with noise nuisance by serving a Noise Abatement Notice.

Circular 10/73 also gives guidance where a new residential or noise-sensitive development could be affected by noise from an industrial or other fixed installation. The maximum Corrected Noise Level within dwellings should be 55 dBA during the day and 45 dBA at night. Similarly, a 'good standard' would be a Corrected Noise Level of 45 dBA during the day and 35 dBA at night. This latter requirement relates to an external Corrected Noise Level of 55–60 dBA during the day and 45–50 dBA at night if windows were openable for ventilation. However, higher standards would be preferable.

Table 1.10 *Suggested models of acceptable conditions relating to noise for use in appropriate circumstances. Extracted from Appendix A of Department of Environment Circular 1/85* [10]

5. . . . .[*activities*] shall not take place anywhere on the site except within . . . . building[s].

The condition should describe precisely the activities to be controlled as well as the particular building(s) in which they are to take place.

6. The building shall be so [constructed/adapted] as to provide sound attenuation against internally generated noise of not less than . . . . dB averaged over the frequency range 100 to 3150 Hz.

7. Noise emitted from the site shall not exceed [A]dB expressed as a [B] minute/hour L_{Aeq} between [C] and [C] hours Monday to Friday and [A]dB expressed as a [B] minute/hour L_{Aeq} at any other time, as measured on the [D] boundary [boundaries] of the site/at point[s] [E].

Specify: A – noise level
B – period
C – times
D – boundary (boundaries)
E – points.

8. [No [*specified machinery*] shall be operated on the premises] before . . . . am on weekdays and . . . . am on Saturdays nor after . . . . pm on weekdays and . . . . pm on Saturdays [nor at any time on Sundays or bank holidays].

9. Before [any] [*specified*] plant and machinery is used on the premises, it shall be enclosed with sound-insulating material in accordance with a scheme to be agreed with the local planning authority.

This condition might be varied where the need was to secure the satisfactory mounting of the machinery to prevent conducted noise and vibration. Advice should be appended to the permission, indicating the attenuation aimed at.

10. Development shall not begin until a scheme for protecting the proposed dwellings from noise from the . . . . road has been submitted to and approved by the local planning authority; and all works which form part of the scheme shall be completed before any of the permitted dwellings is occupied.

Authorities should give applicants guidance on the extent of noise attenuation to be aimed at within or around the dwellings, so as to provide precise guidelines for the scheme to be submitted.

11. Aircraft shall not take off or land except between the hours of . . . . and . . . .

NB. Latest draft version of Circular 10/73 [10] may amend these conditions.

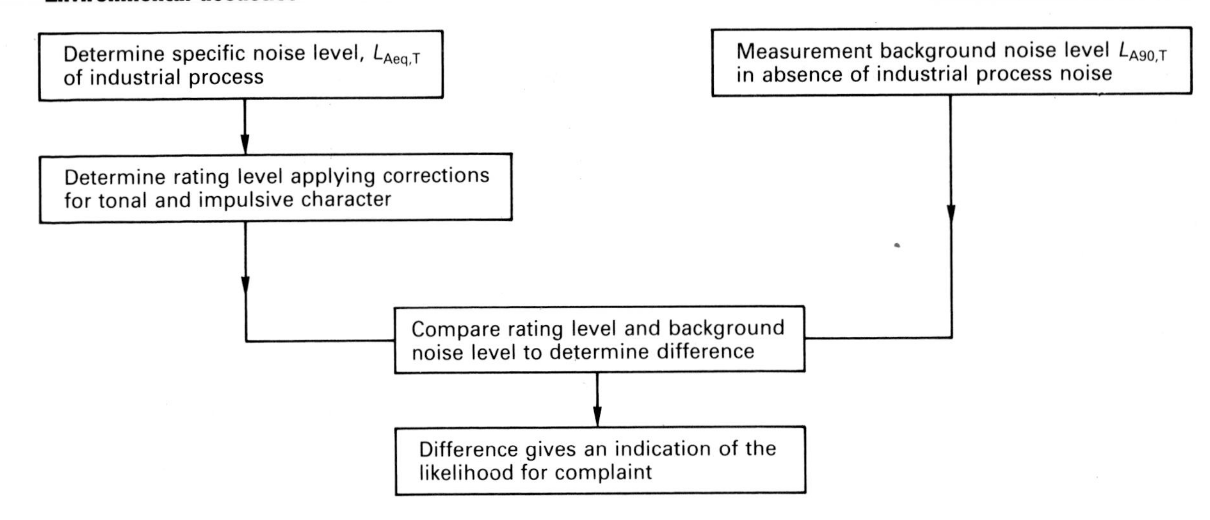

Diagram 1.8 *Industrial noise assessment procedure*

BS 4142: 1990 'Method for rating industrial noise affecting mixed residential and industrial areas' [20] describes methods for determining noise levels from factories, industrial premises, or fixed installations and sources of an industrial nature in commercial premises. The noise level determined in terms of $L_{Aeq,T}$ is corrected for tonal and impulsive character to establish the rating level. This rating level is then compared with the measured background noise level. Even where the noise climate is always affected by the industrial noise, it is possible to measure the background noise level at another location where it is presumed to be equivalent. The process is shown in Diagram 1.8.

The difference between the rating level and the background noise level indicates the likelihood of complaints. A difference of around 10 dB or more in $L_{Aeq,T}$ indicates that complaints are likely. A difference of around 5 dB is of marginal significance. At a difference below 5 dB, the lower the value the less the likelihood that complaints will occur. A difference of −10 dB is a positive indication that complaints are not at all likely. In assessing whether a particular process is causing a noise nuisance, the local authority would normally use BS 4142: 1990.

Prediction

As stated earlier, industrial noise sources can be of various forms and prediction of industrial installation noise requires a clear understanding of the processes involved. Ideally, noise levels should be measured during the operation of similar processes elsewhere. It may be necessary to take account of an increase in size or capacity of an operation. Only in the case of mineral extraction are there suitable data to enable reasonably accurate predictions to take place and these are provided in BS 5228, particularly Part 4.

Data obtained in this way can make the prediction exercise easier since it will normally only require a distance correction to be applied. Where a large number of sources are identified it may be necessary to prepare a computer model. This will require data relating to sound power levels and directivity and is likely to necessitate a three-dimensional model taking into account different heights above a ground level datum.

The sound power level data may need to be estimated by converting anticipated internal levels within a building to those radiating from an aperture or from wall elements or obtained for external sources such as transportation from references.

Noise control

There are a number of alternative methods of controlling noise from an industrial development including:

- attenuators in ductwork of ventilation or extraction systems
- enclosures around mechanical plant like fans and motors, selection of low-noise components
- cladding on ductwork
- silencers in pipework serving valves or engine exhausts
- building orientation during design to avoid openings like doors facing sensitive areas
- building construction
- hours of operation to avoid night-time/evening periods if possible
- methods of operation to avoid high levels of impact noise
- lining bins collecting metallic waste material

Implementation of any noise control needs to take account of other parameters, for example cooling needs and fire ratings, so requirements need to be checked for overall practicality with the manufacturers and/or the suppliers.

Leisure noise

Noise due to leisure activities can be considered as an extension to industrial noise. Leisure activities include discotheques, night clubs, cinemas, ten-pin bowling and clay pigeon shooting. In addition, certain apparently quiet activities such as ice skating can be a potential noise problem due to the accompanying use of a sound system.

Units are as those described for industrial noise, but due to the nature of leisure activity noise, it can be the maximum level which is the most important. Legislation and criteria too are primarily as given for industrial noise for the majority of sources, but there is additional guidance given for certain leisure activities in the form of draft codes of practice, some of which are being considered by the Department of Environment. Examples include:

- clay pigeon shooting [23]
- discotheques [24]
- power boat operation and water skiing [25]

In addition, a draft code for off-road motorcycle sports has been prepared and the Noise Council is preparing a code on pop concerts [26]. With regard to discotheques and pop concerts, the former Greater London Council [27, 28] produced guidelines in a Code of Practice for Pop Concerts which set out means of minimizing annoyance to occupiers near sites. These included controls as follows:

- The L_{Aeq} noise level during any 15-min period of the concert or rehearsals outside the window of their premises should not exceed the L_{Aeq} noise level measured during a comparable period when no concert or rehearsal is in progress by more than 10 dB between 07:00 and 20:00 hours and by more than 6 dB between 20:00 and 23:00 hours where outdoor concerts are held, on no more than 3 days/year, or by 1 dB between 07:00 and 23:00 hours where indoor or outdoor concerts are held on more than 3 days/year.
- No sound should be audible between 23:00 and 07:00 hours.

Prediction of noise due to various leisure activities is best dealt with by using data from existing activities and making correlations to take account of site-specific factors. As well as those methods identified for industrial noise control, including building construction and orientation to avoid doors and windows facing sensitive neighbours, it may be possible to electronically control the sound system output.

Groundborne vibration

Sources

In addition to the airborne noise levels caused by transportation, construction and industrial sources discussed above, there is also the generation of groundborne vibration to consider; this can lead to structureborne noise, perceived vibration or, in rare instances, building damage.

Typical causes of vibration are trains, vibratory rollers, piling equipment and possibly industrial presses or blasting. Unless there are any irregularities in the road surface (for example, 'sleeping policemen' or expansion joints), road vehicles cause relatively low levels of vibration. The ground conditions are important since the vibration levels can be amplified significantly if the soil is marshy/soft.

Units

Groundborne vibration is typically measured in terms of velocity (millimetres per second) or acceleration (metres per second per second). Where sources are impulsive or intermittent it is the peak particle velocity or acceleration which is measured and this is the maximum value recorded during an event.

General advice on the measurement of vibration in buildings is contained in BS 7385: Part 1: 1990 [29]. Vibration can either be considered in terms of the cause of possible building damage, where peak particle velocity is the preferred unit, or the effect on people where either velocity or acceleration can be considered.

In determining the overall vibration value, it is necessary to take account of the vibration levels in the three perpendicular directions, up–down, side-to-side, and front–back, as illustrated in Figure 1.13.

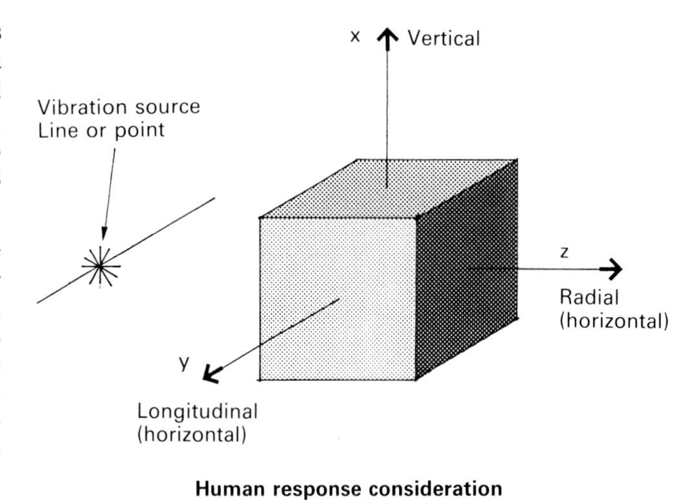

Figure 1.13 *Three directions of vibration measurement*

Legislation and criteria

There are no detailed UK guidelines on acceptable vibration levels to avoid structural damage, although BS 7385: Part 1 does quote general levels. The guidance given in German Standard DIN 4150 Part 3 [30] is therefore often used (refer to Table 1.11). As shown in Figure 1.14, vibration can be felt at levels well below those that could cause structural damage. BS 6472: 1984 [31]

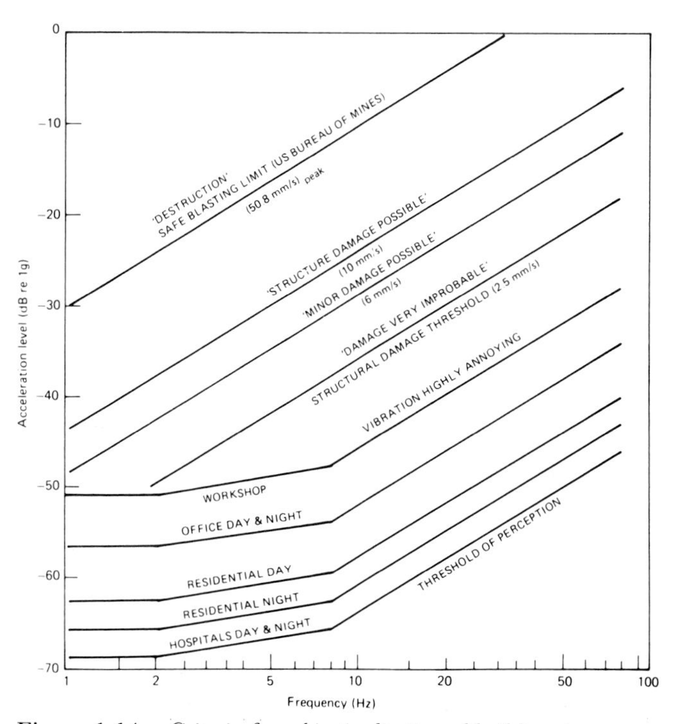

Figure 1.14 *Criteria for subjective limits and building damage*

Table 1.11 *Guideline values of vibration velocity, v_1, for evaluating the effects of short term vibration from DIN 4150 Part 3* [30]

	Type of structure	Vibration velocity, v_i (mm/s)			
		Foundation			Plane of floor of uppermost full storey
		At a frequency of			
		Less than 10 Hz	10–50 Hz	50–100 Hz[a]	Frequency mixture
1	Buildings used for commercial purposes, industrial buildings and buildings of similar design	20	20–40	40–50	40
2	Dwellings and buildings of similar design and/or use	5	5–15	15–20	15
3	Structures that, because of their particular sensitivity to vibration, do not correspond to those listed in lines 1 and 2 and are of great intrinsic value (e.g. buildings that are under a preservation order)	3	3– 8	8–10	8

[a]For frequencies above 100 Hz, at least the values specified in this column shall be applied.

provides guidance on satisfactory magnitudes of building vibration with respect to human response. The factors used to specify satisfactory magnitudes are given in Table 1.12. (Curves relating to these factors are shown in Figure 3.2.)

Complaints from building occupiers about excessive vibration are normally due to the belief that if the vibration can be felt then it is likely to cause damage. Door closure or footfall within buildings often cause levels well above those measured from the source under investigation. Consideration of structureborne noise is only likely to be needed for very sensitive areas such as auditoria, studios and conference meeting rooms. Spaces with windows to outside are unlikely to be of concern since there will be a reasonably high level of low frequency noise breaking into the space via the windows which will mask any structureborne noise. Establishing acceptable levels of structureborne noise in sensitive areas will need to be site specific.

Prediction

In predicting the likely vibration levels from a particular source, consideration needs to be given to:

- the type of source and its interaction with the ground,
- transmission through the ground,
- the likely response of the structure of the building under consideration.

It is very difficult to obtain accurate data on the above and it is essential to undertake measurements on site or in a location representative of site conditions. Reasonable data are available for such activities as blasting from the US Bureau of Mines without measurement but this should be added to by check measurements if possible.

Unless there is a similar building located at the same distance from the vibration source, it is not possible to measure the response of the building. A finite element analysis may be necessary to determine the likely response of the building.

Control

If there is a likelihood of structural damage then, obviously, an alternative form of the source of vibration needs to be found or possibly the structure could be reinforced. However, only rarely is it shown that structural damage is likely to be caused and the method of control normally available is the isolation of the recipient, typically:

- small sensitive equipment, for example optical balances
- entire buildings, for example concert halls
- individual rooms, for example studios.

Only occasionally is it possible to isolate the source, since it is out of the control of the developer, although this has been undertaken with railway tracks and some industrial sources such as presses. It may also be possible to control the operating time of the vibration source to avoid sensitive periods.

New developments as a noise source

In assessing the new development as a noise source, consideration may need to be given to undertaking an Environmental Assessment or simply to meeting a planning condition.

Environmental assessments

There is now a formal requirement in the UK and the rest of Europe for an Assessment of Environmental Effects and the preparation of an Environmental Statement to be undertaken for certain projects. The projects that require an assessment in every case are listed in Schedule 1 of the Town & Country Planning (Assessment of Environmental

Table 1.12 *Multiplying factors used to specify satisfactory magnitudes of building vibration with respect to human response (from BS 6472)* [31]. *Extracts from BS 6472: 1992 are reproduced with the permission of BSI. Complete copies of the standard can be obtained by post from BSI Publications, Linford Wood, Milton Keynes, MK14 6LE*

Place	Time	Multiplying factors	
		Continuous vibration (see note 2)	Intermittent vibration and impulsive vibration excitation with several occurrences per day
Critical working areas (e.g. some hospital operating theatres, some precision laboratoreis, etc.)	Day	1	1
	Night	1	1 (see note 3)
Residential	Day	2 to 4 (see note 4)	60 to 90 (see notes 4, 5, 6)
	Night	1.4	20
Office	Day	4	128 (see note 7)
	Night	4	128 (see note 7)
Workshops	Day	8	128 (see notes 7 and 8)
	Night	8	128 (see notes 7 and 8)

Note 1. This table leads to magnitudes of vibration below which the probability of adverse comment is low (any acoustic noise caused by structural vibration is not considered).
Note 2. Doubling of the suggested vibration magnitudes may result in adverse comment and this may increase significantly if the magnitudes are quadrupled (where available, dose/response curves may be consulted).
Note 3. Magnitudes of impulse shock in hospital operating theatres and critical working places pertain to periods of time when operations are in progress or critical work is being performed. At other times magnitudes as high as those for residences are satisfactory provided there is due agreement and warning.
Note 4. Within residential areas people exhibit wide variations of vibration tolerance. Specific values are dependent upon social and cultural factors, psychological attitudes and expected degree of intrusion.
Note 5. The 'trade off' between number of events per day, their magnitudes and durations is not well established. In the case of blasting, and for more than three events per day, the following provisional relationship can be used to modify the factors for residences in column 4 of this table. This involves further multiplying by the factor $F = 1.7 N^{-0.5} T^{-d}$ where: N is the number of events in a 16-h day; T is the duration of the impulse and decay signal for an event in seconds.

The duration of an event can be estimated from the 10% (−20 dB) points of the motion time histories.

d = zero for T less than 1 s.

for short duration stimuli there is evidence that for wooden floors the human response $d = 0.32$ and for concrete floors $d = 1.22$.

This 'trade off' equation does not apply when values lower than those given by the factors for continuous vibration result.
Note 6. The root mean quad (r.m.q. = $(1/T \int_0^T a^4(t)dt)^{1/4}$) of the weighted acceleration signal $a(t)$ may be used as an alternative method of assessment for impulsive events. The same relation between duration and acceleration may be used to accumulate the exposure to intermittent vibration occurring throughout the day (i.e. accumulated value = $\int_0^t a^4(t)dt$). The value obtained by this method, which should be related to the boundaries for continuous vibration, allows greater magnitudes with shorter and/or less frequent periods of intermittent vibrations.
Note 7. The magnitudes for impulsive shock excitation in offices and workshop areas should not be increased without considering the possibility of significant disruption of working activity.
Note 8. Vibration acting on operators of certain processes such as drop forges or crushers, which vibrate working places, may be in a separate category from the workshop areas considered in this table. The vibration magnitudes specified in relevant standards would then apply to the operators of the exciting processes.

Effects) Regulations 1988 [32] and include oil refineries, power stations, waste disposal and chemical installations, and large transportation schemes. Schedule 2 of the Regulations identifies projects which require an assessment if they are 'likely to have significant effects on the environment by virtue of factors such as their nature, size, and location' and include developments such as mineral extraction, industrial complexes, food industry, infrastructure projects, and holiday villages or hotel complexes. If the project under consideration is deemed by the local planning authorities to be included in either Schedule 1 or 2 then it is likely that an acoustics appraisal would need to be included as part of the assessment. In some cases, the local authority may consider that only an acoustics appraisal is needed. The method of approach is shown in Diagram 1.9; schedules 1 and 2 are summarized in Table 1.13.

The assessment for an Environmental Statement will need to consider the construction and operational phases of the development, including likely transport movements. It will also need to include those elements of acoustic appraisal identified earlier, particularly a site survey. Some site uses in sensitive areas will not be acceptable even with noise control measures, and planning permission for a new development may be refused on environmental grounds.

Planning conditions

Noise emission from the development, or, in the case of affected dwellings, sound insulation measures, will be issues

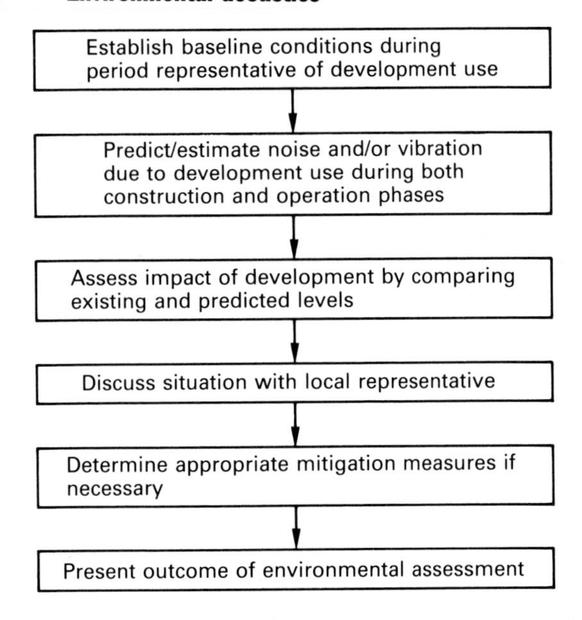

Diagram 1.9 *Environmental Assessment: acoustics appraisal procedure*

Table 1.13 *Descriptions of development*

Schedule 1

Those developments requiring assessment:

Refineries
Power stations
Radioactive waste stores
Works for initial melting of cast iron and steel
Asbestos extraction or processing installations
Integrated chemical installations
Special roads, long-distance railway tracks, or airports
Trading ports
Waste disposal installations
Land-fill sites

Schedule 2

Many projects require assessment in the following categories:

Agricultural processes
Extractive industries
Energy industries
Metal processing
Glass making
Chemical industries
Food industries
Textile, leather, wood and paper industries
Rubber industries

Infrastructure projects including industrial estate development, urban development, and road, harbour, or airport construction not falling into Schedule 1 projects

Other projects including holiday villages or hotel complexes

Modifications to any projects previously in Schedule 1.

covered by conditions imposed as part of the granting of planning permission.

Conditions imposed upon a planning agreement may also refer to the construction and operational phases and will vary from district to district depending on local parameters. Local planning authorities may impose 'standard' conditions as part of the issuing of planning permission. Table 1.10 identifies model conditions which were briefly discussed under 'Industrial Noise'. These should be checked carefully and the questions that need to be asked are: are they clear in intent or too 'catch all'? Is the local authority being reasonable in allowing normal activity and plant noise related to the site use for which permission is granted, or related to existing site noise levels, or is it impossible to operate normally without falling foul of noise limits set arbitrarily too low? Has the design team to demonstrate by a prediction statement that the limits set will be met? Does a Condition apply only to 'fixed plant' or also to traffic movements and activity noise (for example, discotheque sound systems)? Do the same standards apply to intermittent noise sources like standby generators?

The setting of noise control standards at the planning stage and their subsequent observance by the developer or owner does not preclude an individual taking legal action in common law to abate noise nuisance arising from the development. The developer who undertakes a 'shell' major development with subsequent fitting-out by tenants, should be careful in obtaining consents for the total development and agreeing contracts that tenants' plant complies with any planning condition and indeed has been allowed for as a contributing component of noise breakout to the local community.

Construction phase assessment

Reference earlier to dealing with construction activities should provide sufficient information to enable a suitable assessment to be made, However, once the hours of working and the type of machinery to be used are agreed, i.e. the best practicable means have been employed, there could be a requirement for a substantial degree of public relations activity to satisfy potential complainants.

Operational phase

If the development is industrial then adequate guidance may be obtained from earlier guidance dealing with industrial noise. This is also likely to be the case for those developments with mechanical plant for building services as their only noise source – a topic covered in detail in Chapter 3.

Suitable criteria have to be set at the most appropriate locations, normally the site boundary, and discussions with the Environmental Health Departments of the local authorities should lead to an agreement.

In the case of steady noise from mechanical plant, a noise level criterion matching the existing background noise level is normally appropriate. Consideration may need to be given to a criterion 5 dBA below the background noise level in some cases to avoid any significant increase in a particularly sensitive area, but 10 dBA below the background should never be necessary.

Noise from other activities such as leisure, for example night clubs or ten-pin bowling, will need to be carefully considered since music or impulsive noise can be annoying even at noise levels below the existing background. In some cases the only means of control may be a restriction on hours of use.

If there are noise problems once a development is operational, the first line of complaint is frequently the local authority Environmental Health Officer. He may have been asked to comment by the planning authority at planning stage on whether a noise condition should apply, and reacts to the later complaints by carrying out his own checks. If he agrees that the complaint is reasonable and may constitute a Nuisance, he can issue an 'Abatement Notice in respect of Noise Nuisance' under the provisions of the Environmental Protection Act 1990 which gives 28 days to remedy, restrict, or stop the noise. The action to improve matters may vary from 'best practical means', i.e. achieving as much as costs and practicalities allow to provide some amelioration, to radical noise control measures or proscribing activity or plant noise during certain times or even altogether. Once a Notice is served, an Appeal can be made against the basis of the alleged occurrences constituting the statutory nuisance. However, the Notice provisions are not suspended until the appeal court so decides.

Alternatively, a Court Order to restrain a continuation of a wrongful act or omission may be initiated. This is the most common legal action in private nuisance claims. Failure to comply could mean fines and even imprisonment. Unlike a Notice, the restraint requirement is immediate. Further details can be found in *Control of Pollution Encyclopaedia* [33].

References

1. *This Common Inheritance*, Government White Paper on the Environment, 1990
2. Department of the Environment *Report of the Noise Review Working Party 1990*, Batho Report, HMSO, London, 1990
3. Wimpey Laboratories Ltd *Residues and Emissions: Sound and Vibration*, Report No. 15, CTG Channel Tunnel Project – Environmental Impact Assessment, Hayes, Middlesex, September 1985
4. BS 6698: 1986 (amd 1991) *Specification for integrating-averaging sound level meters*, British Standards Institution, Milton Keynes
5. BS 5969: 1981 *Specification for sound level meters*, British Standards Institution, Milton Keynes
6. Organisation for Economic Co-operation and Development, *Fighting Noise in the 1990s*, OECD Publications Service, Paris, 1991
7. The Department of Transport *Calculation of Road Traffic Noise*, HMSO, London, 1988
8. Statutory Instrument 1975 No 1763 *Building and Buildings Noise Insulation Regulations*, HMSO, London, 1975
9. Land Compensation Act 1973, HMSO, London, 1973
10. Department of the Environment *Planning and Noise*, Circular 10/73, 1973
11. The Noise Advisory Council *A Guide to Measurement and Prediction of the Equivalent Continuous Sound Level* L_{eq}, HMSO, London, 1978
12. Shield, B. M. and Zhukov, A. N. *A survey of annoyance caused by noise from the Docklands Light Railway*. Institute of Acoustics Proceedings, **13**(5), pp. 15–23 (1991)
13. Nelson, P. (ed.) *Transportation Noise Reference Book*, Butterworth-Heinemann, Oxford, 1987
14. Greater London Council *GLC Guidelines for Environmental Noise and Vibration*, GLC Scientific Branch, London, Bulletin No. 96, Item 6, June 1976
15. US Environmental Protection Agency Office of Noise Abatement and Control *Information on Levels of Environmental Noise Required to Protect Public Health and Welfare with an Adequate Margin of Safety*, US Government Printing Office, March 1974
16. BS 8233: 1987 *Code of practice for sound insulation and noise reduction for buildings*, British Standards Institution, Milton Keynes
17. Control of Pollution Act 1974
18. BS 5228: 1984/1986/1992 *Noise control on construction and open sites*, British Standards Institution, Milton Keynes
19. Department of the Environment *Control of Pollution Act 1974, Implementation of Part 3: Noise*, Circular 2/76, HMSO, London, 1976
20. BS 4142: 1990 *Method of rating industrial noise, affecting mixed residential and industrial areas*, British Standards Institution, Milton Keynes
21. Environmental Protection Act 1990
22. Department of the Environment *The Use of Conditions in Planning Permissions*, Circular 1/85, HMSO, London, 1985
23. *Code of practice on Noise from Clay Pigeon Shooting*, Midland Joint Advisory Council for Environmental Protection, 1989, revised 1991
24. Noise Advisory Council, *Draft Code of Practice on Sound Levels in Discothéques*, HMSO, 1986
25. Department of the Environment *Draft Code of Practice on Noise from Power Boat and Water Ski Racing*, 1988
26. The Noise Council *Code of Practice for Environmental Noise Control at Open Air Pop Concerts*, London, 1991 (draft document issue)
27. Greater London Council *Code of Practice for Pop Concerts – A Guide to Safety, Health & Welfare at One-Day Events*, Third Edition, London, 1985
28. Greater London Council, *Disco Rules OK?*, London, 1979
29. BS 7385: Part 1: 1990 *Evaluation and measurement for vibration in buildings: guide for measurement of vibrations and evaluations of their effects on buildings*, British Standards Institution, Milton Keynes
30. Deutsches Institut für Normung e.V. 4150: 1986: Part 3 *Structural vibration in buildings: effects on structures*, Berlin
31. BS 6472: 1992 *Guide to evaluation of human exposure to vibration in buildings (1 Hz to 80 Hz)*, British Standards Institution, Milton Keynes
32. Town and Country Planning (Assessment of Environmental Effects) Regulations 1988, SI 1988 No. 1199, HMSO, London, 1988
33. Garner, J. F., Harris, D. J. (eds) *Garner's Environmental Law*, Butterworth-Heinemann, Oxford, first published July 1942, updated three times p.a.

Chapter 2 Design acoustics

Duncan Templeton

Sound insulation

Airborne sound insulation

Airborne sound insulation entails the separation by a physical barrier of the space containing a noise source from an adjacent space requiring protection. The physical barrier may be a partition or wall between rooms or floor for rooms above each other (Diagram 2.1). The term 'partition' implies dry construction (modular panels or clad studded frames) or lightweight, non-loadbearing masonry; 'wall' implies a structure tied into other walls and supporting floors or roof above. There is, however, overlap between the terms. The two systems have differing characteristics acoustically. Partitions may be more subject to edge cracking, flanking and resonance effects. Walls can transmit structureborne sound more than studding-type partitions, or pass re-radiated energy, and without special detailing are more difficult to make discontinuous.

Sound waves in the air on the source side impinge on the partition and 'drive' it (i.e. cause it to vibrate), and in turn the barrier radiates sound into the receiving space. The main issues are the transmission path, the sound-reducing capability of the separating structure, and the components making up this structure. The practical reduction of airborne sound energy is not only dependent on the direct path via the wall, but also on flanking paths.

Flanking paths

Flanking paths occur at the edges of the physical barrier, its junction to floors, other walls, ceiling, or ductwork common to source and receiving rooms (Figure 2.1). The indirect transmission via a flanking path can be reduced by increasing the mass of the flanking walls, increasing the partition mass and bonding it to flanking elements, or introducing discontinuity to side walls in the adjacent rooms, for example by independent wall lining or floating floors.

Sound level difference

The sound level difference between two spaces is dependent on the sound-reducing capability of unit area of the partition, the area of the partition, the acoustic properties of the source and receiving spaces, and flanking effects (Figure 2.2 and Diagram 2.2).

Room-to-room sound level difference. This is defined in Chapter 5. It is often useful in a sequence of measurements for spaces

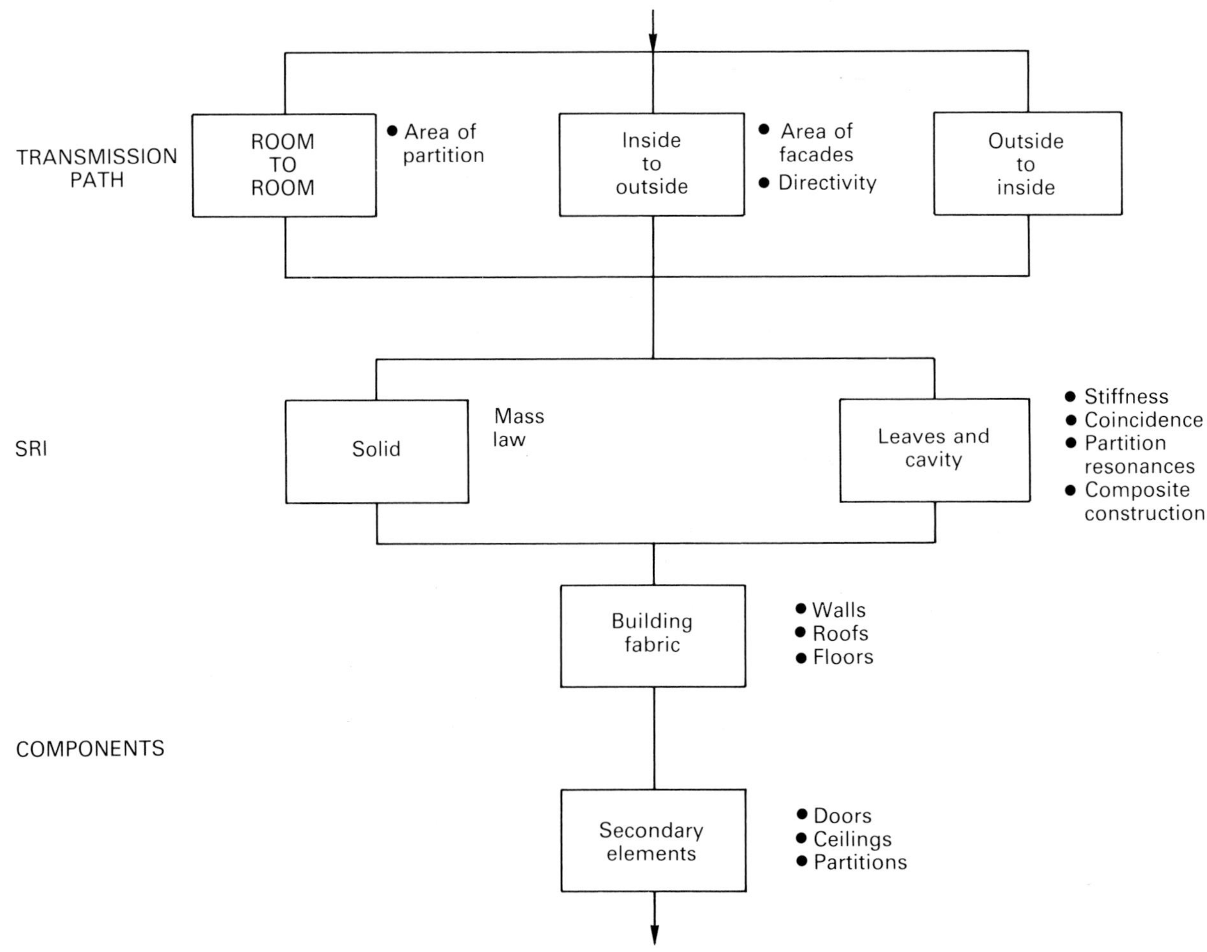

Diagram 2.1 *Sound insulation: considerations in design*

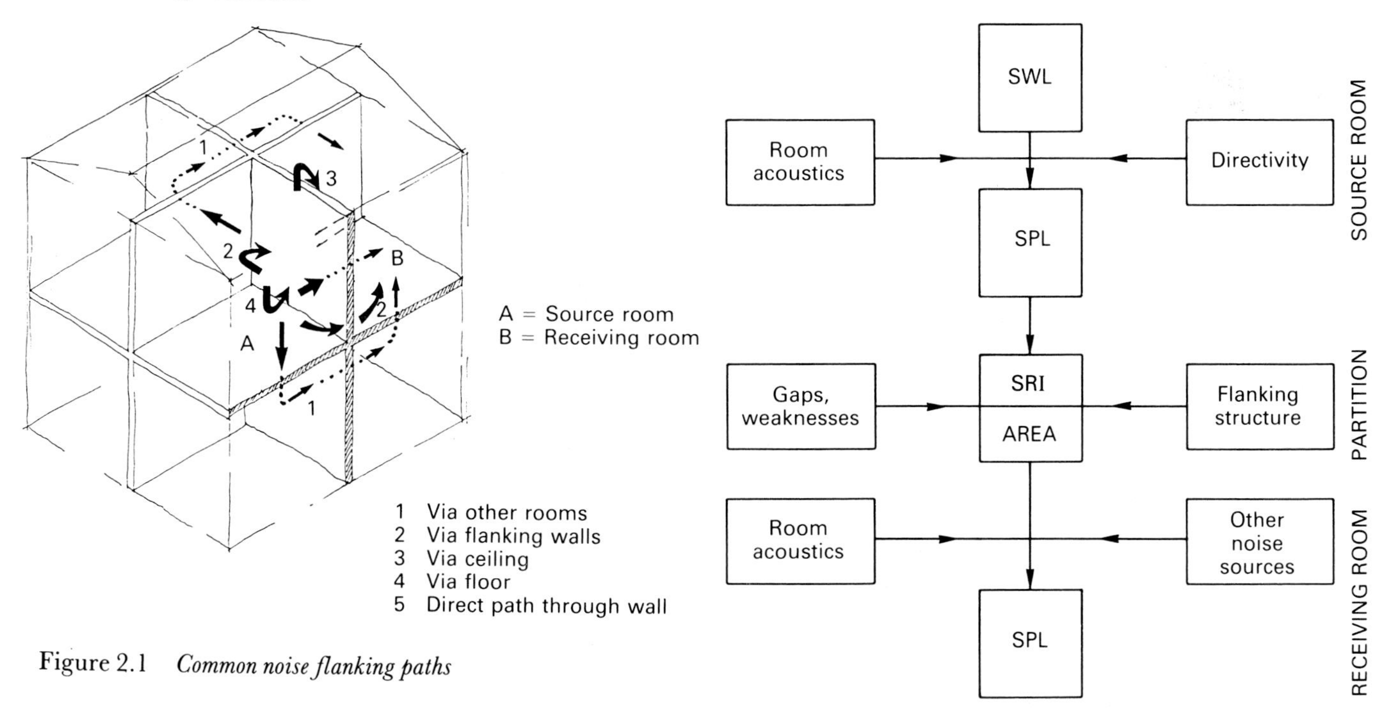

Figure 2.1 *Common noise flanking paths*

Diagram 2.2 *Separation between rooms*

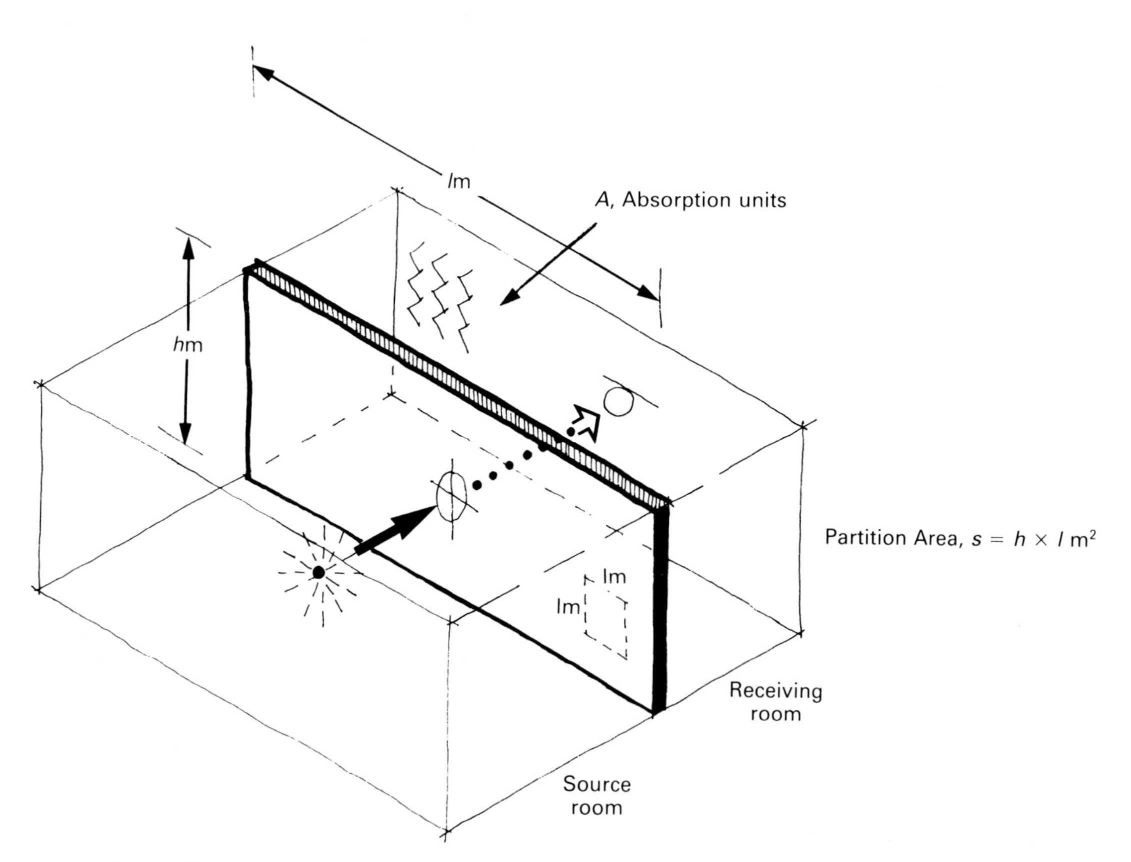

Figure 2.2 *Sound level difference between rooms*

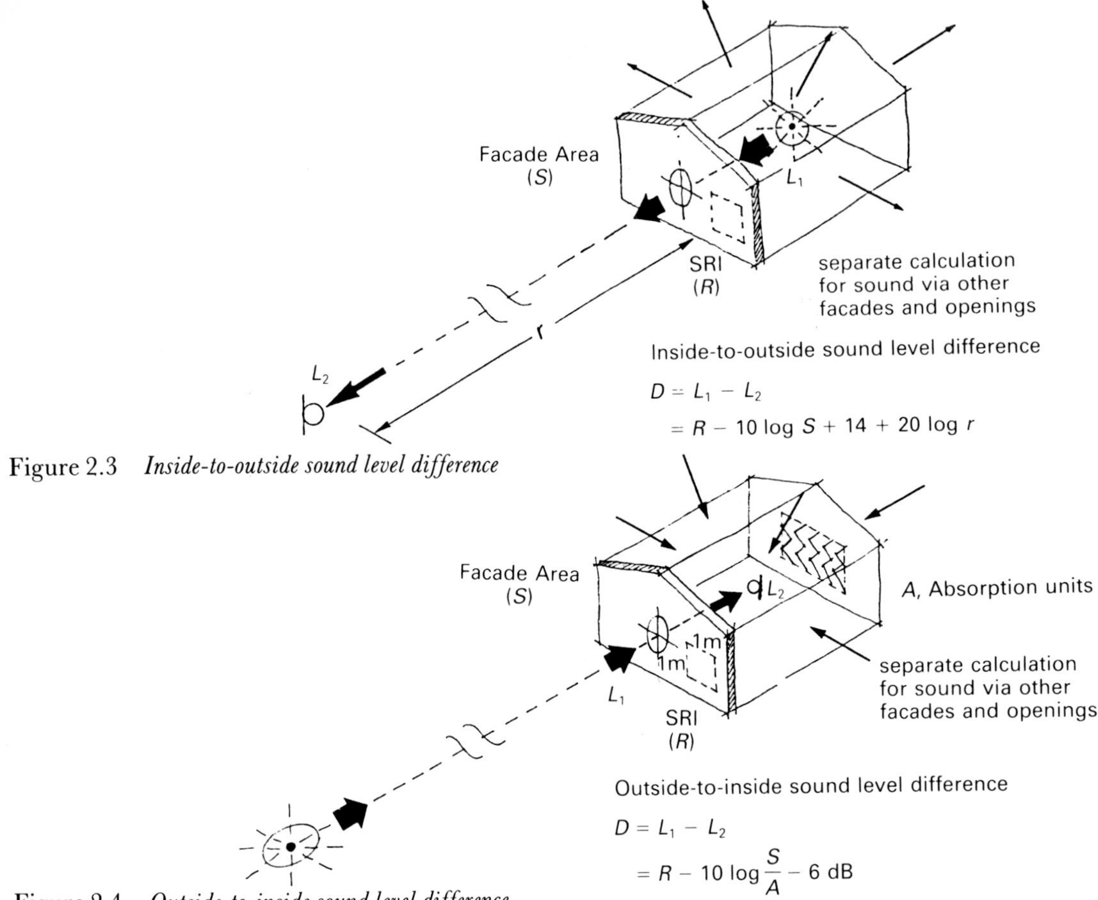

Figure 2.3 *Inside-to-outside sound level difference*

Figure 2.4 *Outside-to-inside sound level difference*

at different stages of fit-out to have a level difference standardized to a reference half second reverberation time. The Weighted Standardized Level Difference ($D_{nT,w}$) is defined in BS 5821: 1984 [1] and referred to in the Building Regulations Part E: 1992 [2].

If two spaces are similar in acoustical character, i.e. have the same amount of absorption in both the source room and receiving room, the measured level difference will be the same whichever is the source room. If one space is much 'deader' than the other, or the spaces are of similar reverberation time but greatly different in volume, the level difference will be greater with the deader (or larger volume) space as the receiving room. This is not an anomaly although a partition apparently more effective for sound in one direction than in the other can require some explanation to a client. In fairly dead spaces, for example cinema auditoria or studios, a true reverberant field in the receiving room may not be generated by the source in the room adjacent and there will be a gradient of sound level away from the separating partition.

Inside-to-outside sound level difference. This is given by:

$$D = R - 10 \log S + 14 + 20 \log r$$

for facade radiation to hemisphere, where S is the outside wall area and r is the radius from the facade centre to the receiving point, or in other situations as defined in Chapter 5 (Figure 2.3). In practice, an assessment of noise break-out from say a factory building requires a sequence of calculations involving the contributions of sound via the roof, other facades, doors and windows. Little useful information exists about sound radiation via angled roofs. In most cases, it is the openings in industrial buildings that determine the noise break-out to adjacent sites.

Outside-to-inside sound level difference. This is given by:

$$D = R - 10 \log S/A - 6 \text{ dB}$$

This assumes that the measuring microphone is well away from the facade. As with break-out, break-in calculations should analyse components of noise from the other faces and roof (Figure 2.4). While separate checks follow for break-in and break-out, both may be of concern on some jobs; for example, a hospital may be considered a noise-sensitive building type but has significant 24-h noise from plant and activity, which may upset nearby housing (Diagrams 2.3 and 2.4).

Sound reduction index

The sound reduction index (SRI) is the basic measure of sound insulation and is the number of decibels that sound power is reduced by transmission through the barrier. The average sound reduction index is usually expressed over 100–3150 Hz one-third octave bands, and will be similar to the single value at 500 Hz (see Chapter 5).

Average sound insulation index rating, R_w

The average sound insulation index rating, R_w, is the weighted single-figure descriptor defined in the Glossary. As with the A-weighting, greater significance to mid and high frequencies is given than for, say, the direct arithmetic

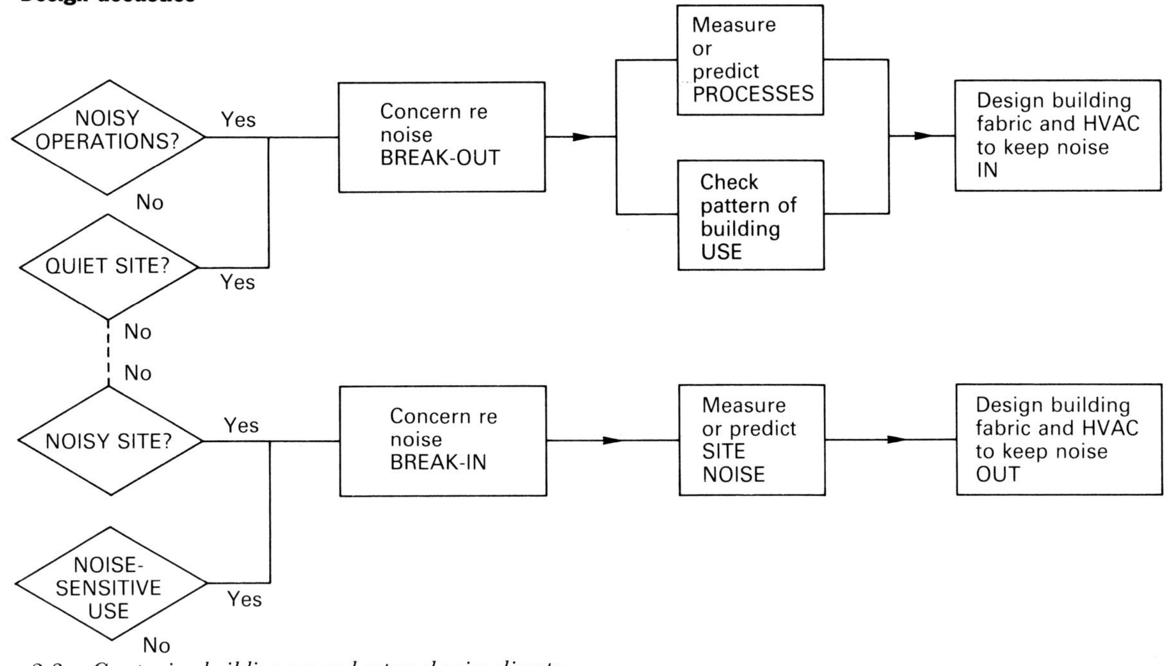

Diagram 2.3 *Comparing building use and external noise climate*

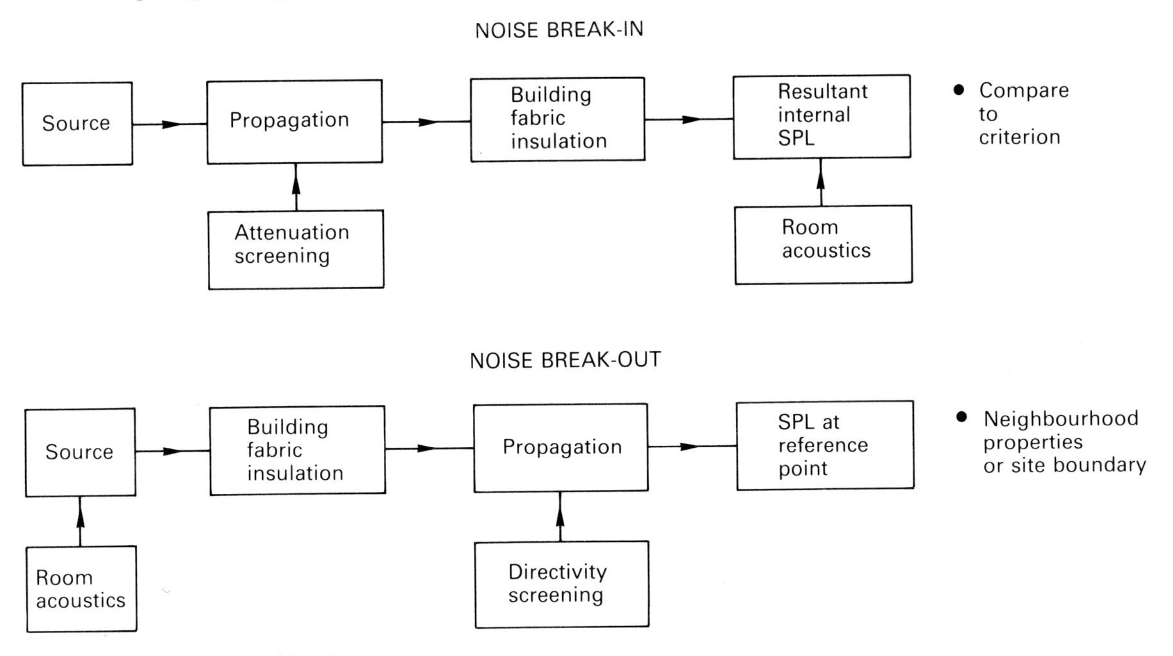

Diagram 2.4 *Noise break-in and break-out*

average of the SRIs. This more accurately reflects the subjective effect of insulation due to the ear's reduced acuity at lower frequencies.

Mass law

The mass law indicates that in theory sound insulation increases by 6 dB for every doubling of weight of dividing element per unit thickness. The sound insulation also theoretically increases by 6 dB per octave. It may be seen from Figure 2.5 that the empirical mass law curve based on results is below the theoretical curve due to coincidence and resonance effects, and approximates to 5 dB per mass doubling. The theoretical sound insulation over the frequency range 100–3150 Hz is given by:

$$R = 7.6 + 20 \log M \text{ dB}$$

where M is in kg/m^2; alternatively, to determine the theoretical performance at a particular frequency:

$$R = 20 \log (fM) - 47 \text{ dB}$$

where f is the frequency of the incident sound.

An example of the effect of increasing weight may be seen by the performance of a brick wall. A single-leaf brick wall may be rated 45 dB average, a 225-mm wall 50 dB, but it takes a thickness of 450 mm to achieve 55 dB, and this performance may well be compromised by edge flanking effects. The performances of typical constructions are scheduled in Table 2.1. A schedule of densities (kg/m^3) of common building materials may be found in BS 648 (1964, much amended since) [3].

Table 2.1 *Sound reduction indices*

	kg/m²	OBCF (Hz) 125	250	500	1000	2000	3000	Mean[a]
Single glazing (mm)								
4-mm glass in aluminium frame, 100-mm opening		10	10	11	12	12	13	11
4 mm	10	20	22	28	34	34	29	28
6 mm	15	18	25	31	36	30	38	29
6.4 mm laminated		22	24	30	36	33	38	30
12 mm	30	26	30	35	34	39	47	35
19 mm	49	25	31	30	32	45	47	35
Double glazing: glass/air space/glass (mm)								
Sealed units								
3/12/3		21	20	22	29	35	25	25
4/12/4		22	17	24	37	41	38	30
6/12/6		20	19	29	38	36	46	30
4/12/12		25	22	33	41	44	44	35
6/12/10		26	26	34	40	39	48	34
6/20/12		26	34	40	42	40	50	39
6.4 lam/12/10		27	29	37	41	42	53	38
Separate panes								
6/150/4		29	35	45	56	52	51	44
6/200/6		37	41	48	54	47	47	46
4/200/4		27	33	39	42	46	44	39
4/200/4, opposite sliders open 25 mm		15	23	34	32	28	32	27
4/200/4, opposite sliders open 100 mm		10	16	27	25	27	27	22
Masonry/blockwork								
102-mm single-leaf fairfaced		36	37	40	46	54	56	45
Single-leaf plastered both sides	240	34	37	41	51	58	60	47
Cavity brickwork with ties	480	34	34	40	56	73	76	52
Double leaf brickwork plastered both sides	480	41	45	48	56	58	60	51
100-mm lightweight blockwork fairfaced	125	32	32	33	41	49	57	41
100-mm blockwork plastered both sides		32	34	37	45	52	57	43
100-mm blockwork with plasterboard on dabs both sides		28	34	45	53	55	52	45
200-mm fairfaced light weight blockwork	250	35	38	43	49	54	58	46
200-mm blockwork plastered both sides		37	39	46	53	57	61	49
200-mm blockwork plasterboard on dabs both sides		33	39	50	55	56	60	49
Three-leaf brickwork plastered both sides	720	44	43	49	57	66	70	55
Two leaves of 100-mm dense concrete blocks, 50-mm cavity, 13-mm plaster both sides, cavity ties		35	41	49	58	67	75	52
Stud partitions								
9-mm plasterboard on 50 × 100 mm studs at 400 mm centres		15	31	35	37	45	46	35
13-mm plasterboard on 50 × 100 mm studs at 400 mm centres		25	32	34	47	39	50	38
13-mm plasterboard on 50 × 100 mm studs at 400 mm centres, 25 mm mineral wool between studs		25	37	42	49	46	59	43
6-mm ply on 50 × 50 mm studs at 600 mm centres		10	14	22	28	42	42	26
Double 13-mm plasterboard on 146-mm steel studs at 600 mm centres		32	41	47	49	53	58	47
Sheet materials/boards								
9-mm ply on frame	5	7	13	19	25	19	22	18
25-mm T&G timber boards	14	21	17	22	24	30	36	25

Table 2.1 (cont.)

	kg/m^2	OBCF (Hz) 125	250	500	1000	2000	3000	Mean[b]
5-mm ply/1.5-mm lead/5-mm ply composite sheets	25	26	30	34	38	42	44	36
Two layers of 13-mm plasterboard	22	24	29	31	32	30	35	30
1.2-mm steel sheet, 18 g	10	13	20	24	29	33	39	26
6-mm steel plate	50	27	35	41	39	39	46	38
Profiled metal sheeting		18	20	21	21	25	25	22
0.8-mm steel trapezoidal section, 50-mm deep cladding panels		14	17	18	20	29	31	22
Duct cladding: plaster/mineral wool	30	11	13	12	12	12	21	12[b]
Duct cladding: lead foil/mineral wool	12	7	8	7	7	7	7	7[b]
50-mm woodwool slabs, screeded to source side	28	26	28	30	32	33	36	30
100-mm woodwool slabs, screeded to source side	50	28	28	32	34	33	38	31
Doors								
43-mm flush, hollow-core door, normal hanging	9	12	13	14	16	18	24	16
43-mm solid core door, normal hanging	28	17	21	26	29	31	34	26
50-mm steel door with good seals		21	27	32	34	36	39	32
Acoustic metal doorset, double seals		36	39	44	49	54	57	47
Floors								
235-mm T&G floorboards, floor joists, 13-mm plasterboard and skin	31	18	25	37	39	45	45	35
235-mm T&G floorboards, floor joists with 50-mm sand between, 13-mm plasterboard and skin		35	40	45	50	60	64	49
100-mm reinforced concrete slab	250	37	36	45	52	59	62	49
200-mm reinforced concrete slab	460	42	41	50	57	60	65	53
300-mm reinforced concrete slab	690	40	45	52	59	63	67	54
200-mm o/a: 125-mm concrete slab and screed on 13-mm nominal glass fibre	420	38	43	48	54	61	63	51

[a]Average 125–4000 Hz octaves. SRI (100–3150 Hz) 0–2 dB lower.
[b]+ value on duct performance.

Double leaves
Double leaves with a gap between allow greater sound insulation than a single layer of equivalent weight. There are two main means of transmission:

- radiation from the first panel into the air space excites the second panel, which radiates energy into the receiving room;
- structureborne transmission between the two leaves by mechanical links, the second leaf radiating the transmitted vibrational energy.

Sound-absorbing quilt in the gap improves the sound insulation. A relatively small provision of absorption is effective because it suffices to ensure that the transmission via radiation is less than the structureborne transmission. Although it contributes little to the total surface mass, it soaks up sound crossing the gap and standing waves of sound within the cavity. The improvement due to absorption is typically at lower to mid frequencies, 200–800 Hz. However, there may be a high frequency benefit in practice also, because the quilt has an attenuating effect on sound via weaknesses at partition leaf junctions.

The schematic sound insulation related to frequency is increased from single-panel 6 dB per octave slope to 12 dB by double leaves plus absorption, so the *overall* effect is greater improvement at higher frequencies.

The sound reduction index will never rise to the arithmetic sum of the SRIs of the individual leaves because the two leaves can never become totally isolated. However, it is usual to obtain a higher SRI from a double-skin construction than from the equivalent-weight single skin. The principle applies for horizontal or vertical dividing elements, i.e. roofs and floors as well as partitions. For a partition of two leaves of like mass (in kg/m^2) with separation of leaves d (in metres),

$$f = \frac{85}{\sqrt{md}}$$

The resonant frequency f can be arranged to fall below the frequency range of interest, say below 50 Hz, by choosing a high value of d in relation to m. Some estimate of the average field sound reduction index can be obtained from the expression:

$$R = 34 + 20 \log md$$

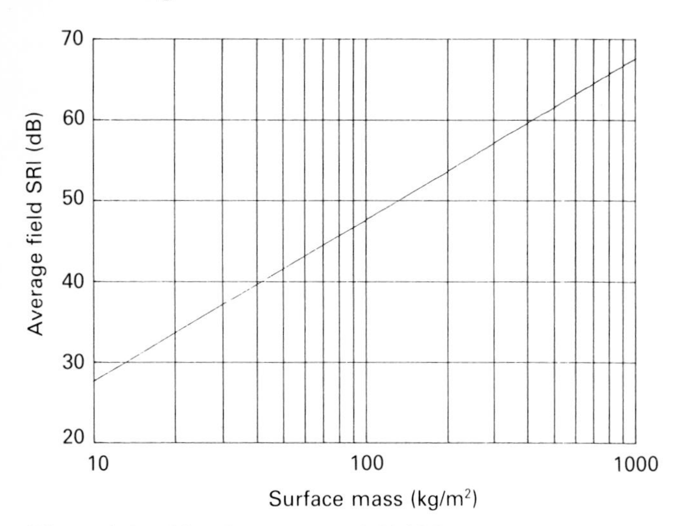

Figure 2.5 *Mass law: average field SRI*

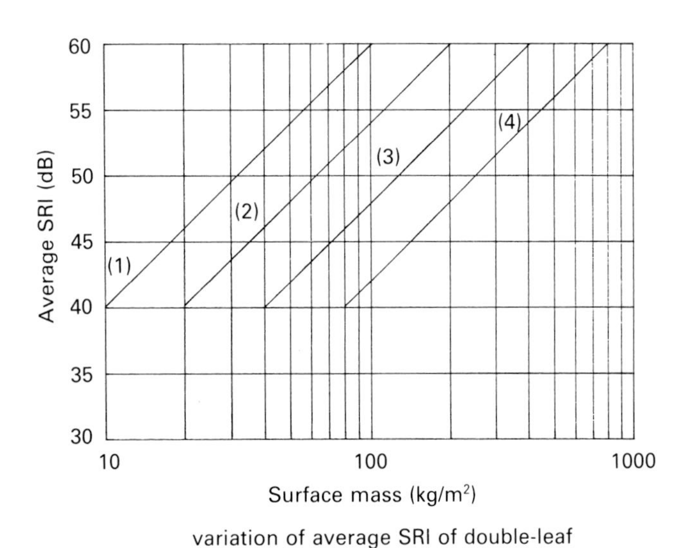

variation of average SRI of double-leaf partition with total surface mass

1 200 mm cavity
2 100 mm cavity
3 50 mm cavity
4 25 mm cavity

Figure 2.6 *Double-leaf construction: effect on SRI*

The effect is shown graphically in Figure 2.6. The minimum useful gap is 50 mm and wide cavities improve the low frequency performance. Different characteristics (weight, thickness) to one leaf offers a further improvement, as the many tests on double glazing combinations demonstrate. Cavity walls in brickwork with ties offer negligible benefit to solid 225-mm construction. A good example of the performance possible is the use of wide-cavity separating walls in multiplex cinemas where double 15-mm plasterboard either side of a 250-mm gap with 100-mm quilt inside consistently achieves 65–70 dB $D_{nT,w}$.

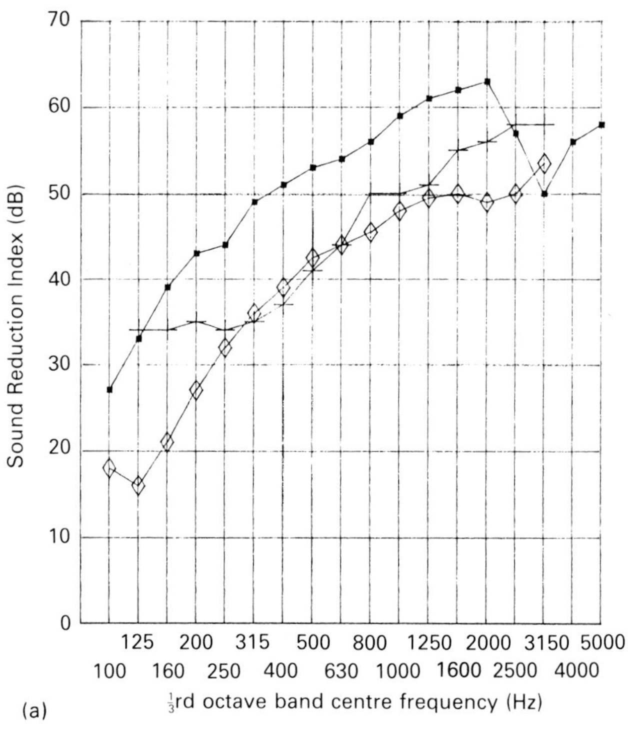

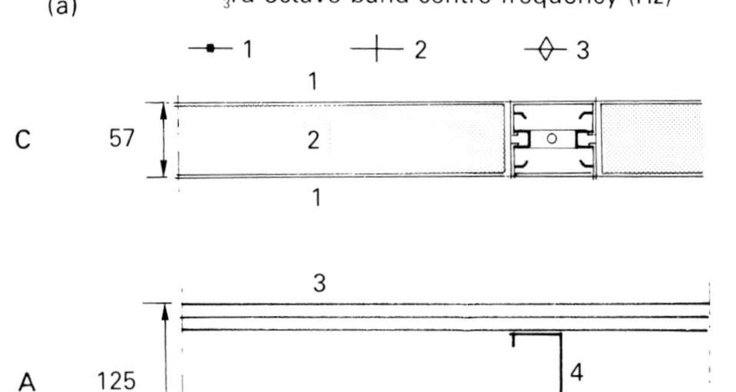

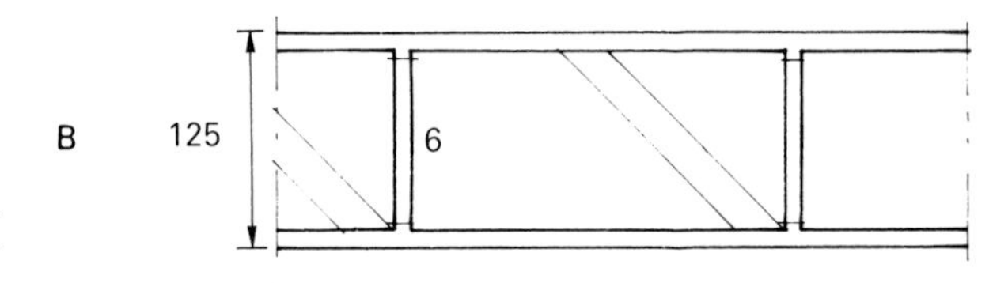

1 0.7 mm steel skin
2 60 mm mineral wool (90 kg/m³)
3 2 × 13 mm plasterboard
4 Metal studs at 600 mm cs
5 25 mm glass fibre
6 Single-leaf common brickwork, plastered both sides (1920 kg/m³)

(b)

Figure 2.7 *(a) Dry lining versus masonry; (b) masonry versus dry construction*

Dry construction versus masonry

Dry construction versus masonry is a frequent design comparison (Figure 2.7). A plastered block wall can be replaced by double-layer plasterboard with quilt inlay at a fraction of the weight, to achieve 50 dB average SRI, but more care is required at the edge and at penetrations. In dry construction, if substantial acoustic doors are used, a structural steel 'H' subframe bolted to the floor and the underside of the floor above should be used to hold the doorset firm; standard partition metal studs allow too much flexing. This has to be borne in mind in any cost comparisons between systems. The acoustic integrity of dry partitioning is more easily compromised by services penetrations, sockets and fixtures.

Stiffness
The stiffness of thin panels is important because of the susceptibility of leaves to be more easily driven by a noise source on one side at certain frequencies. The effect can be seen in duct systems where thin duct walls may easily transmit low frequencies of in-duct sound.

Coincidence effect
The coincidence effect happens when sound waves falling on a panel excite bending waves in it, the velocity of which depends on frequency. Sound transmission is greater at the frequencies where the coincidence effect is greatest and the theoretical R is reduced by as much as 10 dB below the level derived from mass law calculation (see Chapter 5 for full description).

Partition resonances
Partition resonances happen when standing waves are formed within the partition. At the frequencies at which this occurs the resonances will reduce performance. In a single-leaf partition or wall, the fundamental resonant frequency is determined by its stiffness. At higher frequencies, there are other performance 'dips' at harmonics of the resonant frequency. For many partitions the resonances occur at low frequencies outside the range of usual interest and the effect can often be ignored. However, it may be of interest when, for example, checking a specific curtain wall glazing arrangement for low frequency components of traffic noise. The most important is the fundamental resonant frequency, calculated from:

$$f = \frac{\pi t}{2}\sqrt{\frac{E}{12\rho}}\left[\frac{1}{a^2} + \frac{1}{b^2}\right]$$

where a and b are the partition dimensions (m), t is its thickness (m), E its Young's Modulus (Pa), and ρ density (kg/m^3) (Table 2.2).

Discontinuity
The discontinuity of rooms within a building can get complicated at junctions and is most practically implemented on smaller studio spaces than on major auditoria.

Table 2.2 *Values of Young's modulus and density*

Material	Young's modulus, E (Pa)	Density, ρ (kg/m^3)
lead	1.6×10^{10}	11 300
Steel	2×10^{11}	8 000
Aluminium	7×10^{10}	2 700
Glass	4×10^{10}	2 500
Concrete	2.4×10^{10}	2 300
Brick	1.6×10^{10}	1 900
Plasterboard	1.9×10^{9}	750
Plywood	4.3×10^{9}	580

Discontinuity implies the separation of structural elements so that vibrations are not easily transmitted around the main structure to cause intrusive noise in other areas by re-radiation. The ingredients for 'box-in-box' rooms employing a consistent standard of discontinuity are double or even treble walls, a floating floor, and a substantial ceiling and slab above.

Composite construction
Composite construction is that consisting of surface areas of different sound reduction indices, for example a brick wall containing a door and a window. The total sound power through a composite structure is the sum of the components of sound power transmitted by each component separately (see Chapter 5).

Sound leaks
Sound leaks can have a serious deleterious effect on the performance of a partition, wall, floor or roof. The effect is more marked at high frequencies. Figure 2.8 shows that a hole of 0.001 m^2 makes the composite SRI 40 dB for a 45-dB-rated SRI wall of 16 m^2. Gaps at door edges are a typical example of sound leakage.

Building envelope: roofs
Roofs are typically of lighter construction than outer walls and of relatively large area compared with walls and openings. The exposure to road, rail or industrial sources is less, but a building can be vulnerable to aircraft noise or

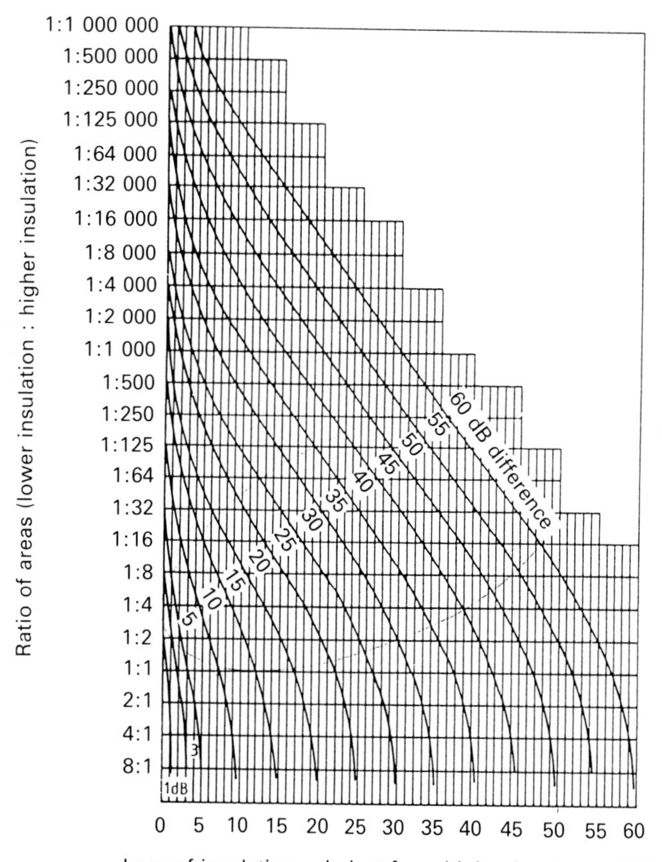

Figure 2.8 *Composite construction sound insulation*

roof-mounted items of plant. Traditional slated roofs have a reasonable surface mass, but gaps up the lapping slates and the need to ventilate under the lapping slates make such roofs poor insulators (27 dB average) on their own. The usual use of a roof void with its thermal insulation and plastered ceiling increases this to 38 dB average.

Flat roofs
Flat roofs of built-up felt on thermal insulation on metal decking or composite construction of profiled metal, insulation and liner tray metal sheeting, achieve only 30–35 dB. Roofs of similar surface mass to floors – screeded topping to precast concrete, for example, with asphalt and insulation above – manage 45–50 dB. For performance exceeding 50 dB, the roof will have to be supplemented by a barrier ceiling below. A conventional lay-in grid mineral tile ceiling will be of little additional value to the roof, particularly if open grilles in the ceiling allow its use as a supply or extract air plenum.

Lightweight roofs
Lightweight roofs with a profiled metal outer face are subject to rain and hail drumming, and can also 'click' and bang during thermal movement. Damping the outer skin by having quilt directly behind it muffles the sound to a degree. Composite metal roofs with a soffit of perforated metal are often used in sports halls – gymnasia, ice rinks and swimming pools – to absorb sound within the space, but the position of the vapour barrier above the perforations needs considering carefully. Too thick a membrane will blank off the absorption capability of the quilt above; some systems have the membrane embedded in the thermal/absorption quilt, but special fixings through this arrangement are needed.

Ceilings

To uprate the sound reduction capability of a roof, the suspension of a barrier ceiling can be included. The performance of a timber floor can be altered from 42 dB to 58 dB by the addition of a British Gypsum M/F ceiling as shown in Figure 2.9. The system uses straightforward metal straps; some other specialist systems use resilient hangers, and any design will have to address problems of suspending ductwork or further decorative ceilings below the barrier ceiling.

Ceiling voids
Ceiling voids are familiar transmission routes for sound between rooms, where partitions are not carried through to the roof or floor above. Carrying the partitions up not only breaks the ceiling but inhibits moving them and affects ventilation arrangements – ducted supplies need the cross-talk attenuation discussed in Chapter 3. Ceiling manufacturers should be able to quote room-to-room transmission characteristics as measured in BS 2750: Part 9: 1980 Laboratory testing [4].

Suspended ceilings
Suspended ceilings tend to be selected for their light weight (hence economy) and for absorption, rather than sound insulation: the level difference through a ceiling is limited to 10–15 dB. For greater performance a closed plasterboard ceiling can be used but recessed lights must be cased and absorption is reduced compared to the proprietary tiled grid.

Walls

Blockwork
Blockwork performs reliably if well constructed and of adequate mass. Lightweight thermal blockwork (350–700 kg/m^3) frequently used in the absence of advice otherwise, is poor. Unplastered blockwork loses sound insulation by its fissures and movement cracks: plastering can improve this. The best blockwork is 2000 kg/m^3 solid no-voids dense concrete masonry (dcm), a thickness of 190 mm achieving 50 dB SRI.

Brickwork
Brickwork is usually better than blockwork; the smaller units can be built around partitions more easily and movement cracking is less. The heaviest (2300 kg/m^3) construction is obtained by using solid engineering bricks; an acceptable everyday use is commons laid with frogs up (1700–2000 kg/m^3). Mortar density is typically 1800 kg/m^3.

Partitions
With care, lightweight construction can outperform masonry, certainly mass-for-mass, and sometimes even for similar thicknesses. In plasterboarded partitions, metal studding has largely replaced timber studding and gives better SRI performance because the leaves are coupled across the studs more resiliently. Plasterboard itself is used less often as the range of metal-skinned modular panel partitioning diversifies and becomes more competitive. The panels take the form of 50- or 100-mm-thick elements with absorption material in a core, for offices. For studios, noise havens or music practice rooms, panels can be perforated on the inner face for absorption and fixed to isolated floor and ceiling panels.

A strong combination is masonry plus independent lining, with quilt in the cavity. Some examples are shown in Figure 2.10. Care must be taken not to make the cavity too small, otherwise low frequency resonances can render the dry lining disadvantageous rather than advantageous to insulation.

Folding partitions. These should be avoided if possible. They are frequently installed and then complained of in places where low background noise levels and need for confidentiality exist, e.g. solicitors' offices and boardrooms. Because of gaps around the suspension gear, the level difference either side is similar to a door's: 15–20 dB average. Folding panel rather than concertina types are marginally better, but beware of the claims of suppliers who quote high sound insulation values, even supported by tests, that relate only to the body of the panels and not to the total assembly. The best types have some closure seal – pneumatic or mechanical – which can lock the panels in place. A robust ceiling at head and a division of ceiling void are necessary. By care, separation in the order of around 35 dB can be obtained: this can be related to speech privacy needs as shown in Table 2.3. An alternative method is to add the background noise level and the partition SRI and work to a total exceeding 65 (Table 2.4). It may be seen from Figure 2.11 that sound insulation performance is particularly important for speech privacy in the frequency range 500–2000 Hz.

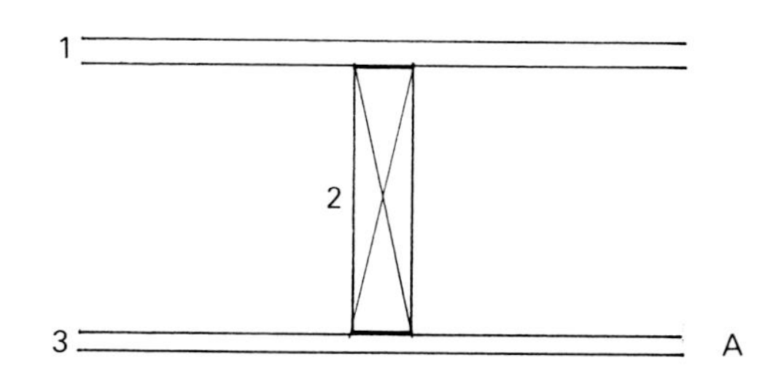

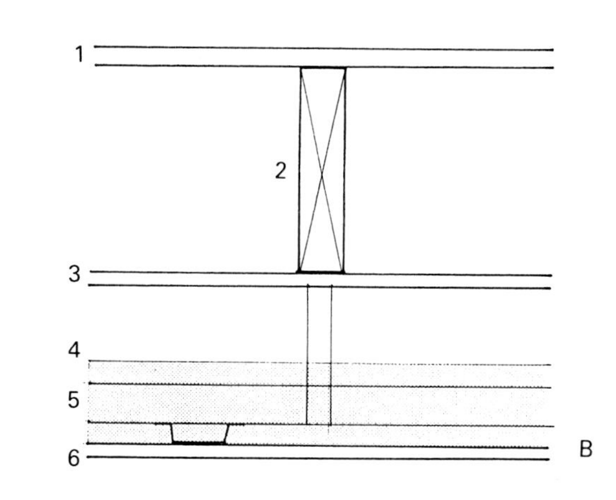

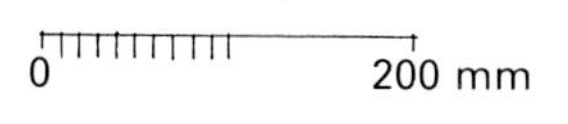

1 18 mm chipboard
2 195 × 45 mm timber joists 600 cs
3 13 mm gyproc
4 150 mm min. spacing/ceiling void
5 80 mm gypglas 1000
6 13 mm gyproc on M/F suspension system

(a)

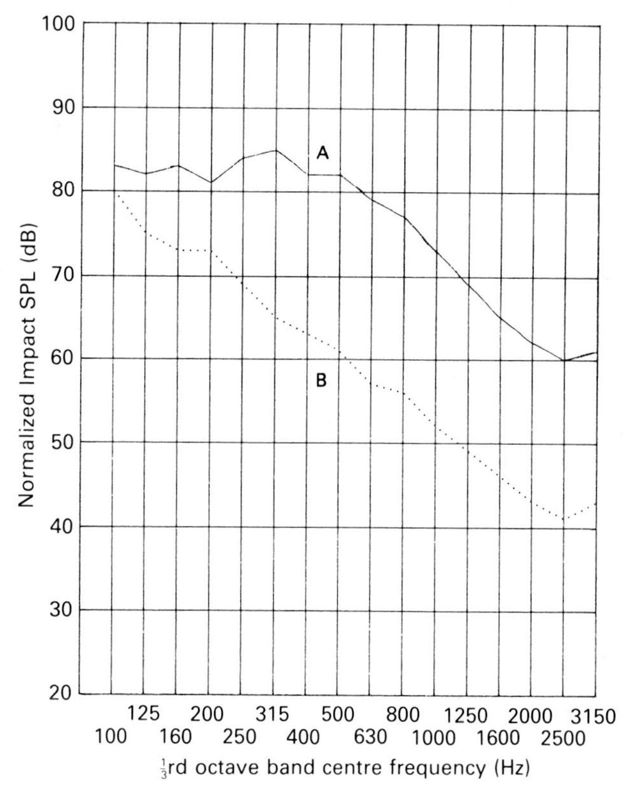

1 Timber floor
2 Timber floor and suspended ceiling

(b)

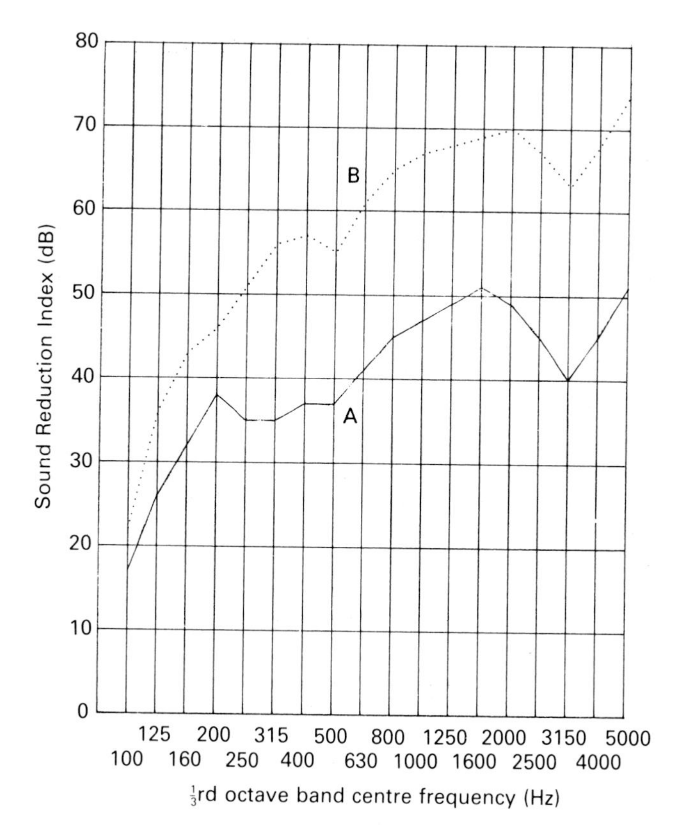

1 Timber floor
2 Timber floor and suspended ceiling

(c)

Figure 2.9 *Sound insulating suspended ceiling. (Courtesy of British Gypsum)*

Table 2.3 *Speech privacy*

Whether conversation overheard other side of division	SRI[a] of dividing element (dB)
Normal speech easily overhead	20
Loud speech clearly heard	25
Loud speech distinguished during normal activity	30
Loud speech heard but not intelligible	35
Loud speech can be heard faintly but not understood	40
Loud speech or shouting heard with great difficulty	45

[a]Average sound reduction index, 100–3150 Hz.

Table 2.4 *Conversation privacy taking account of background noise*

Sound as heard by occupant	SRI + background noise	
	dBA	NR
Intelligible	70	65
Occasionally intelligible	75–80	65–70
Audible but not intelligible	80–90	75–85
Inaudible	90	85

Doors

The typical domestic door, hollow-cored with a loose fit, achieves 15–20 dB average. A solid-core door with fire rating rebates to the frame improves this marginally. The fitting of integral blade or compression seals to edges and the threshold help the value up to about 30 dB. Purpose-made timber or metal doorsets can be selected in the range 35–45 dB average. The weight of acoustic doorsets is substantial and they must either be well fixed directly into masonry or into high-performance partitioning; a metal subframe bolted to the primary structure must hold the assembly in place.

Particular care is required in selecting doorsets on the basis of manufacturers' quoted values, as these are frequently 'composite' values for the doorset in a higher

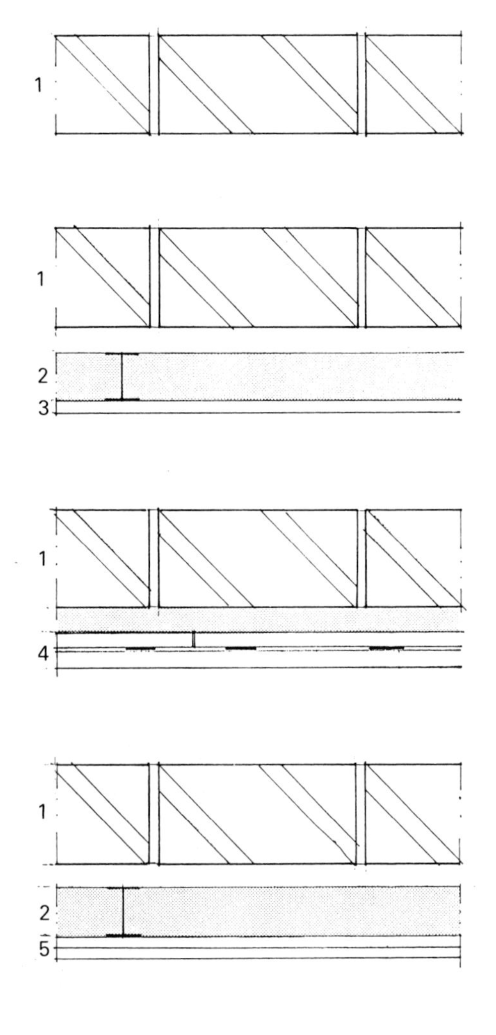

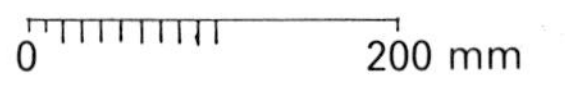

1 single-leaf common brickwork
2 48 mm glass fibre (24 kg/m³)
3 13 mm plasterboard (wallboard) on 48 mm. 'I' section metal studs at 600 mm cs
4 2 × 19 mm 'Gyproc' plank, adhesive between layers
5 2 × 13 mm plasterboard

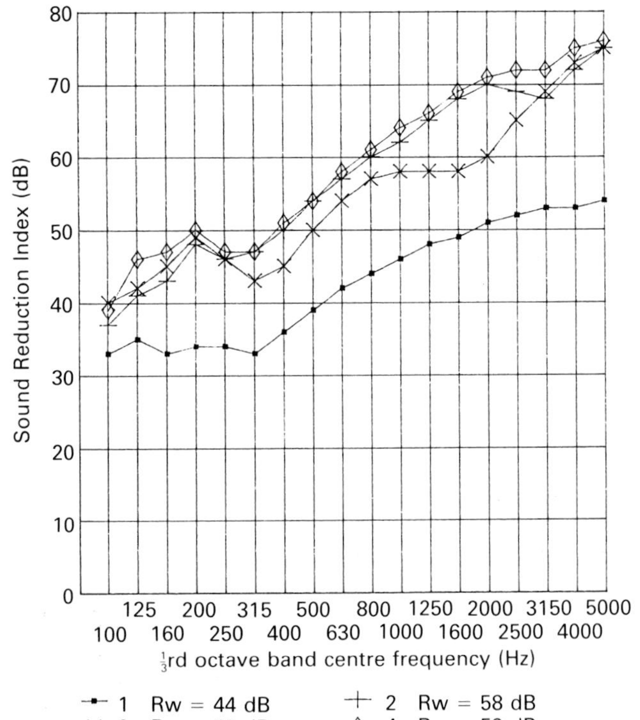

masonry/dry-lining combinations

Figure 2.10 *Masonry – dry lining combinations. (Courtesy of British Gypsum)*

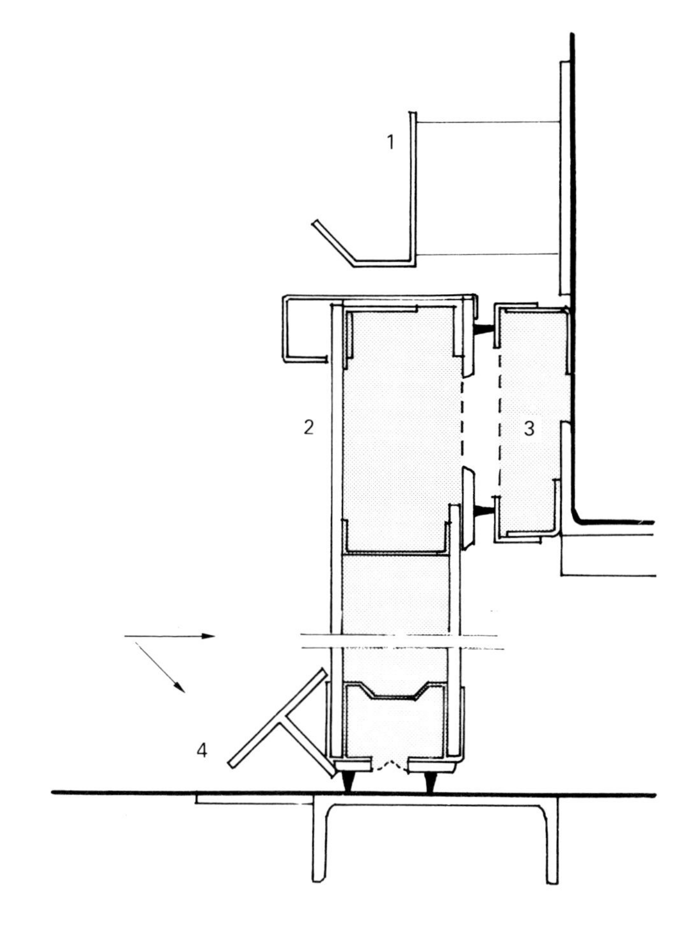

1 Suspension gear
2 Acoustic door: 0.6 mm steel skin/12 mm 'fermacel'/1 mm steel faces to mineral wool core
3 Top and sides attenuating seal strips
4 In-and-down 45° closure action

Figure 2.11 *Sliding acoustic door for stage/scene dock or production studio use. (Courtesy of Envirodoor Markus)*

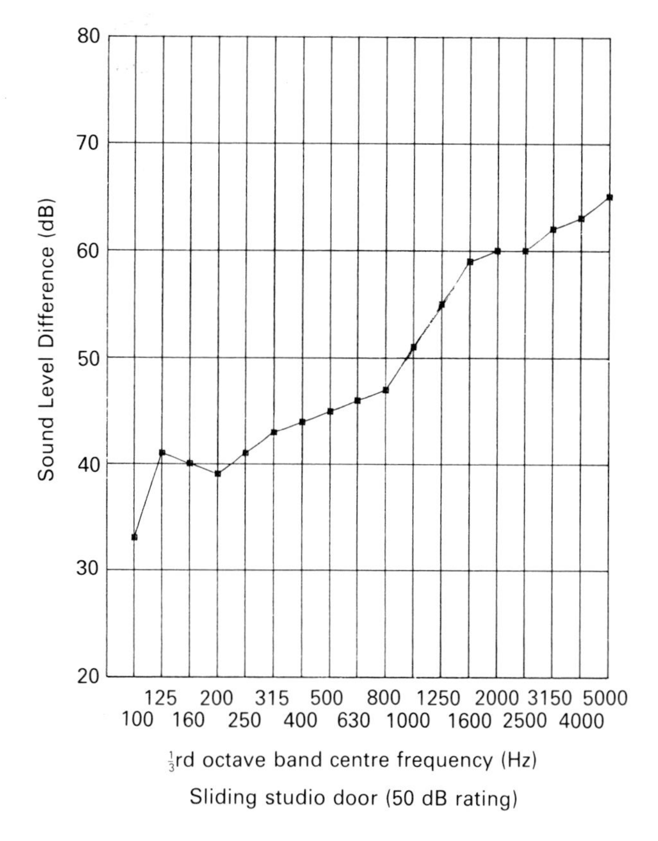

performance wall. Test data should be sought from a suitable (e.g. NAMAS qualified) laboratory to check claimed values.

A sound lobby may be advantageous if space can be afforded: the performance of the individual doorsets can be less and the sudden sound break-out on opening is less. Delayed action final closure of self closers will ensure doors do not bang. Undue pressure should not be required to compress seals, otherwise the door will remain ajar as the self closer will be unable to overcome the seal's resistance. The most highly rated doorsets by virtue of their weight and seals may need significant opening pressure. This may rule them out for panic bar operated fire escape doors in public buildings. Magnetic, compression and blade type seals are alternative edge treatments.

Special sliding doors of greater weight and size can be used, e.g. between studio and workshop or theatre stage and scene dock (Figures 2.12 and 2.13).

Windows

Windows are the weak link element in the building envelope for shielding interiors from intrusive external noise. If noise levels are high, $\geqslant 65\,\mathrm{dB}(L_{\mathrm{Aeq}})$, opening windows may have to be avoided by the use of full mechanical ventilation. The various factors in the performance of the total assembly are as follows.

Figure 2.12 *Sliding acoustic door*

1 Floor boards fixed through Gyproc to flange
2 19 mm, Gyproc
3 Overlapping ledger channels on foam strip
4 Floor joists
5 100 mm Gypglas 1000
6 13 + 19 mm Gyproc
7 Resilient bar
8 Sealant
9 Perimeter seal strip

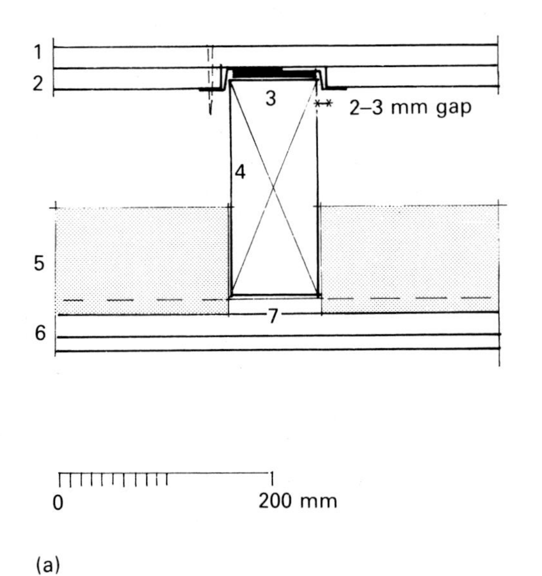

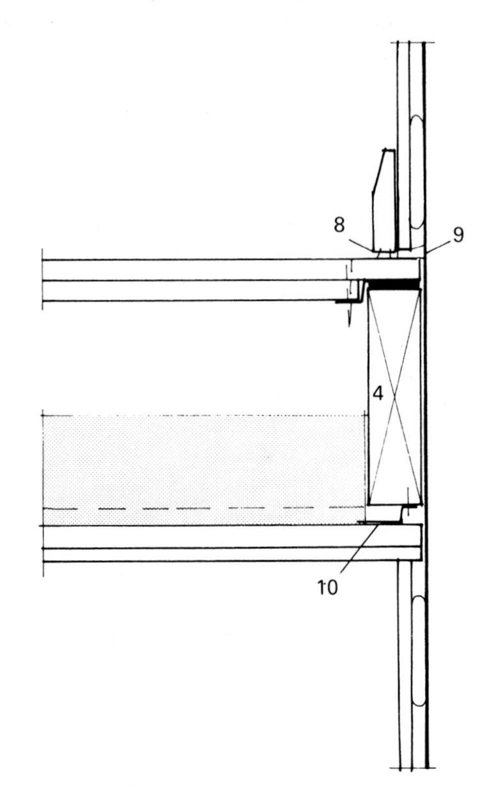

(a)

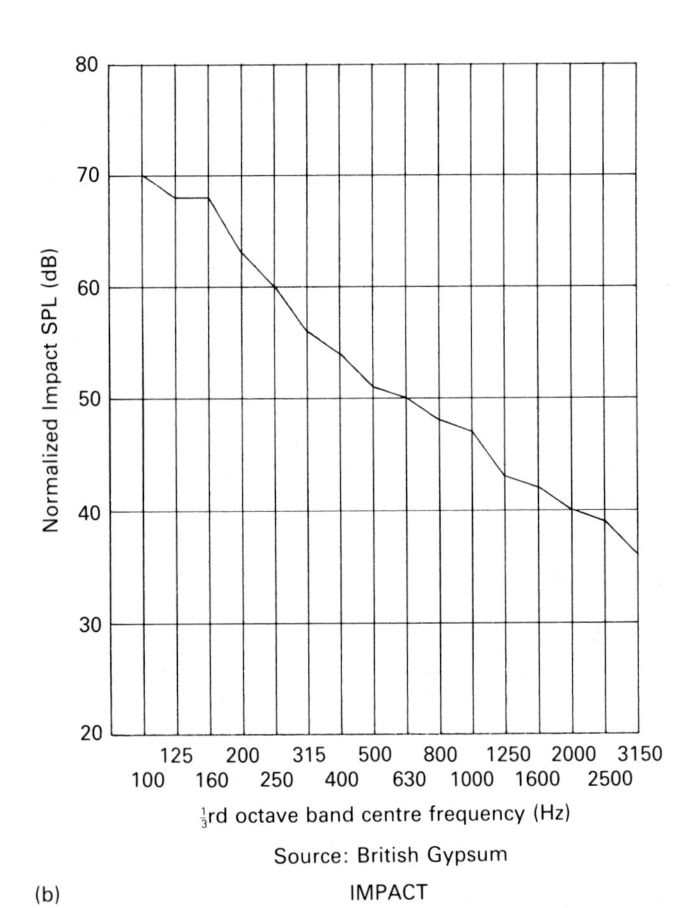

(b)

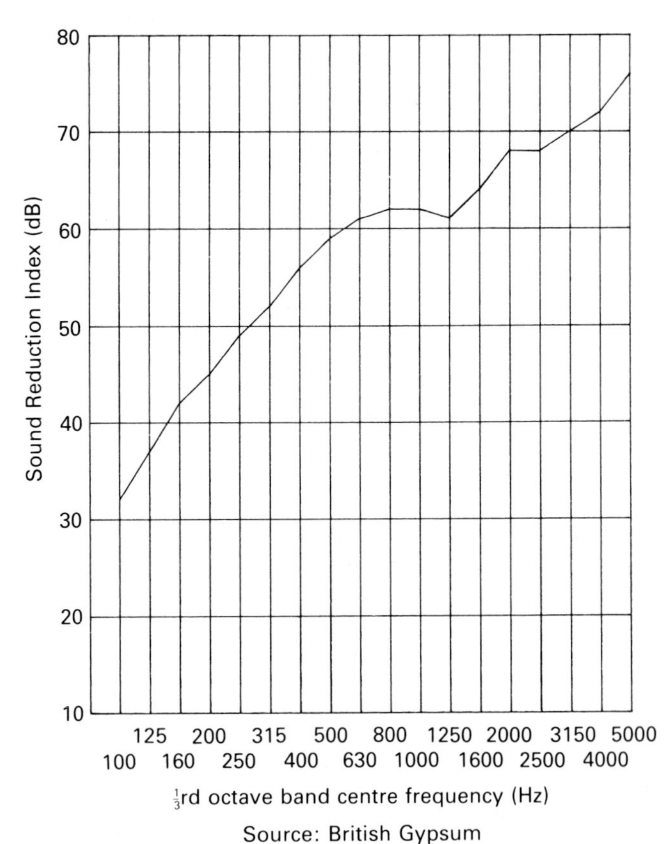

(c)

Figure 2.13 *Timber floors: Gyproc SI floor system. (Courtesy of British Gypsum)*

- *Thickness.* Increasing the glazing thickness increases the mass and stiffness, improving performance and changing the coincidence 'dip'.

- *Stiffness.* Toughening the glass does not affect the bending stiffness and so has no effect on its sound insulation properties.

- *Air space.* Very small air spaces do not help appreciably (compare 6-mm glass with 6/12/6 in Table 2.1, for example); larger spacing with differing glass thicknesses improves the insulation.

- *Lamination.* 'Plate damping' reduces the transmission of sound through a window by transforming resonant vibratory motion in the glass, excited by sound on the incident side of the window, into heat energy. Laminated glass comprises two thin layers of glass bonded by a clear viscoelastic material with high damping characteristics. When laminated glass is combined with air space and a second glass layer in a double-glazed unit, a significant improvement in performance is achieved over a single layer of equivalent mass.

- *Edge damping.* The size of the glass panel and how well it is framed has a bearing on the performance of a curtain wall. Assuming the glazing is well gasketted, mechanical interaction between the glass panels and muntins leads to an improved sound insulation effect for an assembly of many individual panes and muntins.

- *Gas filling.* Some insulating double glazed units are filled with argon, sulphur hexafluoride or xenon. These improve the sound insulation at higher frequencies, but below 250 Hz the reduction in performance outweighs this. As traffic noise has a strong low-frequency component of noise, gas-filled glass is not beneficial compared to conventional double glazing for acoustic protection.

- *Inner windows.* Wide air spaces and decoupled frames allow good performance, although maintenance access and cleaning is a disadvantage. Separate windows can also incorporate off-set opening lights or trickle vents, without a total loss of insulation for opened lights.

- *Frame to masonry.* Frames should be on continuous grounds and well edge-sealed to inside and outside reveals, in order to avoid water as well as noise ingress. A weakness often occurs at the frame head – eaves closure detail, because of relative movement.

Floors

Timber floors as used in dwellings perform as shown in Figure 2.14. Between flats, Building Regulation Part E recommended details are a reference source[2]. The new 1992 version increases new party floor standards, e.g. a base floor slab in a composite system is increased from 220 kg/m^2 to 300 kg/m^2, and brings in onerous provisions for converted residential properties. There are several proprietary isolation systems. Concrete floors can be disappointing if the lightest-weight precast units and nominal-only topping are used. Solid rather than cored units with a generous structural topping are better, achieving 48–50 dB. With floors the impact sound insulation as well as airborne will be of concern. Impact sound happens when a short impulsive blow to a structure 'drives' it and the sound is carried and re-radiated elsewhere. Isolation can be given by either a resilient surface layer, floating screed or a floating slab. Examples of the different types are shown in Figures 2.15 and 2.16. The isolating layer and the slab or screed form a mass-spring system with a low resonant frequency. Effective isolation is only possible at frequencies about two or three times higher than this frequency.

Care in application is required as isolation battens like those illustrated have impact isolation geared to meet Building Regulations, primarily damping footfall and normal domestic activity. However, there is little static deflection inherent in such systems (otherwise there might be too much 'give' in the floor in normal use). In a recent instance of a school dance floor above other teaching rooms, the heavy impact of group rhythmical movements make the total floor system act as one, overcoming the limited deflection isolation system.

The type of floor shown in Figure 2.16 is a proprietary floor by Sound Attenuators Ltd, Sunbury-on-Thames, Middlesex. Usually floor-to-floor heights are at a premium so a wide floor zone cannot be afforded. The system consists of 13-mm ply with a grid of resilient pads 50 mm thick and absorption quilt between the pads. The ply serves as permanent formwork for 100 mm of concrete to be laid as the upper floor slab layer, isolated at the edges. At this 50-mm spacing of slabs, the air gap determines the spring-mass resonance at around 15 Hz. The floor is not intended as a substitute for adequate antivibration mounting of plant, as the floating floor system is only effective for frequencies above 30 Hz. Another proprietary system avoids ply by having isolators jackable so that the upper slab is initially laid directly on top of the lower and then raised and levelled.

The improvement given by a floating floor is such that the nett separation will be determined by flanking effects (for example, via walls or columns common at both levels) rather than by airborne performance as shown in Figure 2.16. Without attention to the flanking routes, the average improvement in the sound reduction index will be limited to around 8 dB.

Sound absorption

Absorption and insulation

Absorption and insulation are not to be confused. The application of sound-absorbing finish to a separating wall will not have any discernible effect on its sound insulation properties at all. All surfaces absorb sound to a greater or lesser extent: bare concrete or marble have a low sound absorption coefficient, and hence absorb little sound and reflect back almost all incident energy.

Absorption coefficients

Absorption coefficients are not considered dependent on the angle of incidence of sound striking the medium: random

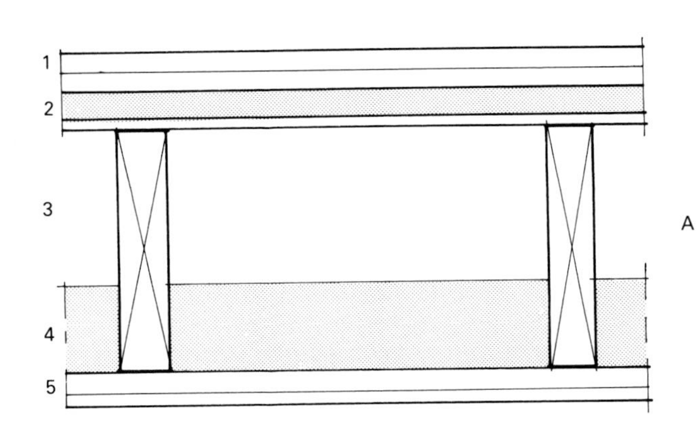

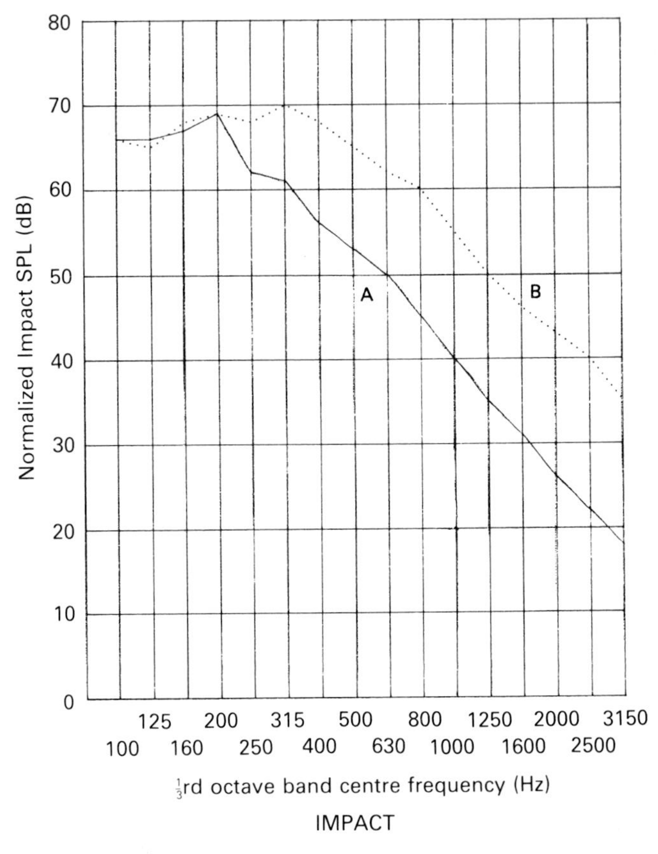

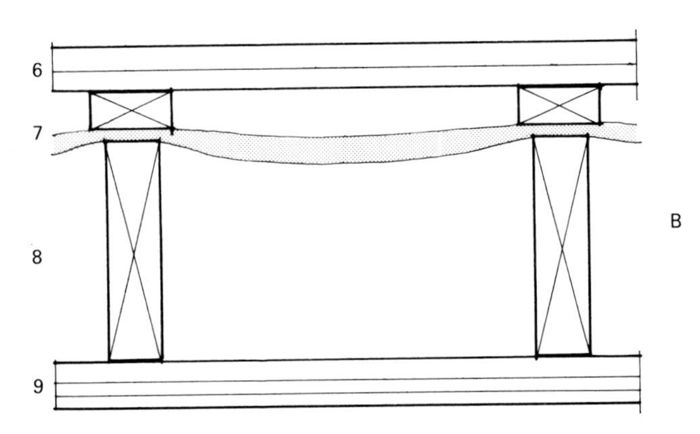

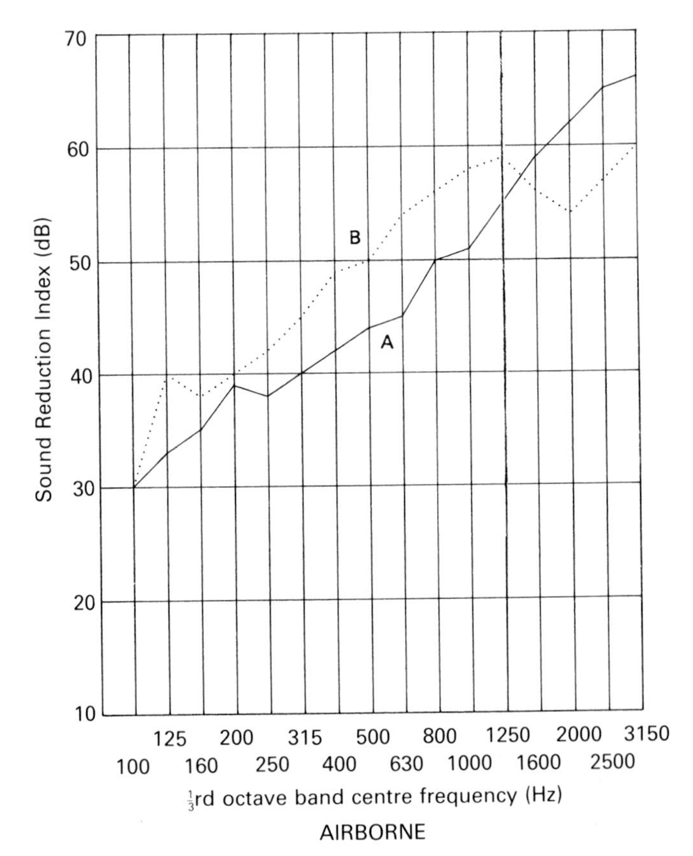

1 18 mm chipboard on 19 mm plasterboard
2 25 mm glass fibre (65 kg/m^3) isolation quilt on 10 mm plywood
3 219 × 44 mm floor joists at 400 mm cs
4 80 mm glass fibre (23 kg/m^3) quilt
5 13 mm and 19 mm plasterboard
6 22 mm chipboard on 19 mm plasterboard
7 75 × 32 mm battens on 25 mm glass fibre (65 kg/m^3)
8 200 × 50 mm floor joists at 400 mm cs
9 19 mm plasterboard and 2 × 13 mm plasterboard

Figure 2.14 *Timber floors. (Courtesy of TRADA)*

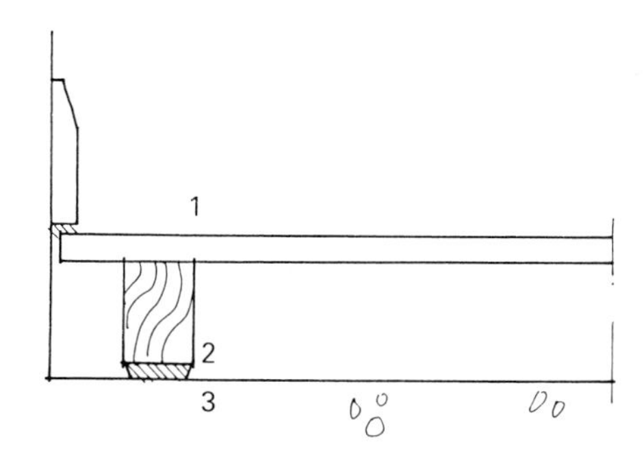

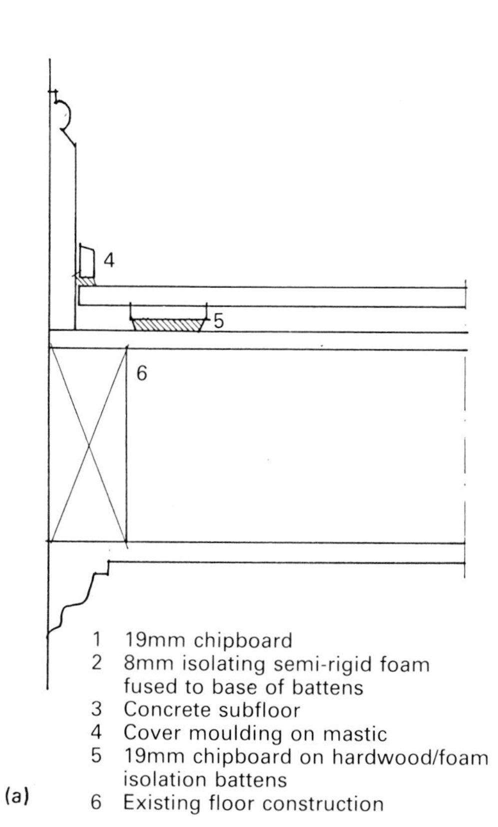

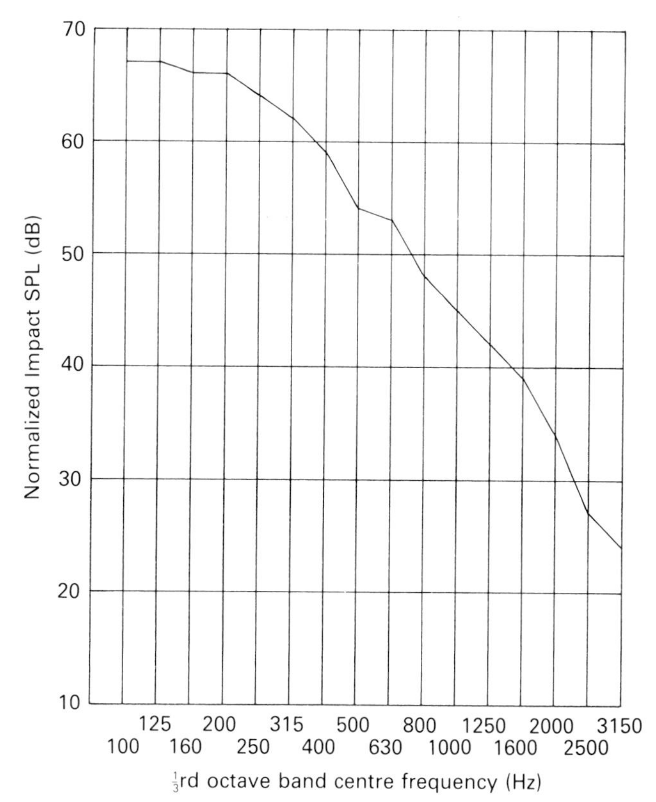

Figure 2.15 *Floating timber floors. ((a) Courtesy of Contiwood (Durabella) Ltd; (b) courtesy of Pheonix Floors Ltd)*

incidence is assumed. Absorption coefficients are normally given for the frequency range 125–4000 Hz (Table 2.5). Third octave band values of absorption coefficients will not differ much from octave band values (octave bands *average* the third octave values, as opposed to NR values where one-third values are *additive* in producing octave bands); NRC and dBA are correspondingly different in derivation.

Types of absorber

Types of absorber include fibrous absorbers, fibrous absorbers with impervious membrane facing, and fibrous absorbers covered with perforated panelling. An example of the first type is quilt batts mounted directly on a wall surface, or carpet on a floor. An example of the third type is metal perforated suspended ceiling tiles with quilt inlay above; provided the open area of the perforations exceeds

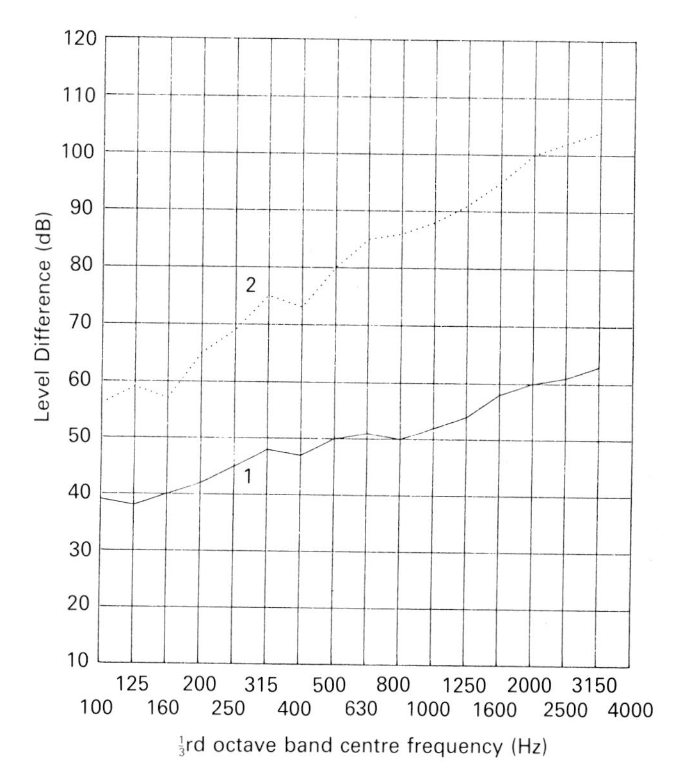

Figure 2.16 *Increase of sound insulation by use of concrete floating floors. (Courtesy of Sounds Attenuators Ltd)*

Table 2.5 *Absorption coefficients*

	OBCF (Hz)					
	125	250	500	1000	2000	4000
'Hard' finishes						
Water or ice	0.01	0.01	0.01	0.01	0.02	0.02
Smooth concrete, unpainted	0.01	0.01	0.02	0.02	0.02	0.05
Smooth concrete, sealed or painted	0.01	0.01	0.01	0.02	0.02	0.02
Concrete blocks, fairfaced	0.05	0.05	0.05	0.08	0.14	0.20
Rough concrete	0.02	0.03	0.03	0.03	0.04	0.07
Brickwork, flush-pointed	0.02	0.03	0.03	0.04	0.05	0.07
Brickwork, 10-mm-deep pointing	0.08	0.09	0.12	0.16	0.22	0.24
Plastered walls	0.02	0.02	0.03	0.04	0.05	0.05
Painted plaster	0.02	0.02	0.02	0.02	0.02	0.02
Ceramic tiles	0.01	0.01	0.01	0.02	0.02	0.02
Marble, terrazzo	0.01	0.01	0.01	0.01	0.02	0.02
Glazing (4 mm)	0.30	0.20	0.10	0.07	0.05	0.02
Double glazing	0.15	0.05	0.03	0.03	0.02	0.02
Glazing (6 mm)	0.10	0.06	0.04	0.03	0.02	0.02
Ceilings						
13-mm mineral tile, direct to floor slab	0.10	0.25	0.70	0.85	0.70	0.60
13-mm mineral tile, suspended 500 mm below ceiling	0.75	0.70	0.65	0.85	0.85	0.80
Metal planks, slots 14% free area, mineral wool overlay and void	0.50	0.70	0.80	1.0	1.0	1.0
Metal tiles 5% perforated, 20-mm quilt overlay and void	0.13	0.27	0.55	0.79	0.90	1.0
Woodwool slabs	0.40	0.40	0.70	0.70	0.70	0.80
Panels						
Solid timber door	0.14	0.10	0.06	0.08	0.10	0.10
9-mm plasterboard on battens, 18-mm air space with glass fibre	0.30	0.20	0.15	0.05	0.05	0.05
5-mm ply on battens, 50-mm air space with glass fibre	0.40	0.35	0.20	0.15	0.05	0.05
Suspended plasterboard ceiling	0.20	0.15	0.10	0.05	0.05	0.05
Steel decking	0.13	0.09	0.08	0.09	0.11	0.11
Ventilation grille (per m^2)	0.60	0.60	0.60	0.60	0.60	0.60
13-mm plasterboard on frame, 100-mm air space with glass fibre	0.30	0.12	0.08	0.06	0.06	0.05
13-mm plasterboard on frame, 100-mm air space	0.08	0.11	0.05	0.03	0.02	0.03
2 × 13-mm plasterboard on frame, 50-mm air space with mineral wool	0.15	0.10	0.06	0.04	0.04	0.05
22-mm chipboard on frame, 50-mm air space with mineral wool	0.12	0.04	0.06	0.05	0.05	0.05
16-mm T&G on frame, 50-mm air space with mineral wool	0.25	0.15	0.10	0.09	0.08	0.07
22-mm timber boards 100-mm-wide, 10-mm gaps 500-mm air space with mineral wool	0.05	0.25	0.60	0.15	0.05	0.10
Treatments						
Curtains in folds against wall	0.05	0.15	0.35	0.40	0.50	0.50
25-mm glass fibre, 16 kg/m^3	0.12	0.28	0.55	0.71	0.74	0.83
50-mm glass fibre, 16 kg/m^3	0.17	0.45	0.80	0.89	0.97	0.94
75-mm glass fibre, 16 kg/m^3	0.30	0.69	0.94	1.0	1.0	1.0
100-mm glass fibre, 16 kg/m^3	0.43	0.86	1.0	1.0	1.0	1.0
25-mm glass fibre, 24 kg/m^3	0.11	0.32	0.56	0.77	0.89	0.91
50-mm glass fibre, 24 kg/m^3	0.27	0.54	0.94	1.0	0.96	0.96
75-mm glass fibre, 24 kg/m^3	0.28	0.79	1.0	1.0	1.0	1.0
100-mm glass fibre, 24 kg/m^3	0.46	1.0	1.0	1.0	1.0	1.0
50-mm glass fibre, 33 kg/m^3	0.20	0.55	1.0	1.0	1.0	1.0
75-mm glass fibre, 33 kg/m^3	0.37	0.85	1.0	1.0	1.0	1.0
100-mm glass fibre, 33 kg/m^3	0.53	0.92	1.0	1.0	1.0	1.0
50-mm glass fibre, 48 kg/m^3	0.30	0.80	1.0	1.0	1.0	1.0
75-mm glass fibre, 48 kg/m^3	0.43	0.97	1.0	1.0	1.0	1.0
100-mm glass fibre, 48 kg/m^3	0.65	1.0	1.0	1.0	1.0	1.0
25-mm acoustic plaster to solid backing	0.03	0.15	0.50	0.80	0.85	0.80
9-mm acoustic plastic to solid backing	0.02	0.08	0.30	0.60	0.80	0.90

Table 2.5 (cont.)

	OBCF (Hz)					
	125	250	500	1000	2000	4000
9-mm acoustic plaster on plasterboard, 75-mm air space	0.30	0.30	0.60	0.80	0.75	0.75
50-mm mineral wool, 33 kg/m^3	0.15	0.60	0.90	0.90	0.90	0.85
75-mm mineral wool, 33 kg/m^3	0.30	0.85	0.95	0.85	0.90	0.85
100-mm mineral wool, 33 kg/m^3	0.35	0.95	1.0	0.92	0.90	0.85
50-mm mineral wool, 60 kg/m^3	0.11	0.60	0.96	0.94	0.92	0.82
75-mm mineral wool, 60 kg/m^3	0.34	0.95	1.0	0.82	0.87	0.86
25-mm mineral wool, 25-mm air space	0.10	0.40	0.70	1.0	1.0	1.0
50-mm mineral wool, 50-mm air space	0.50	0.70	0.90	0.90	0.90	0.80
50-mm mineral wool (96 kg/m^3) behind 25% open area perforated steel	0.20	0.35	0.65	0.85	0.90	0.80
Floor finishes						
Cord carpet	0.05	0.05	0.10	0.20	0.45	0.65
Thin (6-mm) carpet on underlay	0.03	0.09	0.20	0.54	0.70	0.72
Thick (9-mm) carpet on underlay	0.08	0.08	0.30	0.60	0.75	0.80
Wooden floor boards on joists	0.15	0.11	0.10	0.07	0.06	0.07
Parquet floor on timber joists and deck	0.20	0.15	0.10	0.10	0.05	0.10
Parquet laid concrete	0.04	0.04	0.07	0.06	0.06	0.07
Vinyl or linoleum on concrete	0.02	0.02	0.03	0.04	0.04	0.05
Vinyl and resilient backing on concrete	0.02	0.02	0.04	0.05	0.05	0.10
Miscellaneous						
Audience on timber seats (1/m^2)	0.16	0.24	0.56	0.69	0.81	0.78
Audience on timber seats (2/m^2)	0.24	0.40	0.78	0.98	0.96	0.87
Audience per person, seated	0.33	0.40	0.44	0.45	0.45	0.45
Audience per person, standing	0.15	0.38	0.42	0.43	0.45	0.45
Seats, leather covers (per m^2)	0.40	0.50	0.58	0.61	0.58	0.50
Upholstered seats (per m^2)	0.44	0.60	0.77	0.89	0.82	0.70
Floor and upholstered seats (per m^2)	0.49	0.66	0.80	0.88	0.82	0.70
Areas with audience, orchestra, or seats, including narrow aisles	0.60	0.74	0.88	0.96	0.93	0.85
Orchestra with instruments on podium, 1.5 m^2/person	0.27	0.53	0.67	0.93	0.87	0.80
Shading factor (apply to finishes under seats, x coefficient)	0.80	0.70	0.60	0.50	0.40	0.20
Air 30% RH (per m^3 at 20°C)	—	—	—	0.005	0.01	0.04
Air 50% RH (per m^3 at 20°C)	—	—	—	0.005	0.009	0.03
Air 70% RH (per m^3 at 20°C)	—	—	—	0.005	0.009	0.02
Office furniture (per desk)	0.50	0.40	0.45	0.45	0.60	0.70

Values exceeding 1.0 have been rounded down to 1.0.

say 20%, the quilt and air cavity behind the metal tiles is almost as efficient at soaking up sound as if the tiles were not present. In studios, deep boxes with thin membranes can be purpose-designed or selected to even out the reverberation characteristics at different frequencies. Resonance absorption in the range 70–1000 Hz can be produced by selecting appropriate perforation and air space depth. If the perforation rate is 5% or less, the panels are reflecting except for myriad Helmholz resonators formed by the holes. 'Bass traps' are used in studios to provide broadband absorption right down to very low frequencies. They consist of a lined labyrinth air space within which negligible reflection results.

An extreme case of absorptive materials installation is the semi-anechoic and anechoic chambers in acoustic laboratories (Figure 2.17): deep wedges of foam above, below and

Figure 2.17 *Anechoic chamber*

to all sides reduces the reverberation time to a very low value at all audible frequencies.

A feature of absorption is that the more that is put into a room, the less effective it is, because new absorption is 'competing' with the absorption already present to absorb incident sound. The maximum absorption effect is in a diffuse field, i.e. when sound is incident on the absorptive material from all around. There is also a slight 'drawing in' of sound at the edges. For a specific area of surface absorption in a room, the maximum absorption effect is obtained by distributing small areas all around.

Reverberant sound pressure level

The reverberant sound pressure level is given by:

$$L = \mathrm{SWL} - 10 \log A + 6 + 10 \log N$$

where SWL is the sound power level of a noise source within the space, N is the number of sources, and A is the absorption present. As 10 log 2 is 3, it may be gathered that each doubling of the absorption in a room reduces the reverberant SPL by 3 dB.

Reverberation time

The reverberation time is the finite time it takes for the sound source energy in a space to decay 60 dB when switched off. For enclosures in which a diffuse sound field exists and where the average absorption coefficient is less than 0.1, the reverberation time can be found by the Sabine Equation. Where the average absorption coefficient is greater than 0.1, the Norris-Eyring Equation can be used. Both methods are described in Chapter 5. Air absorption effects must also be accounted for in larger volumes.

Larger spaces

Larger spaces produce sound decay characteristics in poor agreement with Sabine or Norris-Eyring. In 'amorphous space' areas, such as shopping malls or industrial halls, there is not a true reverberant field across the space and the sound character will vary in different parts of the space. On the other hand, such spaces are not 'free-field' and the SPL will decay at something less than 6 dB for each doubling of distance from the source. Empirical data suggest 4 dB/doubling of distance across typical industrial halls. Alternative forms of calculation have been proposed for industrial halls. Complex spaces can sometimes be subdivided into individual coupled volumes: if alcoves off a main space have significant absorption such that significant energy is not returned into the main space, the surface area of the alcove opening can be counted as $\alpha = 1$.

Alternatively, a simplified calculation is given by:

$$A = \frac{Ar \times S}{Ar + S}$$

where A is the absorption contribution from the recess to the main space, Ar is the absorption in the recess, and S is the area of the opening between the recess and the main space.

Finishes

Finishes, then, can be designed to affect room acoustics in three different ways: absorption, reflection and diffusion.

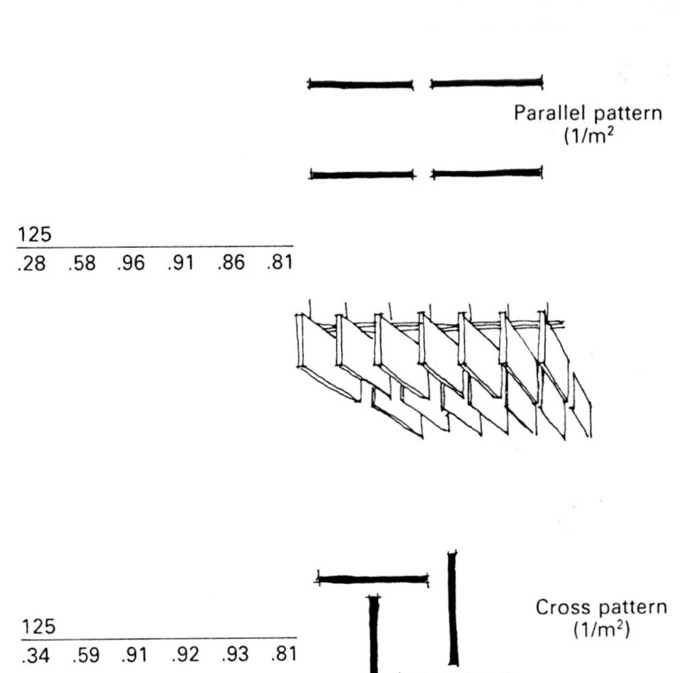

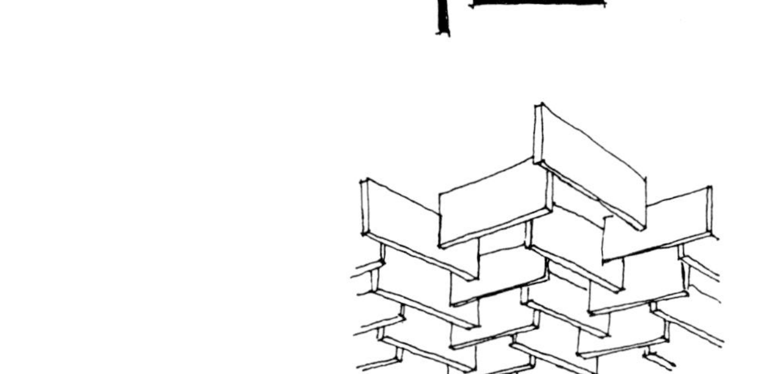

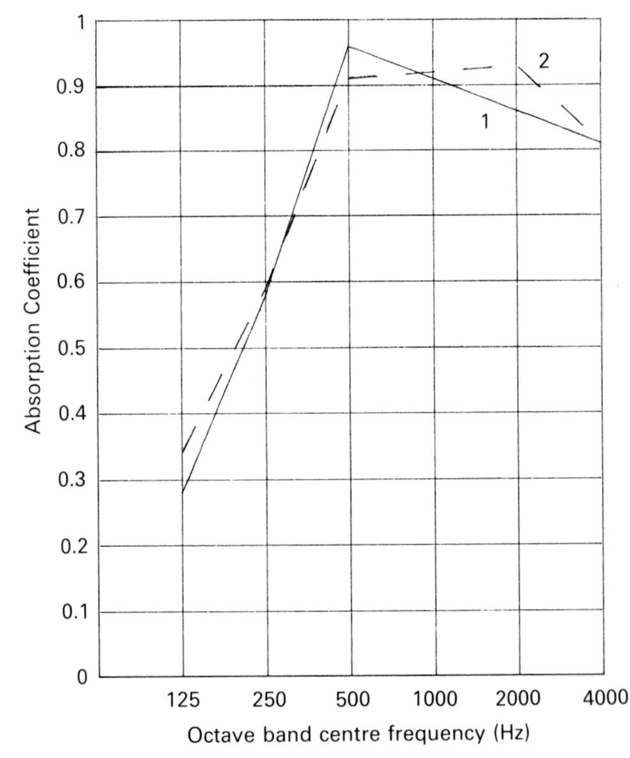

Figure 2.18 *Overhead sound absorbers. (Courtesy of Rockwool)*

Absorption
Absorption can either be integral in a space (e.g. by virtue of blockwork faces or a suspended ceiling) or added as a decorative finish (carpet, wall linings, curtains and furniture), and then, of course, there are the occupants. Absorption is more effective spread around rather than concentrated all in one area. Some elements show the 'perfect' absorption coefficient 1; 'space absorbers', hanging baffles or banners are absorptively particularly effective because they soak up incident sound from both sides. In industrial premises, arrays of overhead absorbers, one per square metre, can be used to soak up high levels of process noise in bottling plants or printing halls (Figure 2.18).

Reflection
Reflection of sound at surfaces not absorptive in nature not only increases the reverberant sound level but also directs the sound to other surfaces. Rooms of particular proportions can give cross reflection effects, impairing speech intelligibility. Curved walls can focus sound, and by so doing starve other parts of a space of sound. Strong directed sound is a feature desired by musical performance audiences in modern halls but this has to be combined with diffusion; the methods of providing adequate surface area are discussed under 'Concert halls'.

Diffusion
Diffusion is an effect whereby the complexity of reflecting surfaces results in an even dispersion of sound in a room: the modelling of reflecting surfaces can ensure this by breaking up incident sound. In order to break up low frequency components of sound as well as middle and upper frequencies, the modelling of surfaces has to be on a large scale, e.g. projections exceeding 300 mm and element areas exceeding $0.5\,\text{m}^2$.

Auditoria modelling

For auditoria modelling of interiors considering the effects of sound-absorbing or sound-reflecting surfaces, there is a choice between physical and computer prediction techniques. Physical models (1:50 or even 1:10) are still the most reliable for the most important projects, e.g. major concert halls, with wavelengths scaled up accordingly, spark sound sources to produce the high frequencies needed, and miniature electret/capacitor microphones. There are a number of computer systems with powerful graphical presentations, emanating from Japan, USA and Europe. Other systems are suited for sound systems design, having been developed by international sound equipment manufacturers.

A room acoustics modelling program should allow basic evaluation of rooms of any shape and complexity. The calculation methods are intended to combine the best of both ray-tracing and image-source methods. Desired features in such a program are:

- fast estimation of room volume and reverberation conditions
- reflectograms with 'sound rose' graphical displays
- 3D tracing of individual reflection paths
- maps of energy parameters over chosen surfaces
- fast recalculation in response to altering receiver position and absorption materials
- easy operation: menu-driven, warnings, and data displays
- project file management which allows consistent analysis and thorough records of the design approach
- multiple source capability
- compatibility with CAD systems
- source directivity factors
- link to audible simulation of acoustic conditions
- RASTI map calculation

ODEON/ODSEE, a system derived for acousticians by the University of Copenhagen, is an example of a program that includes a number of the above features (Figure 2.19), although there is no one ideal system for all purposes. The geometry of a room is typically defined by coordinates for its corners, plus corners to surfaces within. Any system should be validated to 'stock' auditoria which have known characteristics, for example the Royal Festival Hall, or to physical models tested in the laboratory, a technique which has proved comparatively reliable in the past.

Criteria for different building types

Cinemas

Multiplexes
Multiplexes are a new form imported from the USA, involving typically a group of auditoria with a common projection room and concourse/foyer. Often new developments are 8- or 10-plex but versions ranging from 3- to 24-plex have been attempted. The key design issues are sound insulation between auditoria, isolation to outside, good sound systems, 'dead' room acoustics, and moderate ventilation noise. Adequate separation can be provided ($D_{nT,w}$ 65–70+ dB has been achieved on 10 such projects to date by BDP Acoustics for UCI (UK) Ltd, Manchester) by as low a specification as two layers of 15-mm plasterboard on separate studs, significant cavity with 100-mm quilt inlay, and careful head, base and edge detailing (Figures 2.20 and 2.21). The head detail to a lightweight roof is a problem because of roof deflection effects either buckling the partition or causing a gap; this is overcome by a closure angle movement joint.

Outside noise will come in via escape doors (these should be 40+ dB rated, light- and sound-proof), or via the lightweight roof. The surface mass of the roof may have to be uprated or a barrier ceiling included if the multiplex is directly under a light path or close to an elevated motorway. Individual entrances into auditoria should be by acoustically rated (30+ dB) doors in lobby configuration. Some ventilation noise (NR 35) is welcomed as it masks residual intrusive noise from the adjacent auditoria sound tracks and audience sounds. The standard system is a dedicated unit to each auditorium with ducted supply and extract via the ceiling void. In some leisure centres, cinemas are located above or below noisy facilities like ten-pin bowling and discotheques; tenancy agreements on maximum sound levels and double floors are required to avoid operational difficulties.

There are more sophisticated sound systems, for example THX, which are more demanding on sound insulation

Royal Albert Hall
Bare Hall
(Acoustic discs removed)

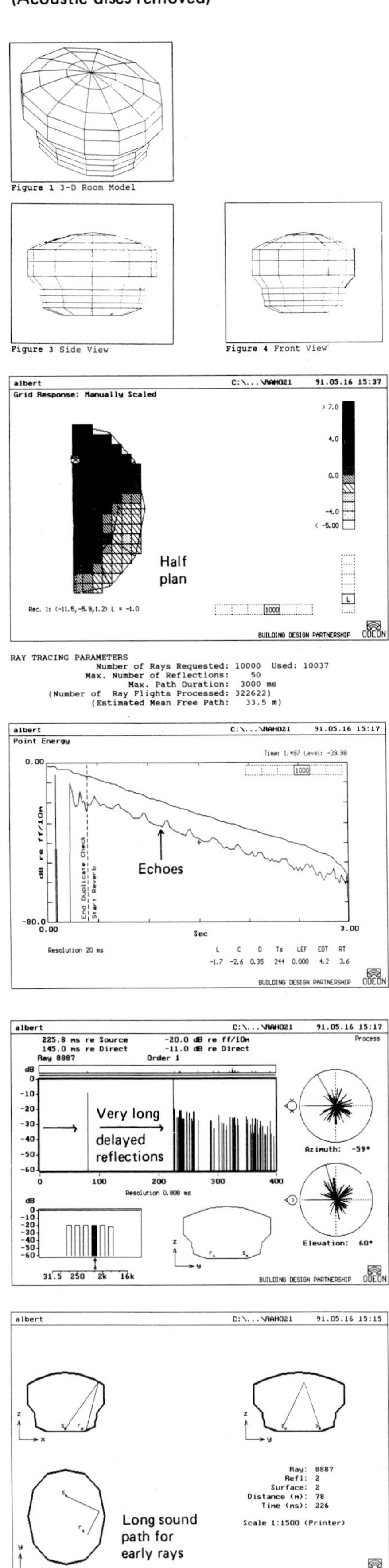

Hall & New Reflectors

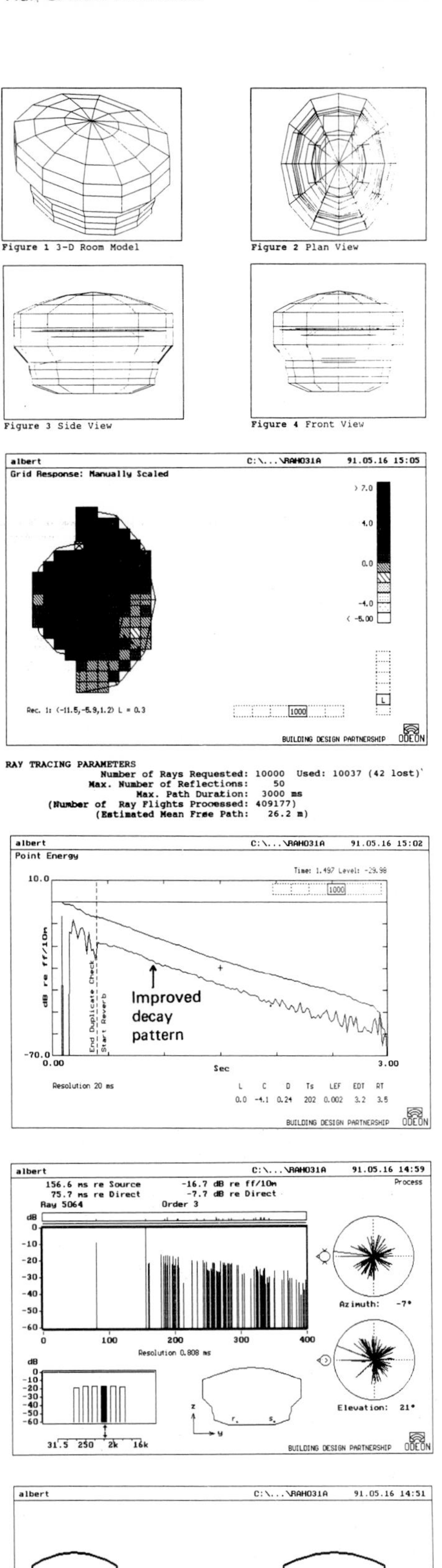

Figure 2.19 *Sound decay analysis: ODEON*

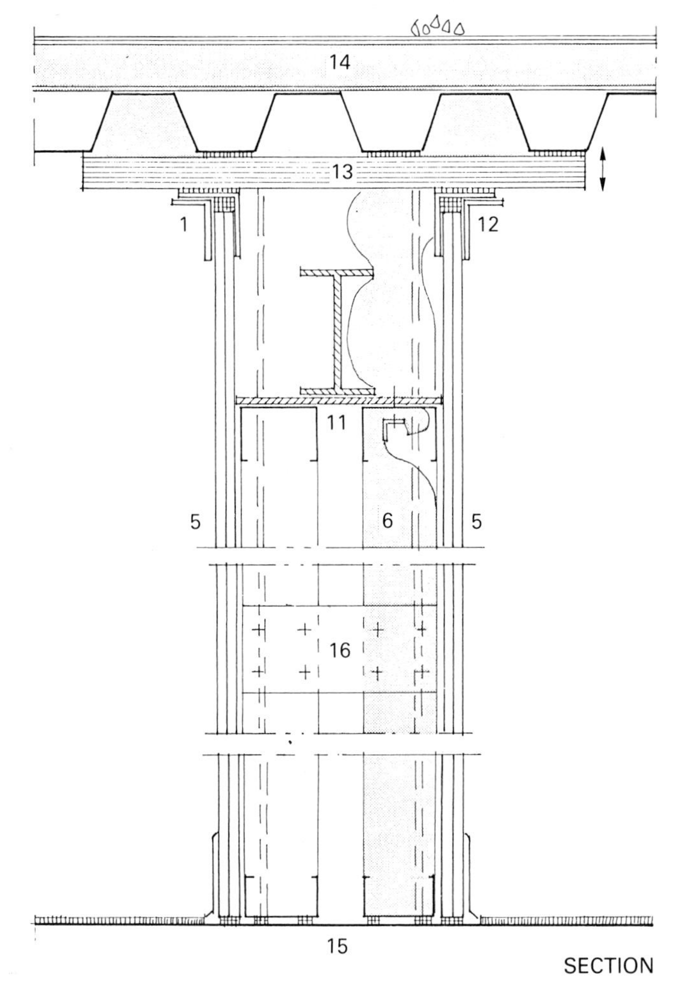

Figure 2.20(a) *Multiplex cinemas: Separating walls — section. (Courtesy of UCI (UK) Ltd/BDP)*

1. 13-mm plasterboard on 13-mm plywood (for ease of cable fixing in projection room)
2. 146-mm metal studs, bedded in acoustic mastic at wall boarding
3. 50-mm glass-fibre quilt cavity inlay
4. 2 × 13-mm plasterboard, acoustic mastic and joint taping at corners
5. 2 × 15-mm plasterboard, lapped joints, face joints taped and filled
6. 100-mm glass-fibre quilt
7. 2 × 92-mm metal studs at 600-mm cs
8. 50-mm rock fibre batts, black tissue faced, as absorption behind screens
9. 50 mm × 50 mm timber batten bedded in acoustic caulk
10. Blockwork inner leaf to outside walls: structural break at separating walls where possible
11. 2 × 92-mm metal head studs fixed to metal plate at underside of tie beam
12. Double angles at head to allow up to 25-mm roof deflection/uplift without losing acoustic integrity
13. 40-mm Vicuclad bedded in acoustic mastic and with profiles packed with mineral wool
14. Built-up chippings/felt roofing/ thermal insulation on metal decking
15. 2 × 92-mm metal base studs and plasterboards bedded in acoustic mastic
16. Bracing to separate studs at max. 3500 cs. support to quilt by 25 × 25 metal angles running between studs

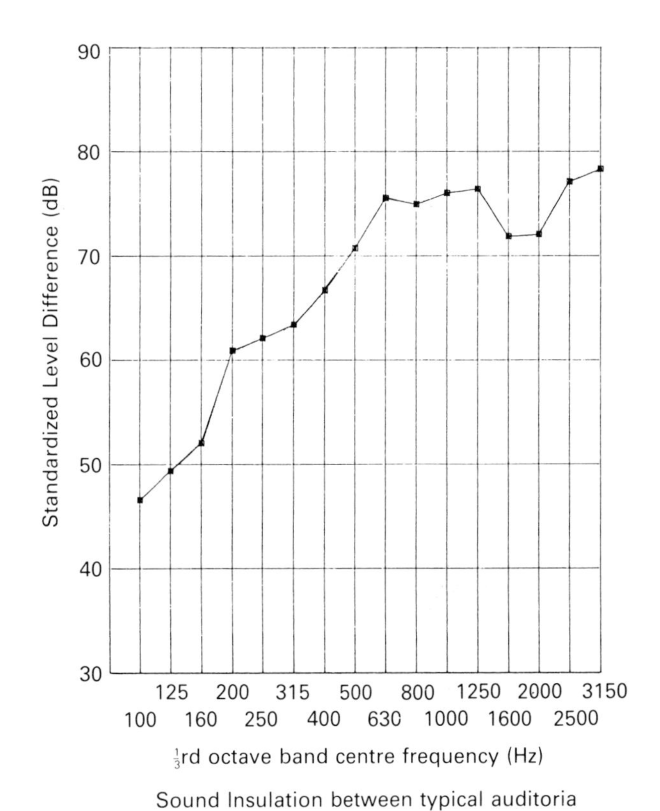

Figure 2.20(b) *Multiplex cinemas: sound insulation between typical auditoria*

between cinemas (masonry wall added between the separate studs) and low ventilation noise levels (NR25). IMax, OMNI and MotionMaster cinemas require specialized attention including vibration control to some of the cinematic effects.

Conversions

Conversion of older theatres and cinemas results in satisfactory results, although it is less common now. Multi-board dry construction linings to form, as far as possible, separate auditorium 'shells' is important; concrete rather than timber floors should be used. Results will not be as good as for a purpose-built multiplex because of flanking via the existing walls; $D_{nT,w}$ achievable may be around 55 dB.

Concert halls

Concert halls require specialist design advice so only a few principles are offered: the acoustics can be made to work for a number of different generic hall types.

Hall shape

Hall shapes can vary from the traditional 'shoebox' and coffin-shape halls, to geometric halls. 'Shoebox' halls have enjoyed a renaissance, as offering good cross reflection characteristics between side walls. They put the audience centre-front to the orchestra as far as possible for the best sound blend. Recent examples at Dallas and Birmingham are attracting publicity. The problem is that the admired nineteenth century shoebox halls like Musikvereinsaal,

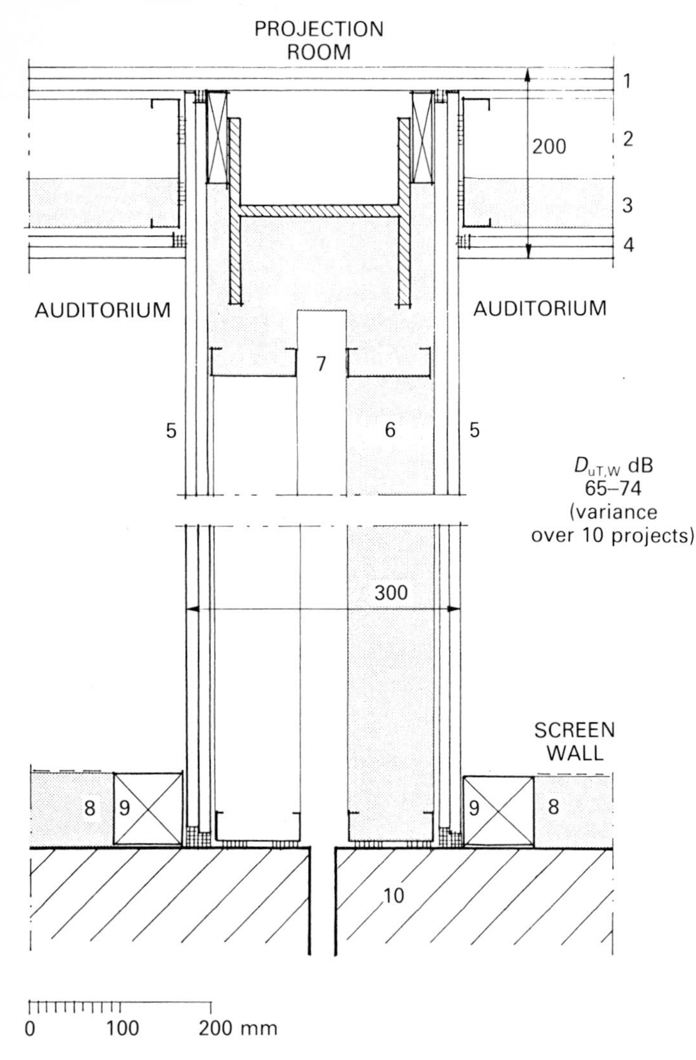

Figure 2.21 *Multiplex cinemas: separating walls — plan (see Figure 2.20 for key). (Courtesy of UCI (UK) Ltd/BDP)*

Vienna, are by modern standards only recital hall size. Even a hall like the 2206-seat Concertgebauw, Amsterdam, would be much larger if replanned to current standards of safety and seating. For a full-sized concert hall (2000–2500 seats) to suit a full orchestra and choir (up to 120 musicians plus 250 singers) the arrangement can lead to a large template – Birmingham is over 60 m long and has a very generously-sized platform focussed in amphitheatre style on the conductor's rostrum. Shoebox halls do however still work well for lesser occupancies, as Jordan Akustik's 1377-seat Odense hall testifies (Figure 2.22). Some concert halls are not even axial to the platform: Aalto's concert halls put more seating on the 'keyboard' side of the piano soloist; Segestrom Hall, Orange Free State, California, interlocks two narrow halls as a means of avoiding the disadvantages of fan-shaped auditoria. Figure 2.23 shows CAD views of a 2500-seat study, with vineyard seating around a semi-surround platform.

Reference can be made to a number of case study collections, for example *Acoustical Survey of Eleven European Concert Halls* [5] and *Halls for Music Performance: 1962–1982* [6].

Figure 2.22 *Odense concert hall, Carl Nielsen Hall*

Volume

The volume should be adequate for a full-bodied sound: the old rule of 'RT × 4 equals volume per person' is if anything on the low side and 10 m^3 per person even in a full-size hall is advisable (Table 2.6).

Table 2.6 *Optimum volumes for performance spaces*

	Optimum volume (m^3/occupant)		
	Minimum	Recommended	Maximum
Theatres	2.5	3	4
Rooms for speech	—	3	5
Opera houses	4	5	6
Concert halls	8	10	12
Churches	6	10	14

Seating

Seating is a key issue as there are so many constraints – sightlines, travel distances, aisle steps, seats per row and balance of seats at different tiers (Figure 2.24). 'Vineyard' seating as pioneered at the Berlin Philharmonie works well in milder form: 'seating trays' of several hundred can be optimally set to face the platform and have local side-reflecting surfaces.

Reverberation time (RT)

Reverberation time is still a fundamental measure in concert hall design although in recent years design concern has extended to not only RT but also to the ratio of early to reverberant energy and to lateral efficiency, i.e. the

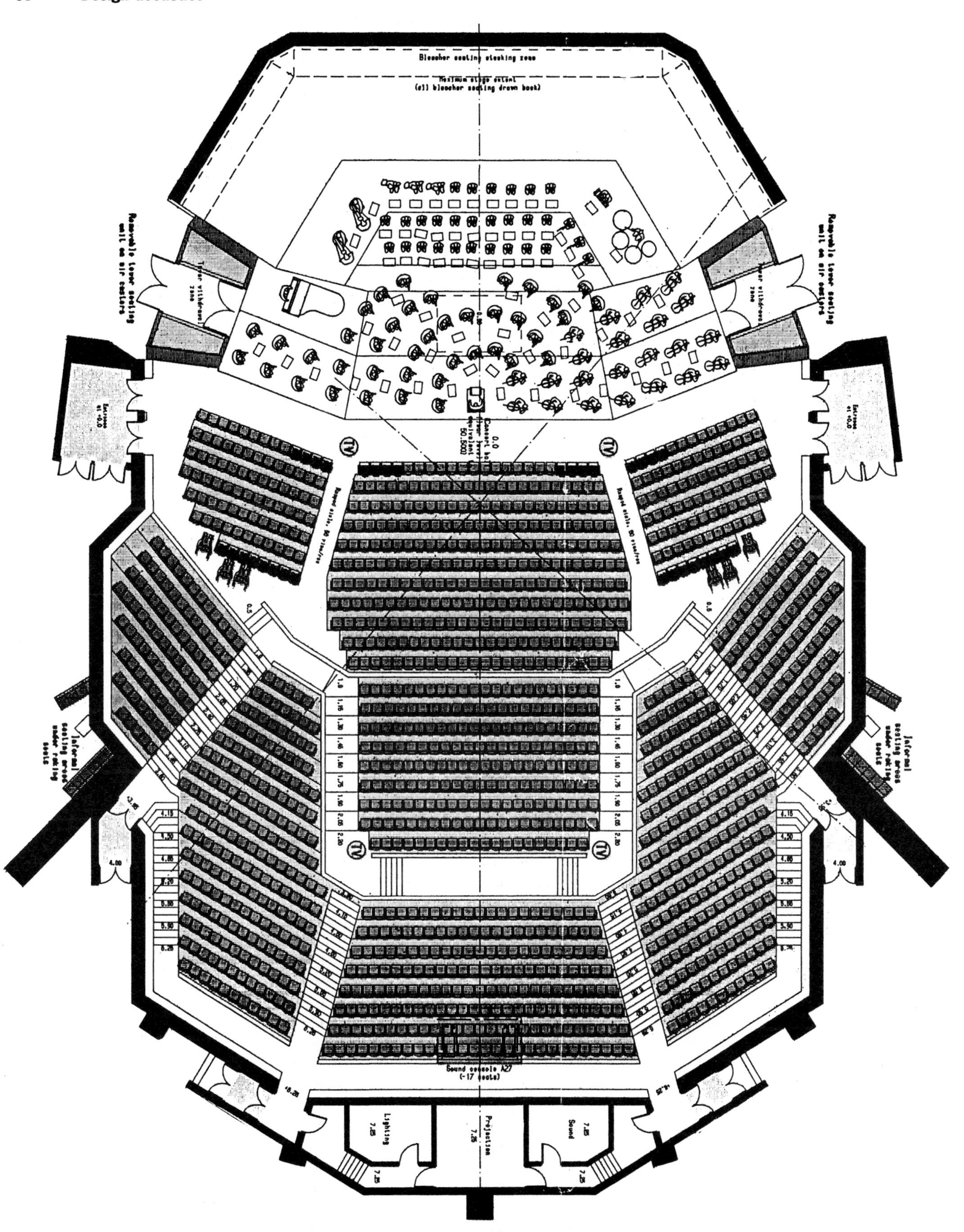

Figure 2.23(a)–(c) *Computer studies: concert hall for 2500*

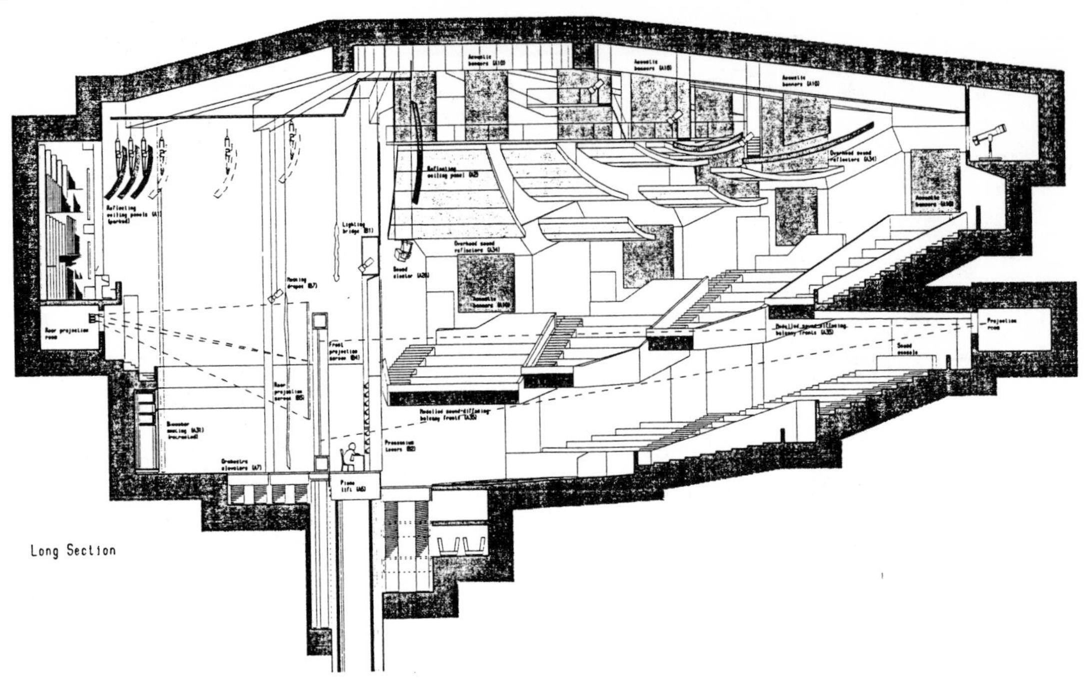
Long Section

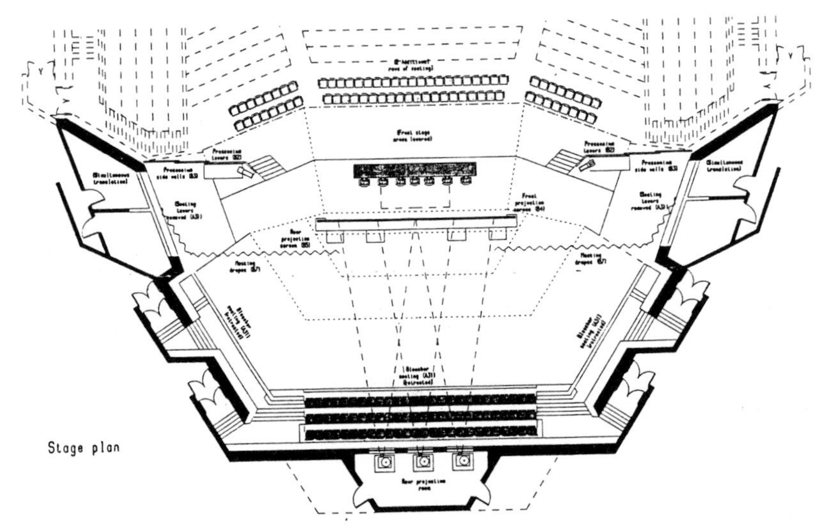
Stage plan

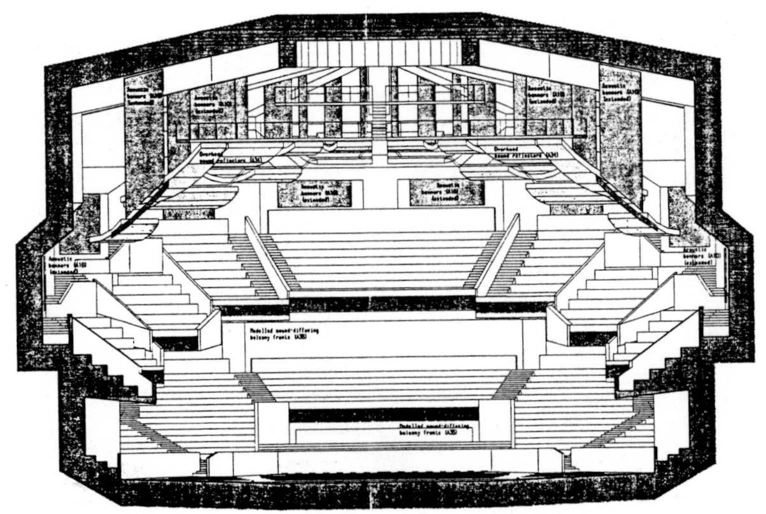
Cross Section

CONCERT HALL
IN CONFERENCE SETTING

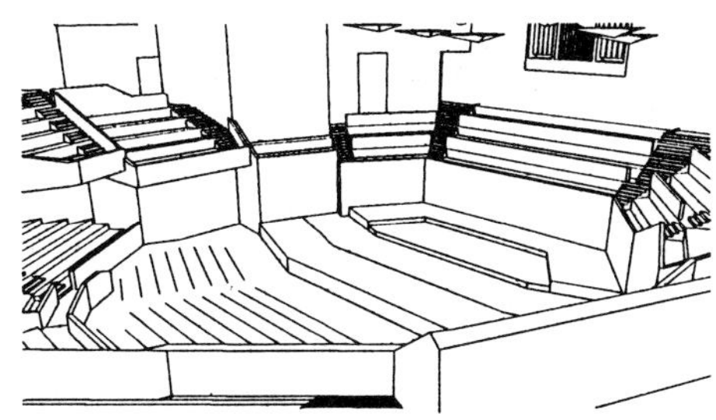
View 2: From a side entrance to the balcony level

View 3: From the balcony seating level with the stage front

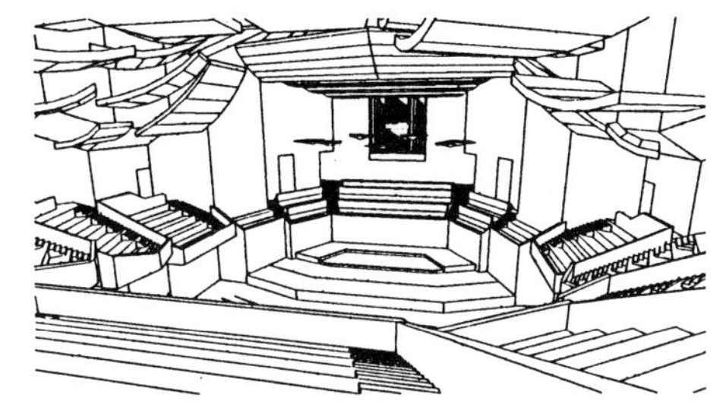
View 1: From the back of the central balcony seating

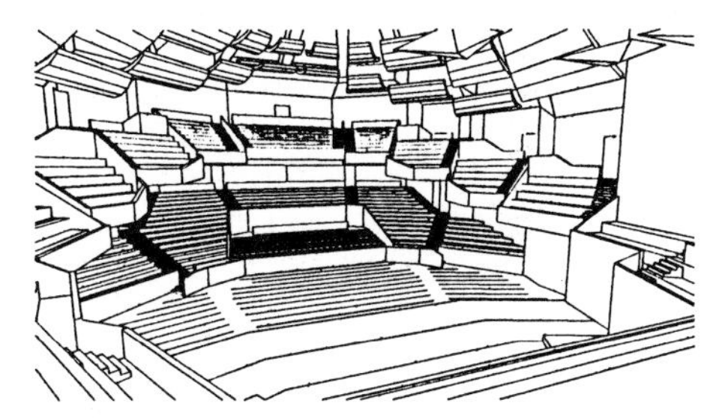
View 4: From the back of the choir seating

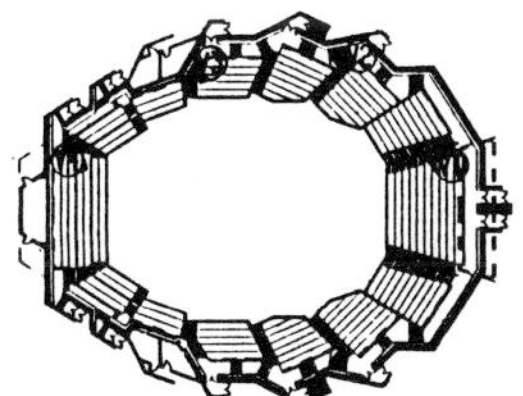
Upper level plan

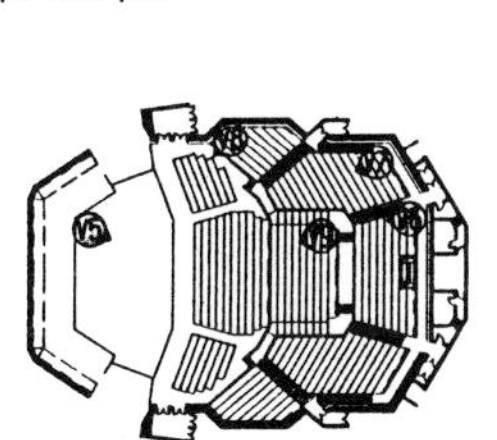
Lower level plan

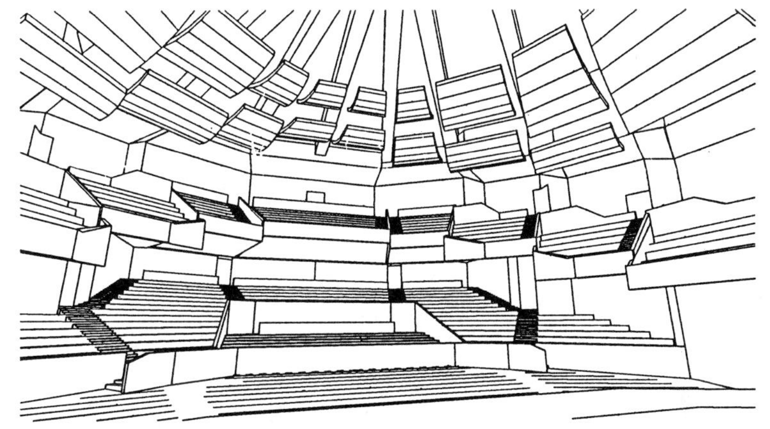
View 5: From the rear of the stage

View 6: From the rear of stalls seating area

View 9: From the central stalls seating area

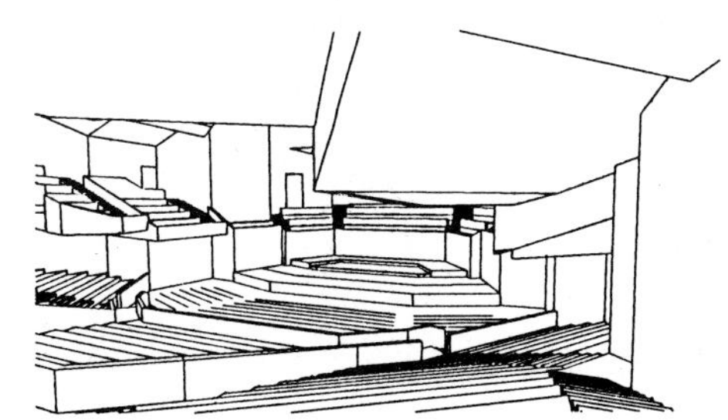
View 7: From the rear corner of stalls seating area

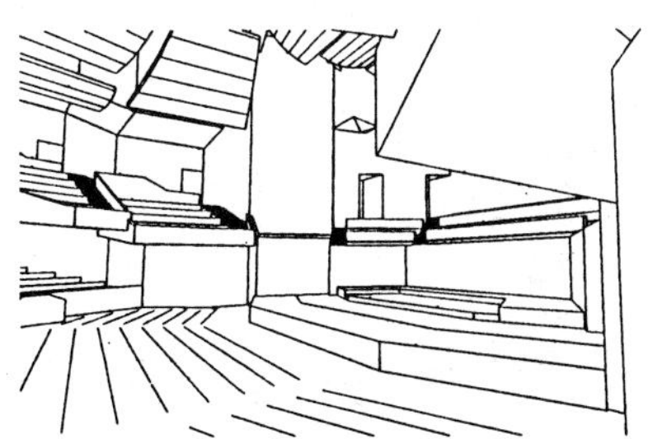
View 8: From the extreme side of stalls seating area

C A D GENERATED
INTERNAL VIEWS

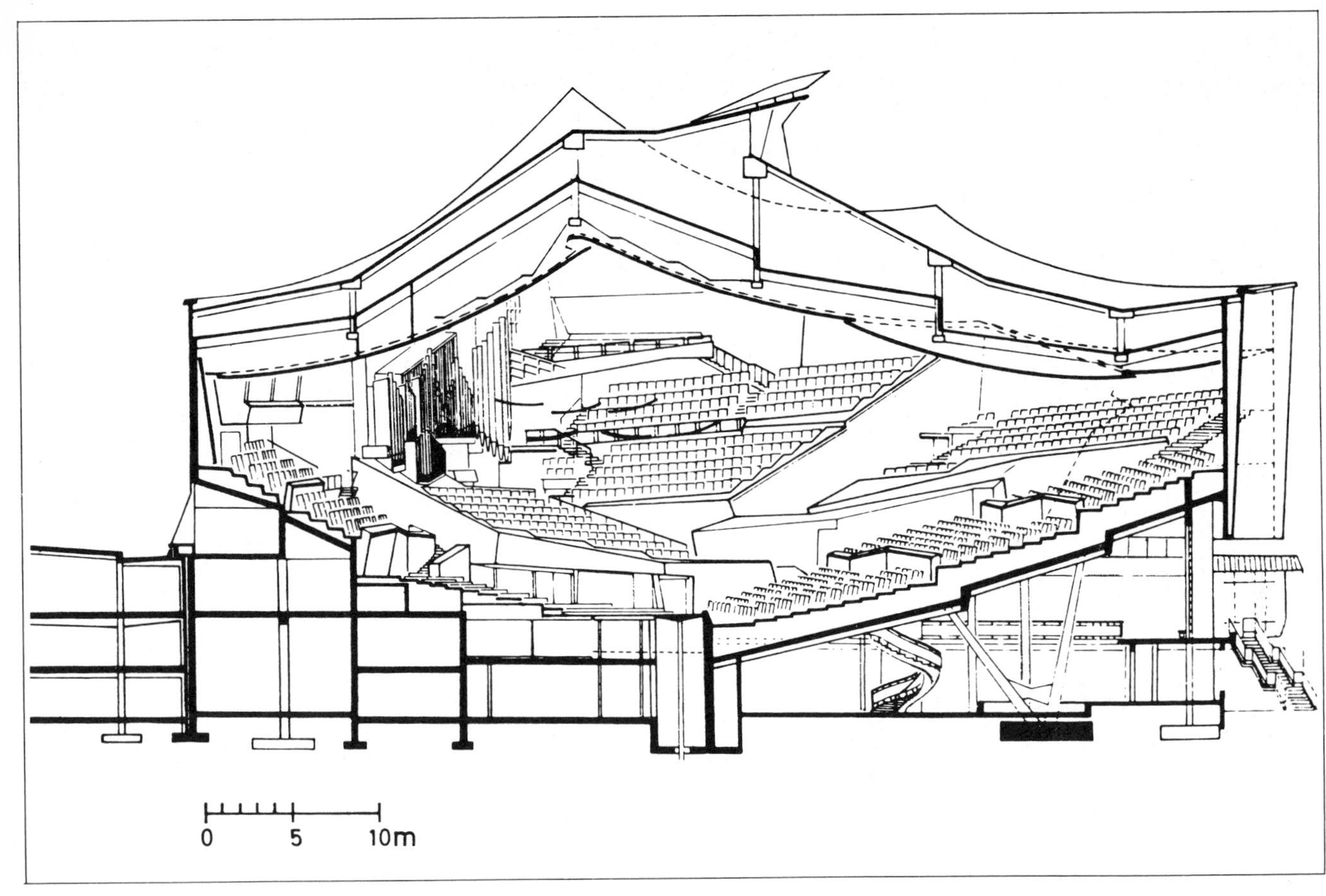

Figure 2.24 *Vineyard seating: Berlin Philharmonie (Sharoun's Main Hall)*

Figure 2.25 *Seating trays: Berlin Philharmonie (recently built Recital Hall)*

Figure 2.26(a) & (b) *Directed sound: Wellington Town Hall, New Zealand*

adequacy of early lateral reflections. A strong, precise and clear sound is the current taste; live music that is expected to match the quality of the living room CD player. The approach needs great care as, if reflections are too strong, tone colourations to sound quality can result. There may even be a loss of reverberance if much sound energy is directed straight to absorbent occupants and seating.

Early decay time

The early decay time (EDT) is the most important criterion in evaluating a hall's acoustics, followed by Lateral Efficiency and Clarity. The EDT should not differ from the Sabine RT more than ±10%; for a concert hall, values between 1.8 and 2.3 s should be sought. It is the decay time measured over the first 10 dB of energy fall-off, equivalent to the slope of energy curve measured within the first 500 ms. Due to the nature of music performance the final part of the sound decay of notes is seldom heard. The later part of the reverberant decay from a specific impulse (transient) is masked by subsequent signals after approximately 10 dB of drop, i.e. peaks in music performance only rise about 10 dB above the average level during that passage. EDT involves measuring the first 10 dB of decay and multiplying by 6 to correspond to RT values. As EDT is sensitive to room geometry, in particular to strong early reflections to reinforce sound in the first 100 ms, the EDT will vary with location around a hall (see Chapter 5).

In huge halls, the EDT can vary spectacularly, highlighting the remoteness of surfaces: the Royal Albert

Hall, London (86 650 m^3), has in its stalls EDTs of 1–1.25 s compared to Sabine RTs of around 2.5 s, across middle frequencies.

Ratio of early-to-late energy
The ratio of early-to-late energy is the measure of the balance between clarity and reverberance in music; different types of music, for example Romantic compared to Mozart, suggest different balance values.

Early lateral energy fraction
The early lateral energy fraction defines the relationship between a sense of spatial impression or envelopment for the listener and the arrival of reflected sound from sidewalls relative to the listener. It is the fraction of lateral energy arriving between 5 and 80 ms after the arrival of direct sound compared to the total sound energy arriving at the listener within the first 80 ms of direct sound arrival. The maximum values found in auditoria are around 0.3.

D50, C50 and C80
Other measures are Deutlichkeit (D50), Clarity Index C50 and Clarity Index C80. D50 derives from the ear's response to consecutive impulses. A sequence of sound impulses delayed more than 50 ms is perceived as discrete impulses, whereas those with less delay combine to enhance the loudness of the impulse before. Good auditoria will have higher D50 values. C50 is slightly differently calculated and is presented in decibel form with values greater or less than 0.5 appearing as C50s with positive or negative values respectively. For a speech-orientated hall, positive values of C50 are desirable. C80 uses the limit of perceptibility 80 ms to suit music uses, again in decibels. Positive values result in a crisp acoustic suitable for classical music and some operatic use but will not provide a suitable setting for romantic and choral works which are enhanced by a greater reverberance (values range from 0 to −3).

Directed sound
Directed sound has become an issue in acoustic design. Marshall incorporated the idea of supplying early lateral energy when he was appointed acoustic designer for a concert hall in Christchurch, New Zealand, in 1972. The lack of lateral reflections at centre seats due to the hall width is compensated for by an array of large reflector panels suspended overhead angled to direct sound into central seats deficient of lateral sound (Figure 2.26). Schroeder-designed reflector panels used in the 1983 Wellington Concert Hall by Marshall are based on a repeated 'Quadratic Residue Sequence' of different depth wells along the panel surface. Interference due to the pattern selected contributes to a wide diffusing area (Figure 2.27).

The concept of using other reflecting surfaces within the auditorium was also used by Cremer in the innovative Berlin Philharmonie; other halls in which Cremer has been involved have incorporated 'stepped hexagons' in which seating is split into hexagonal stepped terraces which provide all seats with localized directing surfaces. Even seat backs to rearmost seats can play their part (Figure 2.27).

Musicians
Musicians as users have specific needs and the priorities are slightly different to those of an audience. The Musicians' Union stresses the following needs: reverberation time (full-bodied); variability (not altered by occupancy); dynamic range; frequency response (balance of sound, no orchestral section or pitch band receiving over- or under-emphasis); clarity and separation (between solo instruments and sections). In addition, on the platform there should be integration (sound similar to that perceived in the auditorium), ensemble (ease of hearing between sections), and floor response to instruments.

Ventilation
Ventilation in halls can serve both energy-efficiently and quietly by undeerseat supply and overhead extraction (Figure 2.29). As exposed ductwork at a high level in halls presents an unpredictable source of low-frequency absorption and duct-noise breakout, its presence in the hall should be minimized or set behind an acoustic 'shell' ceiling. A sensible approach to adopting a criterion is NR 20 for concerts, and an assured NR 15 maximum for broadcasting/recording.

Multiuse
Multiuse is a fact of life, even for halls with resident orchestras. Choir seating can be on bleachers extending a platform back for conference use, seating to the immediate sides can be on towers to be removed to form 'wings'. Stage lifts can alter the configuration for different music settings, dropping to form an orchestra pit for opera. Acoustically, banners can drop within the hall to reduce the reverberation time up to 15%. This will improve speech intellibility in the hall. Overhead acoustic 'clouds' to help the orchestra maintain ensemble can be flown aside to enable sets or a projection screen to be dropped. Back projection, simultaneous translation, projection room and control rooms will help the operational efficiency. In a number of halls, large acoustic canopies slung over the platform are claimed to adjust the acoustics for different sized events.

Multiuse halls are a design challenge and the most demanding use in design criteria terms may only be a prestigious occasional use, so it is realistic to have the most frequent use as a base setting with means of altering conditions for other events. Reverberant sports or concert halls can be made deader for drama by dropping banners or drapes, small halls or dead halls enlivened acoustically by electronic means. Figure 2.30, Diagram 2.5 and Table 2.7 list some typical events which can crop up: these cluster to particular settings (flat floor, etc.) although inevitably there are varying requirements within headings, for example 'classical music' can mean conditions ranging from 1.2 s for chamber music to 2.5 s for major Romantic and choral works.

Courts
Acoustic attention centres around the courts themselves, as they will have to be isolated from other areas by sound lobbies, acoustic doors, and 50+ dB walls. Ventilation noise levels should be kept to between NR 30 and NR 35. Good conditions are important as close concentration is demanded over long sessions.

To keep out intrusive noise, perimeter accommodation as well as the courts may have to be mechanically ventilated. Floors under courts and over cells should be 50+ dB, e.g.

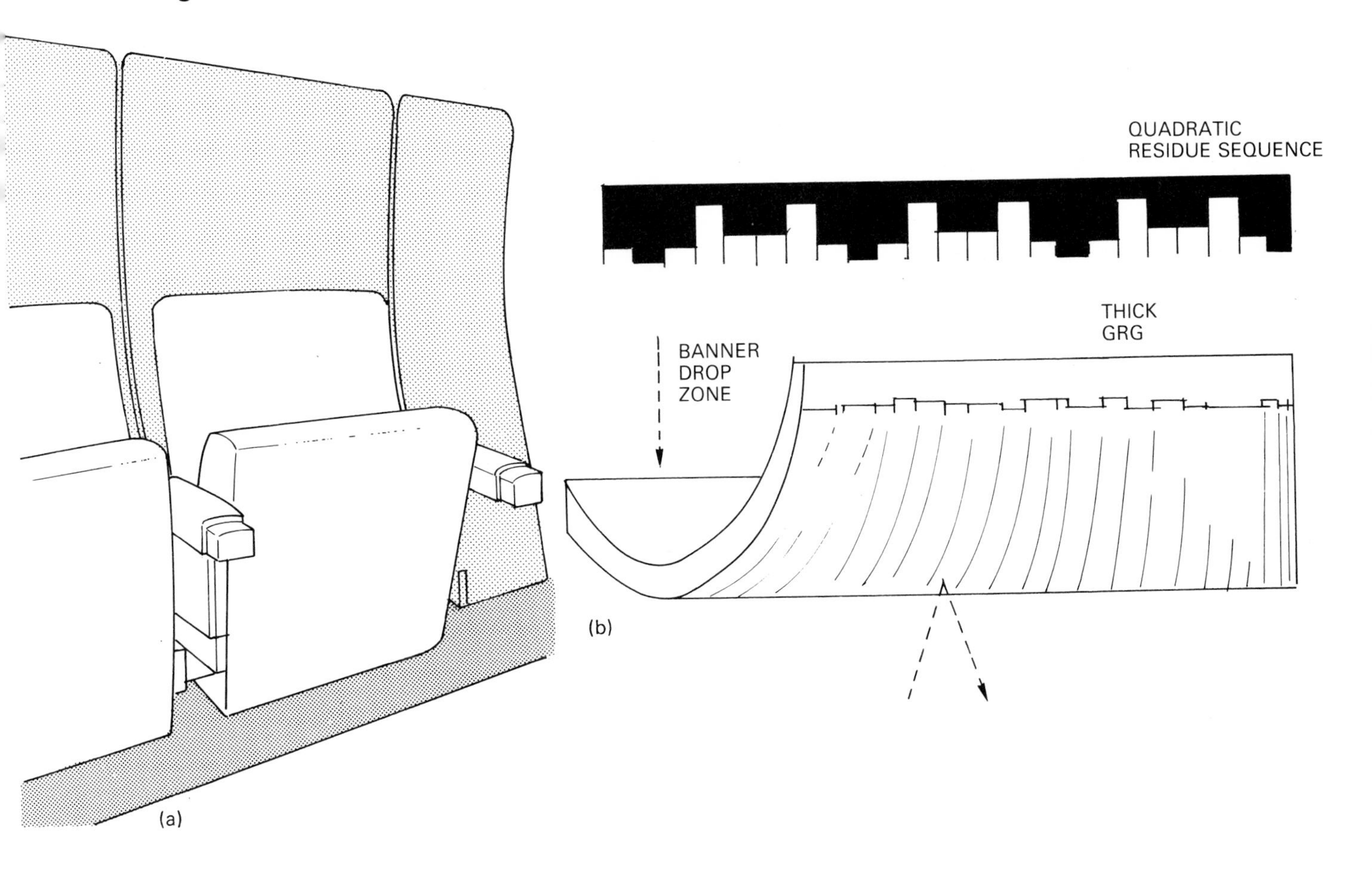

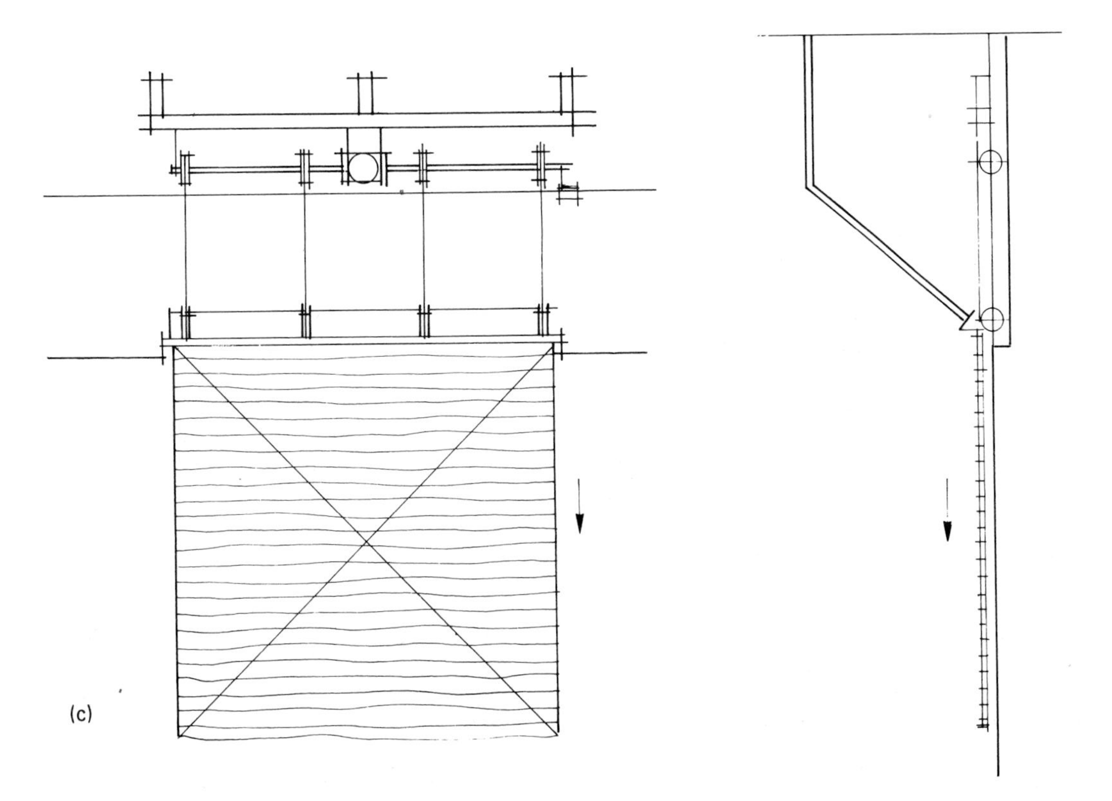

Figure 2.27 *Concert hall details: (a) high-back seats at rear; (b) ceiling reflectors; (c) side-wall absorber banners*

Figure 2.28 *Acoustic model testing: Linköping Concert Hall, Segestrom Hall (California) and Wycombe Entertainments Centre*

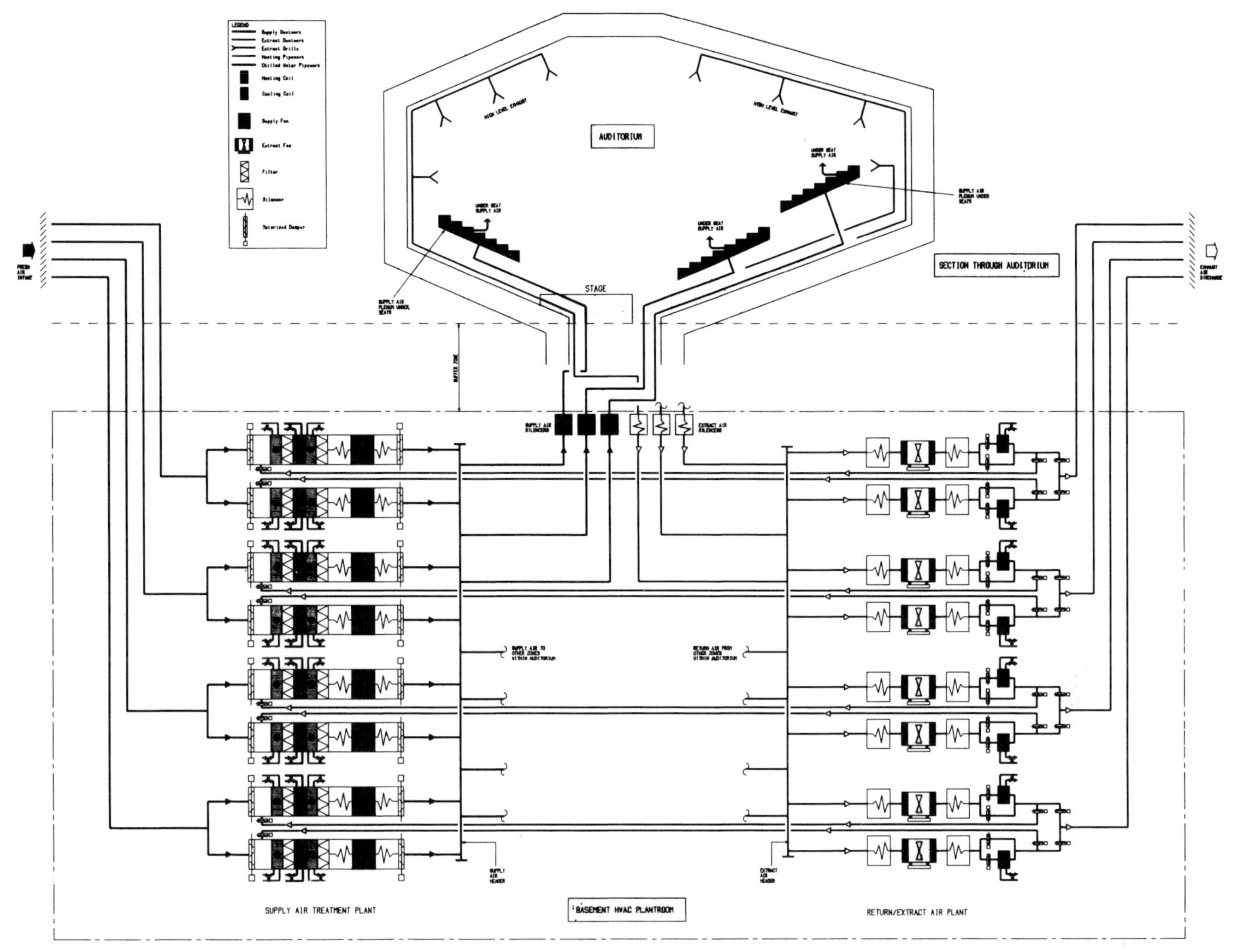

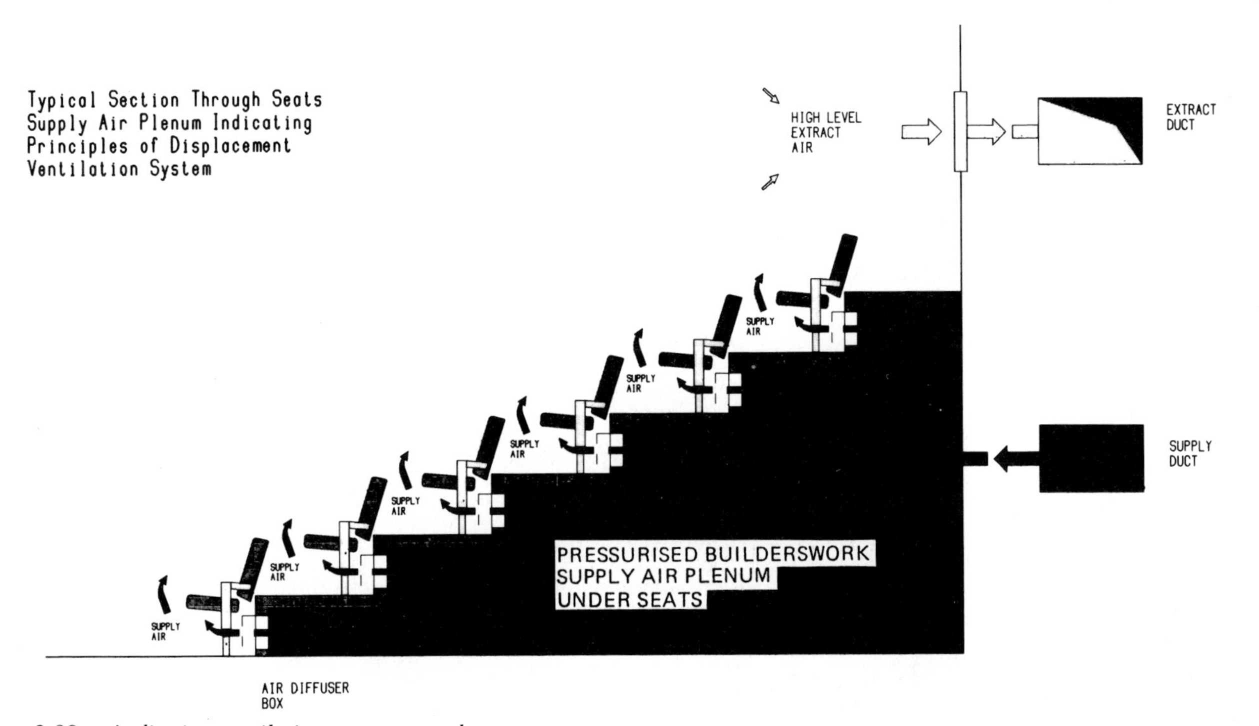

Figure 2.29 *Auditorium ventilation system — study*

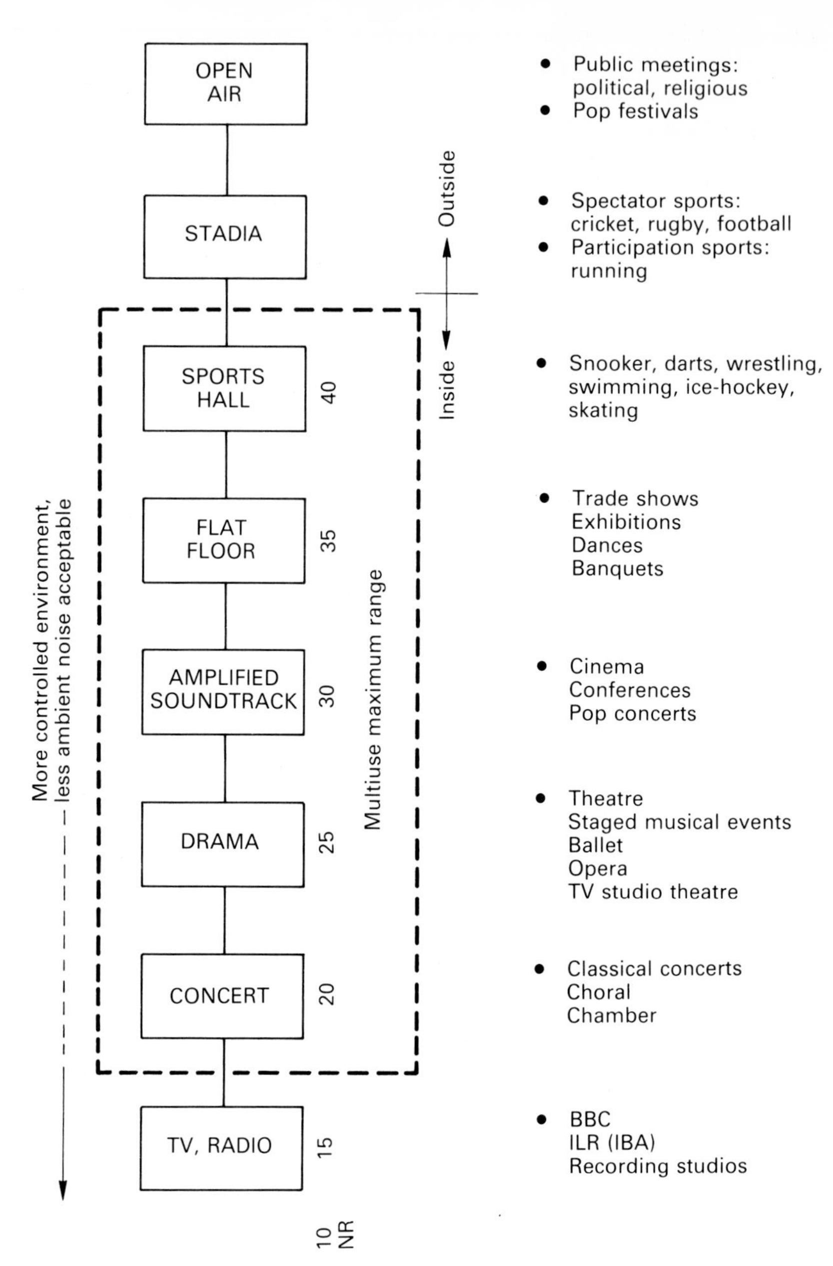

Diagram 2.5 *Public performance spaces*

Table 2.7 *Recommended mid-frequencies reverberation times*

Activity	RT	Building type
Broadcast	0.2–0.25	Sound dubbing, announcer booths
	0.3	Small speech studios
	1.0–2.0	Large classical music studios
Speech	0.6–1.2	Council chambers, law courts, lecture theatres, meeting rooms, conference halls
Drama	0.9–1.4	Theatres, function rooms
Amplified sound	0.5–1.2	Multiplex cinemas, pop concert venues, discotheques, videowall settings
Multiuse	1.0–1.7	School assembly halls, community halls, sports/arts halls
Opera	1.0–1.6	Opera houses, theatres with orchestra pits
Soloists, ensembles	1.2–1.7	Recital halls, orchestra rehearsal halls, chamber music salons
Orchestral music	1.7–2.2	Concert halls
Organ and choir music	2.0–5.0	Ceremonial halls, organ concert halls, churches, cathedrals

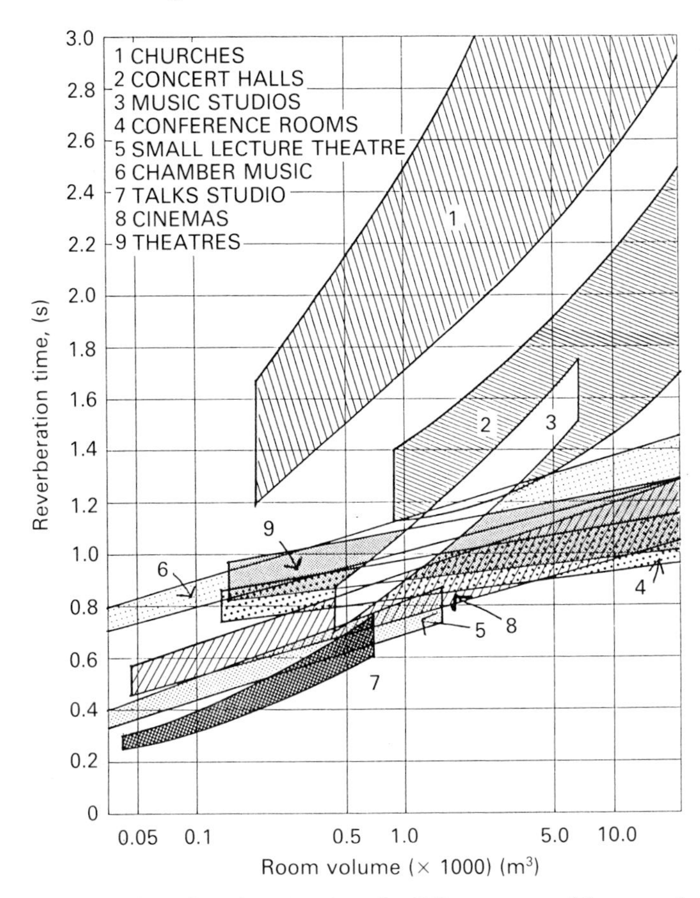

Figure 2.30 *Reverberation times for different spaces. (Courtesy of Sound Research Laboratories, Colchester, Essex)*

Table 2.8 *Specific needs of courts and surrounding rooms*

	Sound insulation	Sound absorption
Main courtrooms	•	•
Juvenile courts	•	•
Sound lobbies		•
Magistrates' retiring rooms	•	•
Court hall		•
Duty solicitor's room	•	
Social workers' offices		•
Holding/waiting rooms		•
Internal circulation		•
Clerk to the justice's office	•	
Deputy clerk's office	•	
Secretary's office		•
Legal/administration/accounts areas		•
Public zone – counter		•
Interview rooms	•	•
Consulting rooms	•	•
Refreshments area		•

200-mm concrete slab plus 50-mm topping (concrete waffle with thin minimum slab thickness or hollow precast units should not be used). Plant rooms should be kept remote from courts.

Some rooms with specific needs, as identified in Home Office Room Data Sheets, are identified in Table 2.8.

The natural acoustics of a courtroom should allow good speech intelligibility. To some extent this is engendered by the reinforcement of direct sound from the speaker by early reflections which combine with the direct sound to provide a strong signal. Late reflections (arriving much after 30 ms) are however counterproductive as they blur the original message. The desired finishes will therefore be limited local sound reflective surfaces combined with sound absorption to other surfaces: floor carpet (this will damp impact/footfall noise as well as provide general absorption), selective wall linings and ceiling treatments.

To achieve these target values local treatment will be required behind and to the side of the public gallery to absorb intrusive visitor noise, and to the side of the bench, where the confidentiality of whispered briefing needs to be maintained. The important parts for the maximum audibility are witness box to jury, witness box to judge, counsel and dock to judge. Proceedings audibility to the public seating area is a slightly lesser priority.

PSA and Home Office guidance advises as a suitable setting for court activity, that the reverberation time calculated for the furnished room shall be 0.8 s at 125 Hz falling to 0.6 s at 250 Hz and continuing at 0.6 s for frequencies up to 4000 Hz, with tolerance on these values ±0.15 s.

As far as possible, good natural acoustics should be relied on for proceedings. The judge and counsel will be used to speaking clearly and addressing the court. Speech reinforcement may be considered, however, to reinforce speech from the witness box, particularly to the public galleries at the rear of the courts. Security enclosure of the dock in some courts may impede reception of speech from this source, so here too speech reinforcement may be considered. The system should be high quality and free from audible hum, noise and distortion.

Discotheques

The attraction of discotheques is precisely their noise and frenetic activity. The problems associated are noise break-out to neighbouring properties, hearing damage to employees, and structureborne re-radiated sound to other facilities in the same building. Average sound pressure levels on the dance floor climb to the range 90–110 dBA as the night goes on, with a high component of low frequency sound. Levels tend to be 8–10 dBA higher by the end of the disco session. Where amplified live music is performed, levels can be even higher. Such sound power levels exceed classical forms of music – orchestras in loud passages may be a more modest 80 dBA or so. The Code of Practice on sound levels in discotheques [7] identifies the hearing risk for staff in particular. The Code recommends an L_{Aeq} not to exceed 100 dB at the nearest point in the premises to operating loudspeakers, the value referred to as the Maximum Permissible Exposure Level. It assumes 25% of the total public area is given to rest areas, otherwise a MPEL of 95 dB would apply. It is not desirable to have direct-to-ear sound, diffuse and reflected sound being better to even out exposure. Previous GLC guidance recommended an $L_{\mathrm{Aeq,8h}}$ not exceeding 93 dB, or for external audience protection 93 dB at 50 m. Discotheques are an obvious candidate for assessment against the recent Noise at Work HSE statutory guidance [8]. In considering an Entertainments License, any authority will consider the following:

- objection at any public hearing
- duration and timing of concert
- frequency of concerts at the same premises
- noise complainants at previous concerts
- location (in relation to noise-sensitive buildings)

Noise-limiting devices on sound systems can be included although commercially it is not realistic to set values below 90 dBA. Premises should have full mechanical ventilation (direct-to-atmosphere extracts with no attenuation will in most cases be unacceptable because of the impact on the environmental noise climate) and lobbied doorset accesses at the entrances. Re-radiated structureborne sound demands great care in design, one approach being to isolate the disco walls by using a drylining 'shell' and effectively a triple floor – a dance floor on a concrete slab on isolators on a second concrete slab – to alleviate the problem on the floor below.

Education buildings

The old but still useful Building Bulletin, *Acoustics in Edcuational Buildings* [9], and Building Bulletin 30, *Secondary School Design: Drama and Music* [10], offer detailed layout advice. DES Design Notes 17 (*Guidelines for Environmental Design and Fuel Conservation in Educational Buildings*) [11] and 25 (*Lighting and Acoustic Criteria for the Visually Handicapped and Hearing Impaired in Schools*) [12] should also be referred to. BS 8233: 1987 [13] classifies four groups with sound insulation requirements varying from 25 dB to 45 dB. Classroom conditions should be controlled to 0.75 s at middle frequencies and 40 dB average separation between reading areas. More importantly, overhead ceiling surfaces can usefully be sound reflective if side and rear walls are panelled in say pin boarding to damp sound-blurring cross reflections. General mechanical ventilation should be designed to within NR 35 in teaching rooms. In primary schools in particular, opening doors and windows from classrooms are expected, so a consideration of facade aspect will avoid distracting external noise levels.

School theatres are no longer exclusively an adaptation of the assembly hall but mimic public theatres. A reduction in scale, including platform height and size allows for the lesser projection of child voices compared to adult.

Health buildings

Noise control rather than room acoustics is important. Hospitals are highly serviced and reference to Building Notes, Technical Memoranda, Health Circulars and Hospital Data Sheets should be made. Recommended design criteria are shown in Table 2.9, as a more detailed interpretation to rating values included in Chapter 3. Separation between rooms has to be carefully considered given the requirement to stop most partitions off at ceiling level.

New major hospital developments have considerable impact on the local community and the services centre with its standby diesels, boilers and chillers needs particular attention. Attenuation in the ventilation systems is by absorptive material protected by plastic membrane and perforated sheet to avoid the risk of fibrous particles release. Sound-absorbing ceilings (cleanable) and cushioned vinyl floor finishes will contribute some noise control within wards.

Hotels

Hotels vary in their standards, most new-built projects being 2- or 4-star. To serve guests, they will usually be near busy roads or airports, or in city centres. The main issues will be noise break-in from outside, privacy between rooms and to public rooms, and ventilation noise.

Windows

Windows in hotels have opening lights even in the noisier situations; good weatherstripping and double glazing are essential. Some protection to road noise can be given by inset balconies.

Privacy

Privacy between rooms will be of a reasonable standard if separating walls and floors are selected with an average SRI of 50 dB (for example, plastered 200-mm dcm, blockwork and solid precast concrete floor units with structural topping). Creation of a lobby outside the en suite bathroom will give isolation to corridor noise. Cross-talk attenuation to bathroom extracts will prevent this being a route for plumbing sounds. If single doors from corridors are used, these should be 35-dB rated, i.e. solid core plus seals, well rebated. Bathroom–corridor walls should have an average SRI of 45 dB. Partitions must extend full height, structural floor-to-floor, and weaknesses like back-to-back electrical sockets must be avoided. Room television and radio sets should not be fixed directly to the room separating walls.

Ventilation noise

Ventilation noise should be kept within NR 35 in any hotel and down to NR 25 in good-standard bedrooms. It is argued that unless a system is audible, guests will think that it is inoperable. Atmosphere connections and chiller plant

Table 2.9 *Recommended noise ratings for health care facilities*

Facilities	NR[a]
Quiet wards, overnight stay rooms, chapel, resuscitation	25–30
Children's wards, treatment and recovery rooms, staff rest rooms, staff bases, offices	35
Operating theatres, circulation, utility rooms, day rooms, pharmacy, reception areas	40
Kitchens, laundry, changing rooms, OT exercise areas, X-ray process areas, clean rooms	45
Utility rooms, stores, cleaners' rooms	50

[a]dBA levels approximate to NR + 6.

should be remote to hotel bedrooms, or well screened and attenuated. Time clocks on, say, kitchen extracts could help by providing a cut-off time so plant is not noisy in the early hours. Plumbing noise, particularly 'water hammer', should be avoided by a 'head' to water pipework (30″/720 mm US practice) or a balloon-type relief valve.

Housing

A BRE Report, *Building Regulations and Health* [14], mentions a 1980 survey where 18% of residents of new houses said they were 'seriously bothered by neighbour noise'. Understandably, statutory controls centre on providing reasonable conditions in peoples' homes. Detailed advice is given in CIRIA Report 114 [15]. The latest Building Regulations Part E (June 1992 updating the 1985 version) [2] extend insulation requirements to the conversion of houses into

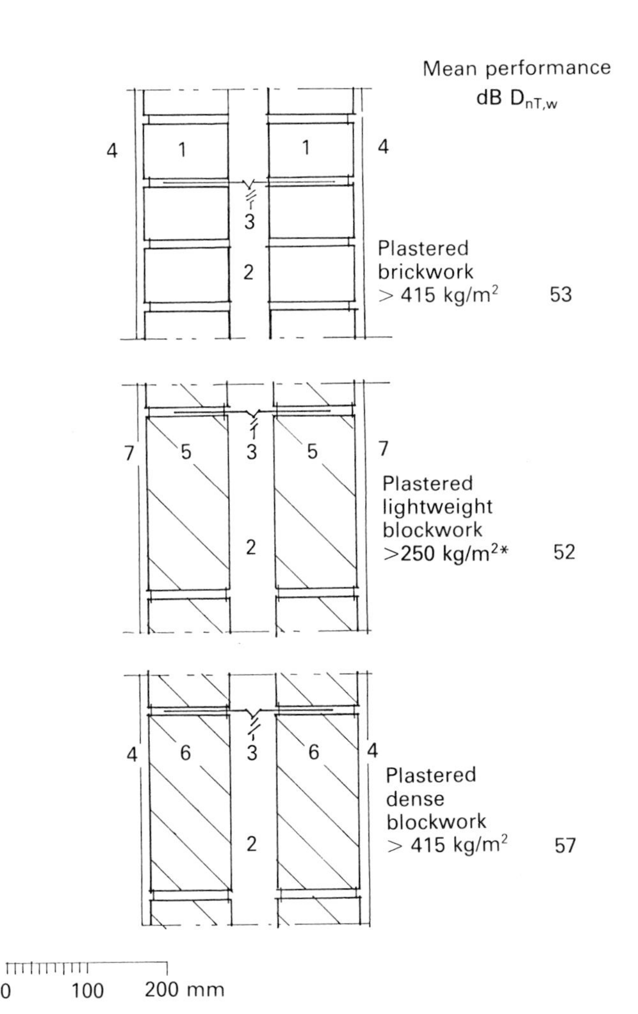

1. Solid common bricks, frogs fully filled
2. Cavity at least 50 mm wide
3. Wall ties spaced at 900 mm horizontally; 450 mm vertically
4. Wall plaster, 13 mm
5. Lightweight blockwork, 100 mm
6. Dense blockwork, 100 mm
7. Plaster or plasterboard dry lining on dabs, 13 mm

*amended to 300 kg/m² in June 1992 Building Regulations (original specification can still be used with a step or stagger)

Figure 2.31 *Selected cavity masonry separating walls. (Source: Approved Document of Building Regulations; CIRIA 114)*

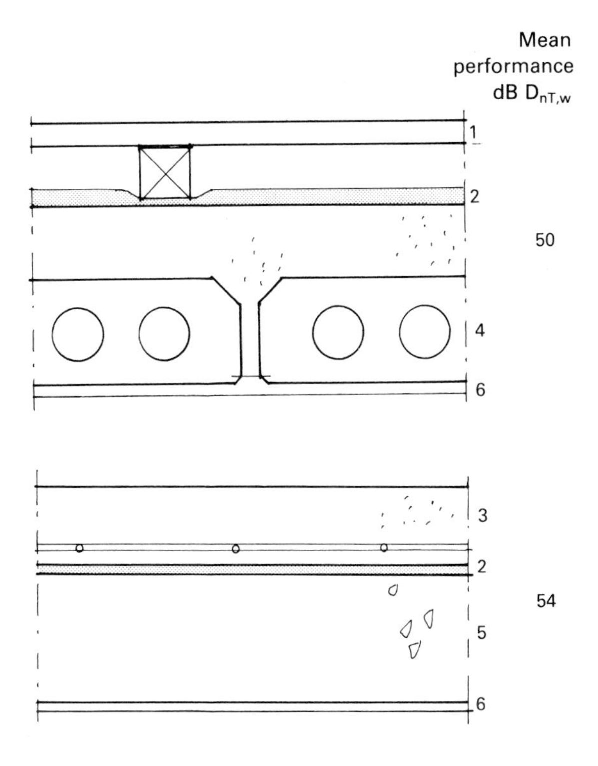

1. 18 mm T&G Floor boarding or 22 mm flooring grade chipboard on 50 × 50 mm battens
2. 13 mm/36 kg/m³ mineral fibre quilt resilient layer
3. 75 mm reinforced concrete screed
4. Cored precast concrete units, at least 220 kg/m²*
5. In-situ concrete on permanent formwork
6. Plastered soffit

*amended to 300 kg/m² in June 1992 Building Regulations

Figure 2.32 *Selected floating floor constructions. (Source: Approved Document of Building Regulations; CIRIA 114)*

flats and increase significantly the surface mass of concrete in party floors. They determine issues like party walls and the surface mass of walls that can be taken through party floors (in flats). The minimum sound insulation for party walls is 52 dB ($D_{nT,w}$). The corresponding figure for party floors is 51 dB, and for impact noised the maximum value is 62 dB $L_{nT,w}$. Some sample Approved Document Constructions are illustrated in Figures 2.31 and 2.32.

A frequent difficulty is maintaining impact isolation between flats where tiled floor kitchens or bathrooms are installed. There have been recent developments in thin isolating screeds. In considering intrusive noise, environmental health officers will bear in mind recommended maxima for steady intrusive noises (BRE Digest 266 [16]/ BS 8233) of $L_{Aeq,T}$ 30–40 dB for bedrooms and 40–50 dB in living rooms.

For proposed residential developments, land where the existing or predicted $L_{Aeq,T}$ within 15 years is 65 dB should be avoided or sound insulation measures should be provided. Near airports, planning permission will be refused for a location subject to 60 NNI (Noise and Number Index) or above, and doubtful even with sound insulation measures for NNI 40 to 59 (see Table 1.7). Developments adjacent to railway tracks will not only experience high airborne noise but also within 30 m may experience ground vibration effects, depending on ground strata. The necessity for special foundation design should be avoided if possible. A control value for road traffic noise is 68 dB ($L_{A10,\,18h}$)/ 65 dB ($L_{Aeq,\,18h}$). For industrial noise BS 4142: 1990 [17] provides guidelines on determining the acceptability of background external noise in an area.

Industrial buildings

The main issues are noise break-out (particularly for 24-h operating buildings like printing works, bakeries and flour mills) and good conditions for workers within. The planning authority may be expected to set a limiting value at the boundary of the industrial premises, particularly if housing is involved. Noise break-out may occur via the body of the building, if airborne noise levels are high inside and the building is constructed of lightweight cladding. Break-out will also occur via atmosphere processes and ventilation plant connections, for example flues, smoke and process extracts, and via goods doorways (particularly roller shutters) and personnel doors.

The traffic flow to industrial buildings may itself be a noise source problem, for example vehicles parked outside a dairy-produce factory with refrigerator plant running continuously.

Ancillary sources like signal klaxons, public address sound leakage or external PA, and occasional sound from tests of emergency procedures – standby generators, smoke shutters, valve releases – can add to the process and ventilation sources as regards noise break-out.

Within industrial premises, the protection of employees is afforded by legislation, in particular the 'Noise at Work Regulations' of the Health and Safety Executive. The effect of the updating of this statute on 1 January 1990 was that three times as many employees became involved. Failure to comply can mean prosecution or even closure for an employer. There is a general duty to reduce risk to employees' hearing, on employers and designers, by reducing exposure to the lowest level reasonably practicable. In known noisy places of work, noise assessments should be made by a Competent Person, and records of assessments kept until new ones are made. Ear protection zones are designated areas where noise levels will trigger the 'Second Action Level' as defined by the regulations. The unit which applies is the Daily Personal Noise Exposure Level which relates the potential for damage to hearing to both level and duration.

$$L_{EP,d} = 10 \log \frac{1}{T_o} \int_0^{T_e} \left[\frac{P_A(t)}{P_o} \right]^2 dt$$

where T_e = duration of exposure, T_o = 8 h, $P_A(t)$ = instantaneous sound pressure (Pa) varying with time, $P_o = 20 \times 10^{-6}$ Pa.

The trade-off has an additional 3 dBA on noise level offset by a halving of duration, 85 dBA $L_{EP,d}$ is the First Action Level and 90 dBA $L_{EP,d}$ is the Second Action Level. Claims of industrial deafness hinge on causation, showing loss of hearing on the balance of probability is due to noise at the place of work. The provision of protection and regard for levels and duration plus a plaintiff's exposure to noise prior to employment by the defendant may show that there has not been a Breach of Duty. Contributory Negligence by the plaintiff may occur if he has not worn ear protection when it was offered or has chosen to ignore directives on duration of noise exposure. The standard method for testing hearing is Pure Tone Audiometry, where the employee's hearing threshold is established at defined frequencies.

In setting noise hazard limits the following should be considered:

1. continuous noise sources (to avoid hearing loss),
2. impulsive noise sources (to avoid temporary threshold shift),
3. perception of danger warnings, signals (fire alarms, machinery start-up),
4. adequacy of speech communication.

For only modest local control (up to 9–10 dBA reduction) a noisy process, for example band resaw machines, can be screened off using flexible loaded PVC hung full-height, of around 5 kg/m^2 surface density lined with 25-mm polyurethane. Solid panelling of 13-mm fibreboard with 25-mm bagged rock fibre lining instead can lower the level at operator position around by 12 dBA. A modular metal panel system can achieve a 25 dBA reduction even with small openings for conveyors to pass through, whilst full enclosure test cells can obtain a 30 dBA reduction. Modular GRG equivalents are not quite as effective, with a reduction potential of 20 dBA. Applications are illustrated in the HMSO/Health and Safety Executive publication, *100 Practical Applications of Noise Reduction Methods* [18]. Another useful reference is SRL's *Noise Control in Industry* [19].

Often noise in industrial premises is a mixture of high constant noise (engines, compressors and conveyors) and high maximum noise events (hammering, grinding and stamping). Pneumatic power processes can have intense high-frequency noise components. Localized activity in a large industrial interior forms noise 'hot spots'. If noisy processes are spread out, general treatments will help because workers will each experience noise as a mix of direct sound from other nearby activity and reverberant sound, as well as near-field noise from their own efforts. If, on the other hand, direct sound from very noisy processes nearby dominate, the reduction of the reverberant component of noise will be of little benefit. Only both a noise survey and an understanding of activities can throw light on this.

A room treatment of ceiling absorbers can be of benefit as follows:

- sound decay away from industrial noise sources will be greater, perhaps becoming −5 dB/doubling of distance rather than −3 dB/doubling of distance;
- reverberant levels will be lower and there will be a slight reduction of continuous noise, for example from extract fans, due to the added room absorption;
- reflected sound will be reduced;
- the 'ringing' character of sound impacts will be lessened.

An 'applied' treatment like hung absorbers can be efficient but a built-in inclusion of absorption is cost-effectively dual purpose. The roof soffit can be rendered absorptive rather than reflective by using a perforated profiled metal deck rather than a plain profiled metal deck, or a lining treatment that is inherently absorptive. The roof offers greater scope than walls because in a large factory, wall surfaces will be of relatively less surface area and perimeters may be remote to working areas. One problem with perforated soffit roof decks is the tendency for a vapour barrier directly behind the perforations to some degree blank off the higher frequency absorption of the quilt behind the vapour barrier. This can be reduced by placing the vapour barrier as an interlayer between absorption and thermal quilt layers.

Lecture and conference rooms

Lecture rooms

Lecture rooms, purpose-built for 50 up to 500 persons, are a feature of many education and business centres. Up-to-date techniques have revolutionized forms of presentation: videoconferencing, simultaneous translation, back projection, video recording (for training use) and CCTV.

The lecture theatre, unlike say the council chamber, has set locations for speaker and listener, so finishes can be tailored to flatter the speaker. The front two-thirds of the ceiling should be sound reflective rather than absorptive, the rear absorbing; better still, reflecting ceiling panels can be optimally tilted to give strong early reflections via the ceiling, reinforcing the direct sound. Good sightlines are essential for adequate sound reception: a dais at the front and tiered seating rows enable this.

Reflecting surfaces at the speaker end can be 'hard' but best modelled to avoid local cross reflections and resultant flutter echoes which will be off-putting to the speaker and will lessen speech intelligibility for the audience. The rear of the lecture room should be sound-absorbing to damp long sound path reflections from the back (Figure 2.33). In considering whether overhead reflecting surfaces usefully reinforce the direct sound, one can consider a limiting

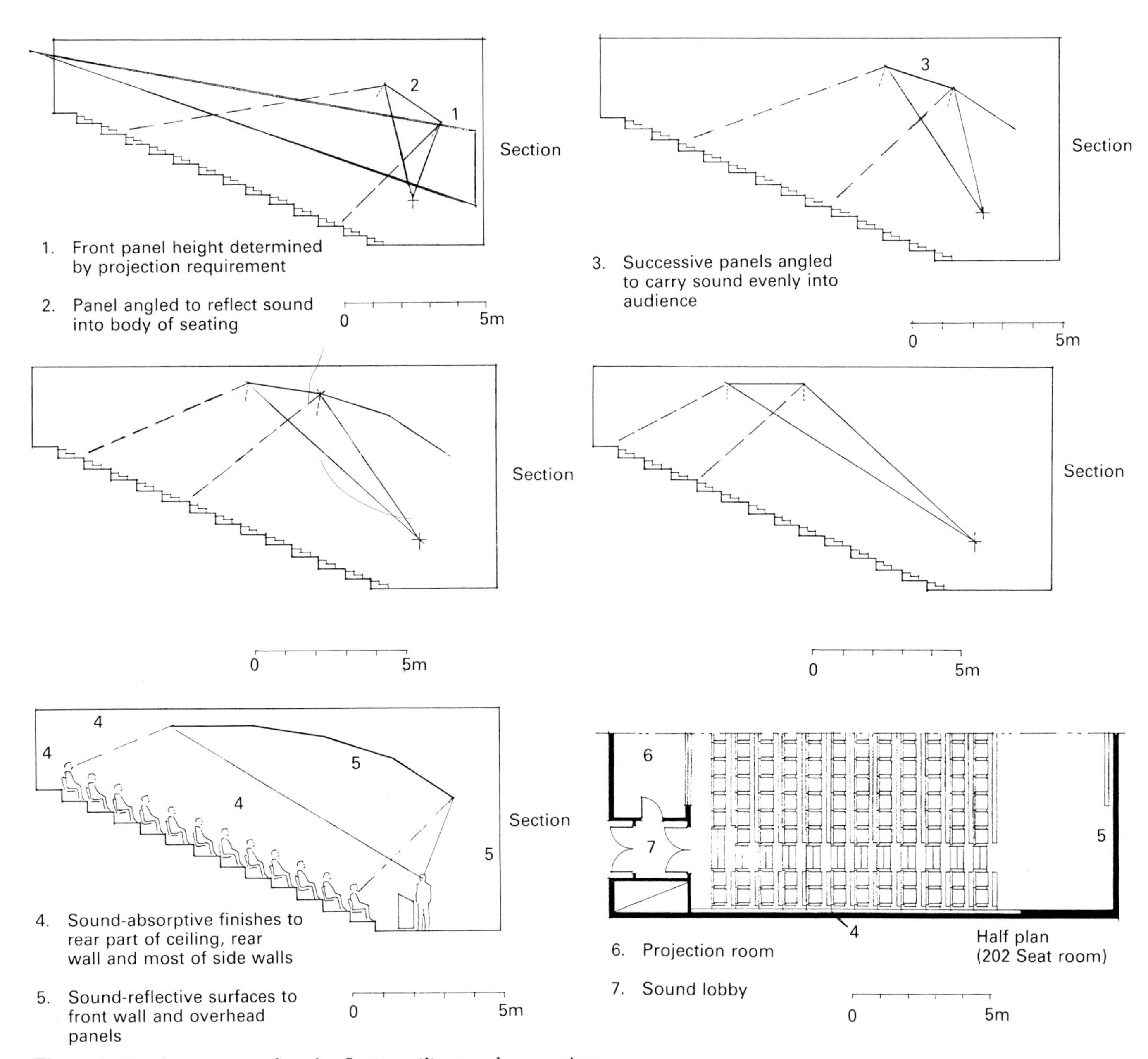

Figure 2.33 *Lecture room: Sound-reflecting ceiling panels — sections*

ellipse within which surfaces can reflect back without echo, but outside of which surfaces should be either absorptive or reflect incident sound so it remains outside the ellipse. A discrete second image of sound is discerned if the reflected sound energy arrives more than 40 ms after the direct sound, i.e. if the distance travelled approaches or exceeds 14 m. The notional limiting ellipse is therefore defined by $AB + AC < BC + 14$ where C is the source point, B the receiver location, and A the surface of reflection.

In a space with good room acoustics for speech (RT 0.5–0.75 s) the following guide applies:

- up to 15 m relaxed listening
- 15–20 m good intellibility
- 20–25 m satisfactory
- 30 m limit of acceptability

Reading lip movements helps intelligibility, which is of assistance up to 15 m. When the distance between the speaker and farthest listener exceeds 10–15 m, a speech reinforcement system may be considered.

Video conferencing

Video conferencing is a new development entailing voice-activated cameras and microphone systems to connect specially-adapted meeting rooms. Ideally the rooms could be designed to talks studio standards but more frequently standard conference rooms are adapted. The wall behind the seated participants should be fully treated with absorptive facing. Intrusive outside and ventilation noise sources should be kept within NR 25.

Conference rooms

Conference rooms vary from meeting rooms to large auditoria where massed delegates can attend a convention. Ideally, office conference rooms should have sound-reflecting ceilings but edge absorption strips and absorptive wall linings to at least half the wall surfaces (Figure 2.34).

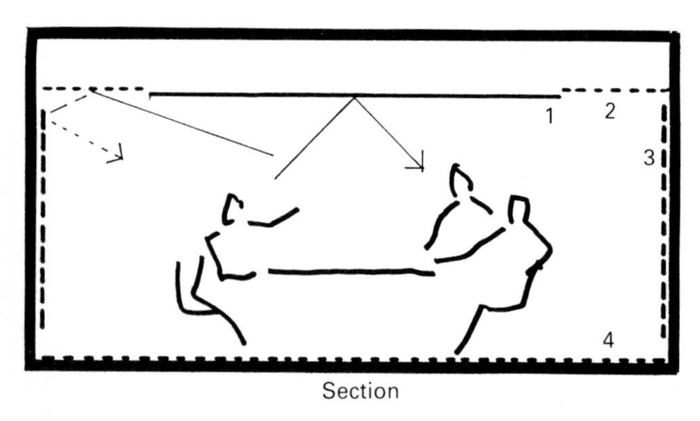

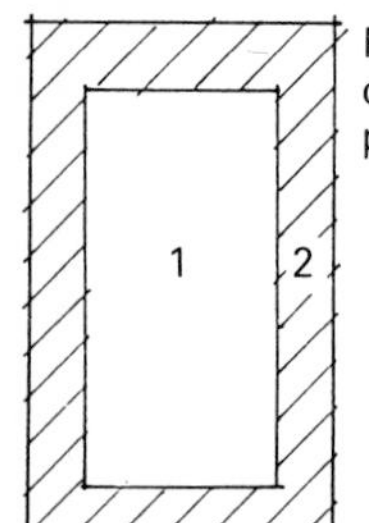

Figure 2.34 *Reflecting and absorbing surfaces in a small meeting room*

PSA and CIBSE recommendations suggest for intrusive/ventilation noise criteria, NR 25 for 'large' conference rooms (>50 persons), NR 30 for rooms holding more than 20 persons, and NR 35 for the smallest rooms. Sound lobbies should be planned into both larger lecture rooms and conference rooms.

Libraries and museums

Library activity varies from busy popular fiction and cassette loan areas to quiet reference areas, so the subdivision of the facility by bookstacks and exhibition screening can allow this. Sound-absorbing carpet, acoustic ceilings and soft furnishings can help keep reverberant sound levels low. Ventilation noise should be controlled to NR 30, intrusive noise from traffic to 45 dB ($L_{Aeq, T}$).

Museums should be lively centres of activity and 'interactive'/participatory exhibits may have to compete noisily with repetitive video presentations. Careful zoning, sound absorption materials, and good-quality directional sound systems can help. 'Theme tours' are a new derivative. The close arrangement of different 'sets' in a tour can allow the effects to be spoiled if noise from one area is distracting and intelligible in other areas.

Music practice rooms

The standard of facilities varies widely between the rooms provided in state school music departments – little different to normal classrooms – and isolated, controlled environments provided for professional, and trainee professional, musicians. The smallest practice rooms hold only two (instructor and soloist); slightly larger ones take a small group. Large rehearsal spaces can hold sections of the orchestra. Surfaces within can usefully be sparse (RTs typically 1 s at 500 Hz for practice rooms, 1.5 s for rehearsal rooms), and non-parallel: offsetting alternate walls in a row of rooms by 7° or more is adequate to prevent distracting cross reflections. Low-frequency absorbers may be useful to balance reverberation characteristics, and velour curtains can allow some user-choice of playing conditions. Ventilation and steady intrusive noise should be controlled to within NR 25 and in all but the quietest settings, rooms should be fully mechanically ventilated. Adequate cross-talk attenuation is essential.

The critical issue is the isolation afforded to the rooms. School rooms tend to be a lower specification both because of costs and the desire of teachers to know pupils are practising in individual rooms. Single acoustic doors rather than using sound lobbies may have to suffice. Figure 2.35 shows the commissioned results for a school. The isolation from separated wall leaves is not fully realized becauses of the flanking effects of roof and floor continuity. The more costly but effective isolating ceiling and floor shown in Figures 2.36–2.38 show a worthwhile gain in performance, again by care with acoustic doors in sound lobbies. Even with this degree of separation, music practice will be discerned in the adjacent practice room. Care in workmanship and supervision is needed, as contractors find it hard to resist tying structures together for stability during building.

An alternative approach, comparable with studio technical facilities, is to use modular dry-construction 'boxes' erected within a building shell. These may well be as expensive for the same acoustic performance, but have the

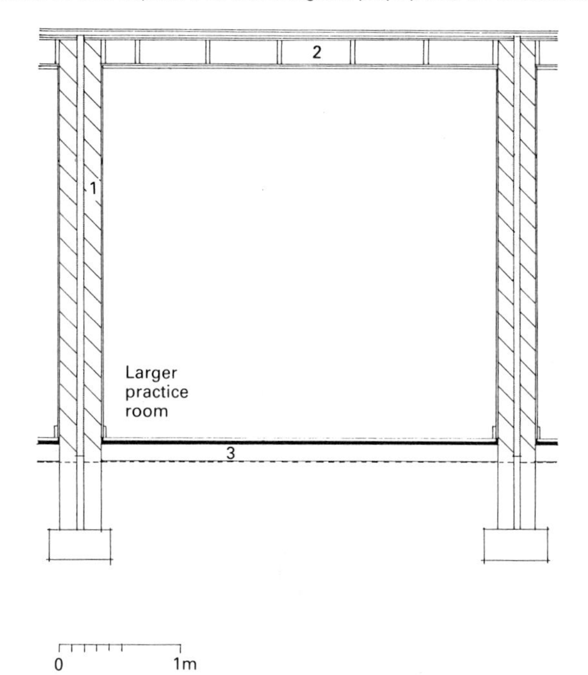

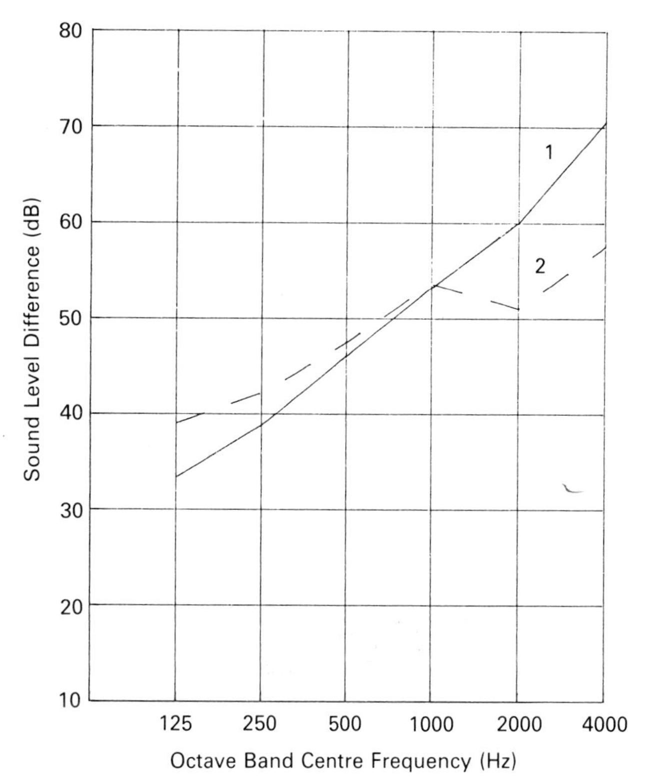

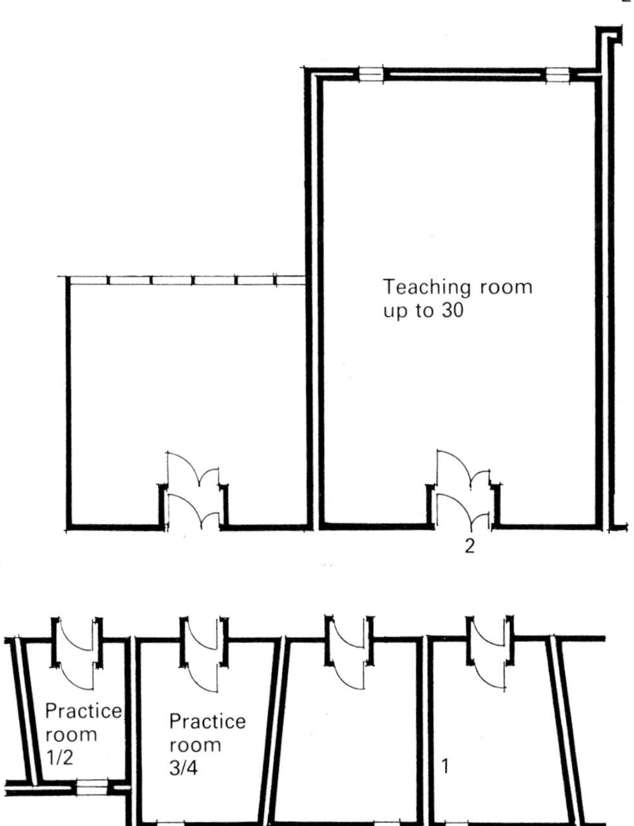

Figure 2.35 *Music practice rooms: Manchester High School for Girls*

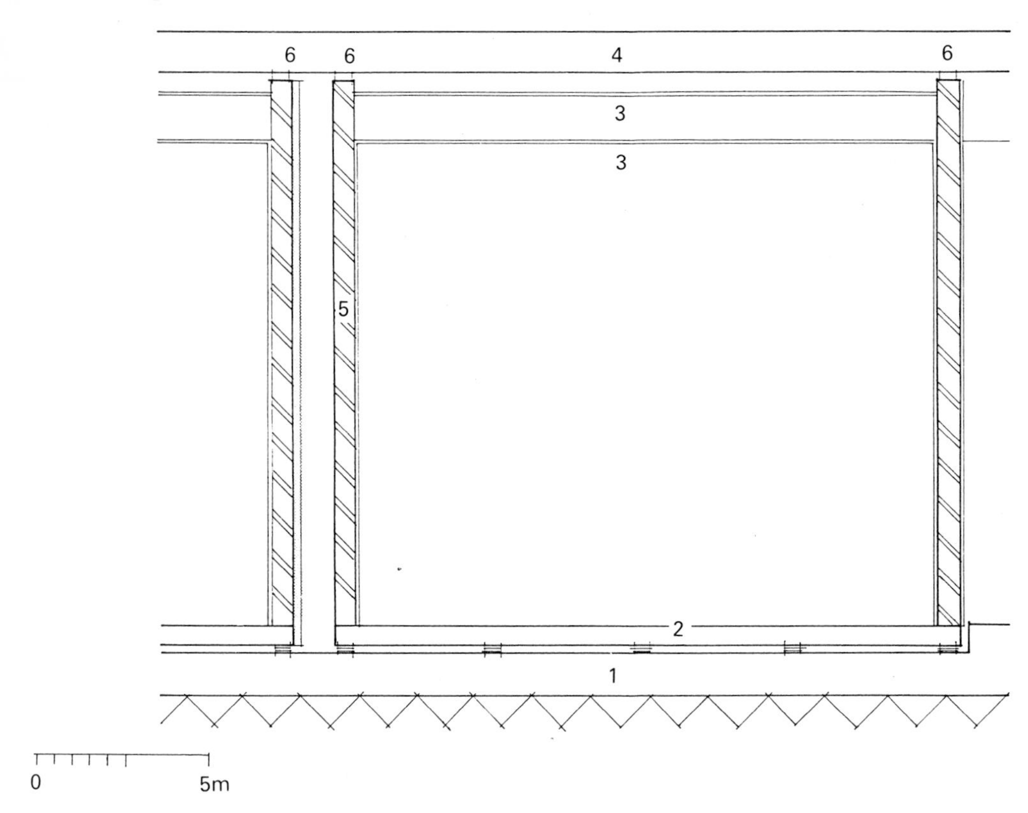

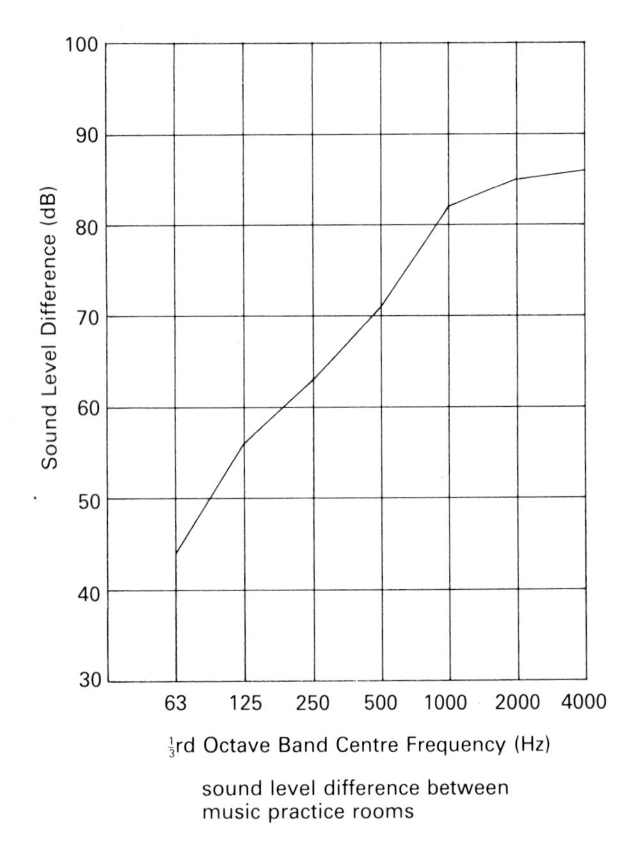

Figure 2.36 *Music practice rooms: Royal Academy of Music, London. (Courtesy of BAP)*

advantage of fast installation on site and a relocation ability.

Offices

Complaints from office workers arise from intrusive outside noise, high noise levels within offices, and poor insulation between cellular offices. BS 8233 recommends $L_{Aeq,\,T}$ values of 40–45 dB for private offices and office conference rooms, and 45–50 dB for open-plan offices. Above a general level of 57 dBA, occupants have to raise their voices to offset the background noise, which further raises internal levels.

Outside noise levels

Outside noise levels can influence the whole form of an office complex: natural ventilation for a 15-m-deep template or natural ventilation plus ventilated core for an 18-m-deep template allows in transportation or industrial noise, but deep-plan sealed fully mechanically-ventilated office buildings offer a more controlled environment; 4/12/6 glazing is usually adequate, but better glazing combinations (6/20/10 or even double windows) may be required in exceptional circumstances.

Atria

Atria form the central features of many large office or mixed development/leisure complexes, where the working spaces are clustered around a glazed central area which provides a controlled internal area with some of the character of an external space. Glazing panels act as low frequency absorbers but are otherwise strongly sound reflective. Combined with hard floor finishes and wall claddings, a

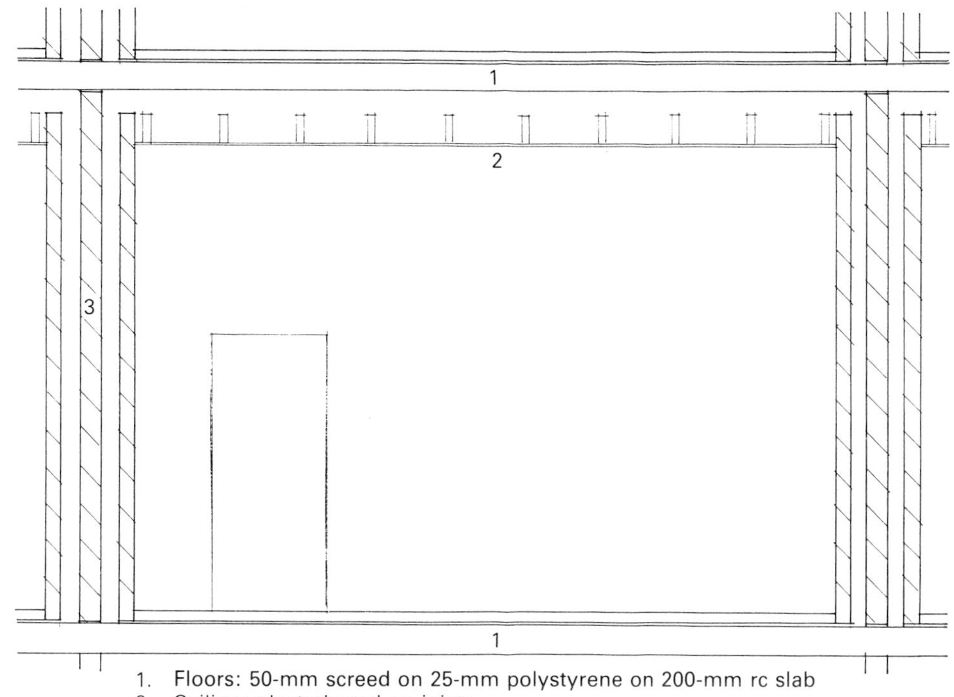

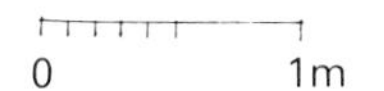

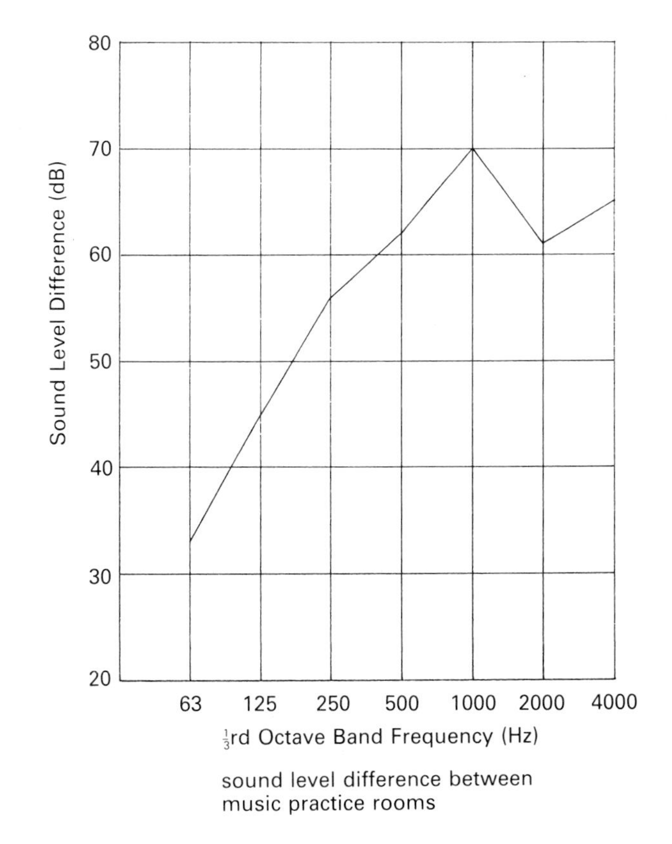

Figure 2.37 *Music practice rooms: Birmingham School of Music. (Courtesy of BAP)*

clattery, reverberant character will result unless a proportion, say 25%, of the wall surface is clad in absorptive panelling. Internal modelling serves to diffuse the sound and features like trees, banners, umbrellas, kiosks and other furniture all serve to soak up sound and reduce reverberant sound pressure level. Water features can provide useful masking sound (70 dBA at close range).

There is little information on atrium acoustics. A paper by de Reuiter [20] compares shopping centre, office and hospital examples. In the UK, Gaughan of the Institute of Environmental Engineering, South Bank Polytechnic, has taken extensive measurements in London atria at the Broadgate Centre (two, each 20 m × 11 m × 18 m) and the Sedgewick Centre (35 m × 20 m × 15 m).

Broadgate's metal, marble and glass finished interior court, four floors high, has reverberation times of 3 s at 125 Hz increasing to 9 s at 500 Hz. The average absorption coefficient is only 0.05.

Sedgewick's seven-floors height by contrast has a mean α of 0.2 and diffuse, almost Sabine, character of RT 3 s at 125 Hz, 500 Hz and 1 kHz. Both centres have ambient noise levels around NC 50 due to ventilation plant and continuous escalators operation. Three-dimensional sound propagation contours show greatly varying and uneven decay rates of sound from source position to position, due to multiple and complex reflection patterns.

Internal noise levels

Internal noise levels can be kept reasonable by including a sound-absorbing ceiling, carpet and screen-based work-

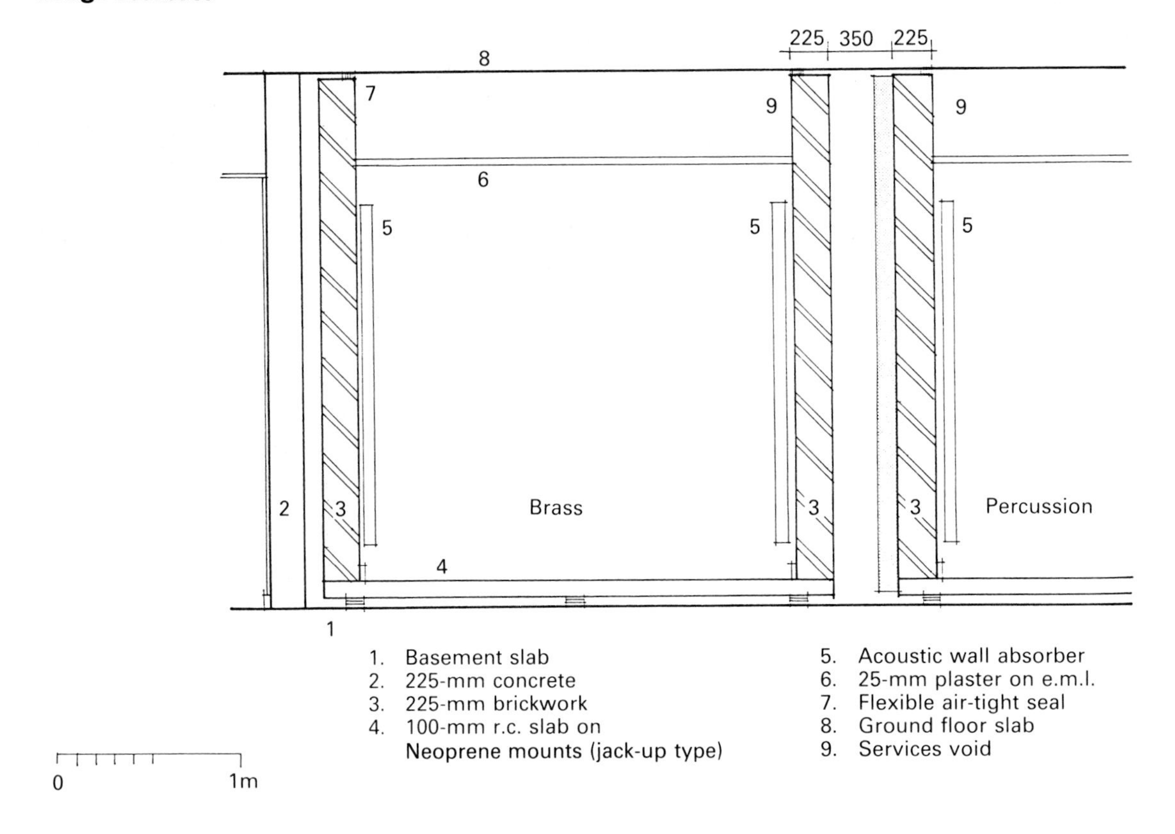

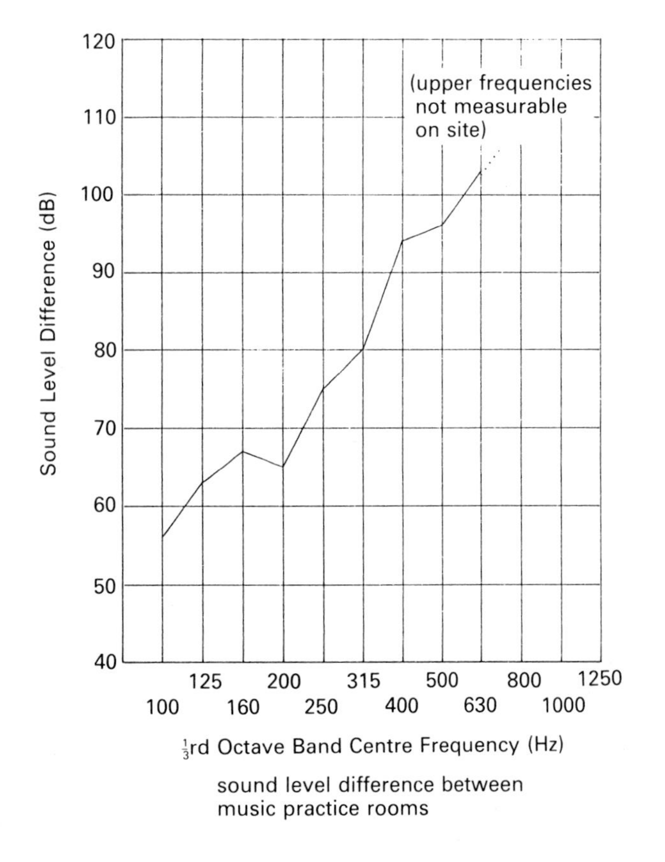

Figure 2.38 *Music practice rooms: Central London Music College. (Courtesy of BAP)*

station arrangement. Reverberation times are not relevant in open-plan offices as the perimeters are remote. Modern office equipment like laser printers and computer keyboards are much quieter than electric typewriters, e.g. laser printers are typically 64 dBA at 1 m compared to 83 dBA for mechanical printers. A recent article by Morland [21] offers a means of evaluating speech privacy in open-plan offices using the Articulation Index. It may be tempting to consider the use of a sound conditioning system. This consists of concealed loudspeakers emitting masking white noise. Care is required as the working sound level for efficient use is narrow: too noisy and the sound is objectionable, or at least draws undue attention to the sources; too quiet and the system is ineffective.

Privacy

Privacy between work places is only in the order of 17–20 dBA between open-plan screen-based work-stations at 12 m^2/work-station, and this may be compared with speech privacy needs as set out in Tables 2.3 and 2.4. These reflect the subjective reactions of office workers, recorded in Table 2.10. Screens can be tested for sound absorption (BS 3638) [22] – a NRC of 0.6 to 0.8 being desirable – and speech privacy noise isolation class (NIC). There is a hierarchy of privacy in offices as shown in Diagram 2.6. Surveys have indicated that the thought interruption due to office noise can amount to significant work 'downtime' for employees, and good privacy arrnagements in open-plan offices can increase productivity by between 3 and 10%. In cellular offices, proprietary 50-mm-thick metal-skinned

Table 2.10 *Relationship of background noise and annoyance in offices*

Activity + ventilation noise (dBA)	Staff in an adjacent work-station annoyed by normal speech (%)
35	65
40	40
45	25
47	16
55	4

SENIOR MANAGEMENT
BOARDROOM
LARGE CONFERENCE ROOMS
MEETING ROOMS
TECHNICAL & ADMINISTRATIVE STAFF
AMENITY
Cellular
Open plan
Increasing privacy
Lower ambient noise

Diagram 2.6 *Offices: zoning for privacy and quiet*

panels with mineral wool core (mass <40 kg/m^2), can achieve 30–35 dB average SRI if well installed or up to 40–45 dB if high performance panels are used. The problem arises with relocatable partitions at ceiling and suspended floor (this is covered earlier under 'Sound insulation'). The privacy obtained depends on:

- the background noise level and its masking effect (care should be taken below NR 35),
- room-to-room sound insulation (aim for 40+ dB on a BS 2750: Part 9: 1980 test),
- 'cross-talk' attenuation (particularly on plenum ceiling or floor void: fully ducted systems preferred),
- the partition–ceiling junction.

Partitions should be carried through the ceiling void to prevent flanking; alternatively a jointless plasterboard ceiling can be installed; the effect of a continuous joint at the ceiling–partition head is shown in Figure 2.39. It is all too easy to drop to around 25 dB separation for demountable systems.

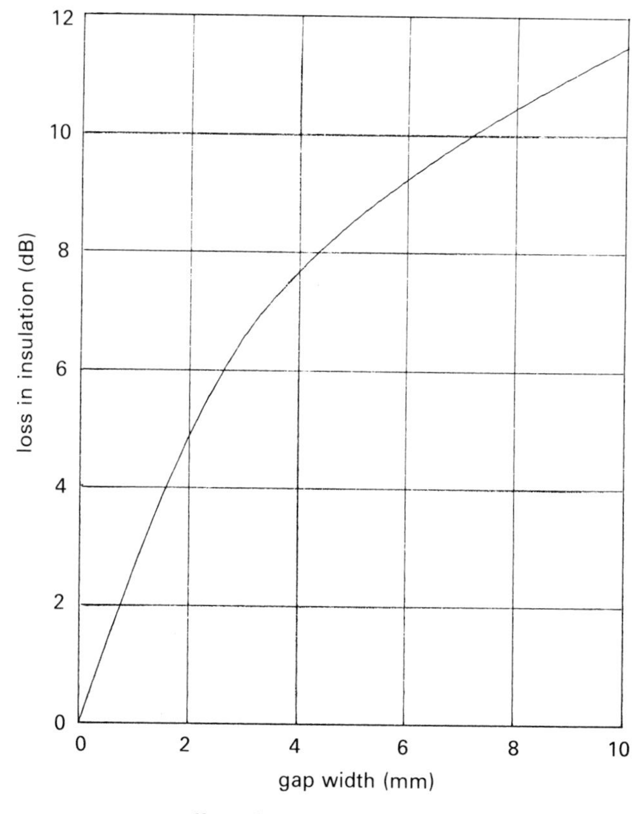

Figure 2.39 *Effect of small gap above an office partition on sound insulation*

Computer rooms

Computer rooms have inherently high noise levels, the acceptability of which will depend on the occupants. If staff work-stations are within, a background level of NR 45–50 should be aimed for; if only intermittent occupancy is involved, for example in machine room areas, NR 60 may be acceptable. Measures to reduce noise in the room airhandling units include lower air velocities, ducted supply and return with silencers, and double-skin casings.

Sports and leisure

Swimming pools are inherently noisy: hard surfaces for wear and hygiene and the reflective qualities of water exacerbate the shouting and splashing. Ice halls have similar problems. Good sound systems and built-in sound absorption by ceiling or banners will give more comfortable conditions to achieve Sports Council recommendations (1.8–3 s mid-frequency values for empty halls). Ten-pin bowling has made something of a come-back; high impulsive noise levels occur close to the racking machines during 'strikes' (typically peaking at 102 dBA close to) and during racking by the machines beyond. Absorbent surfaces and enclosure of the machine room will contain this, and absorptive ceiling finishes improve conditions near the lanes. Slab isolation under the racking machines and under the lanes (to stop ball-rolling thunder) is required to prevent structureborne sound transmission. Indoor or outdoor shooting facilities demand specialist advice.

Television and radio facilities

Video production

Video production demands that the main studio should be reasonably isolated to outside and to other facilities, for example by double-leaf walls, and a roof of reasonable SRI (35+ dB average). Acoustically-rated doorsets should be used to the control room suite but not necessarily to other rooms. Room acoustics aspects will not be exacting. The additional cost to standard accommodation of basic specification is in the order of 25%. Facilities like this are sometimes built as a fit-out of modern industrial estate/business park units or conversion of older buildings to multi-tenancy units. In the case of the latter, transmission through the floor may need to be checked. The studios are suitable for promotional or video films, centring on sets.

Commercial television

Commercial television management inevitably took a short term view on capital expenditure returns because franchises were, until 1991, renewed on a 5-year cycle. Small firms producing programmes on the fringes of the main networks may have standards no better than for video production. Regional franchise holders will have a range of accommodation from property stores to warehouses with larger sets inside, applying standards no better than for outside broadcast, to control rooms and controlled environment studios. Standards in commercial television production improved as the value of good-quality music scores in drama became recognized.

The construction standard will be higher than for smaller facilities. Nevertheless, because the programmes in production are recorded and not 'live', the very occasional intrusion of a loud noise can be accepted. This implies that design is determined by, say, L_{A10} or L_{Aeq} values rather than L_{Amax}. In some live studios now under construction as fast news centres, background activity is put on show so a standard more exacting than NR 20 is not called for. In control rooms, NR 20 for sound dubbing and control facilities will be adequate; the background levels in such rooms may be determined by equipment cooling fans rather than ventilation noise or intrusive noise.

Larger studios may double as theatre audience venues, for example for game shows. Ventilation noise control to NR 25 will be adequate, provided that for lower lighting and occupancy loads a lower ventilation noise value (NR 15 or 20) could be provided for, say, recording drama.

High-quality radio, recording, broadcasting, or live television facilities

Local radio stations met IBA standards by complying with pass/fail tests. The standards are set out in the Engineering Code of Practice for Independent Local Radio, uprated as the IBA Specification for Studio Centres [23]. The document itself should be studied in detail but covers:

- background noise
- reverberation times
- performance characteristics for line path, microphone path, and storage media

Acoustic criteria for transmitting equipment are also specified. Typically double- or triple-leaf walls and room-dedicated floors and 'lids' were coupled with quiet (NR 15) ventilation systems and 'dead' room acoustics (RT 0.16–0.3 s). The European Broadcasting Union's Report R22 [24] is an important reference for standards including preferred volumes of rooms, proportions of spaces (avoid single integral ratios between length, width and height) monitoring positions, reverberation times (over one-third octave bands 200–2500 Hz, the average RT should be 0.3 ± 0.1 s. At low frequency one-third octave band 50 Hz, the RT should not exceed 0.45 s. The RT should not vary by more than 0.04 s between adjacent one-third octave bands in the range 200–10 000 Hz, and ventilation noise should be less than NR 15 with no cyclic variation or pronounced tones.

Recording studios demand similar high standards. The quality criteria here are dynamic range, distortion (attenuation, phase or non-linear: harmonic, intermodulation or amplitude), noise (ambient, system), wow and flutter (short-term speed fluctuations) and electronic cross-talk. Detailed advice is given in Borwick's *Sound Recording Practice* [25]. Recommended reverberation times vary from 0.3 to 0.4 s for small speech studios to 1–2 s for classical music recording. Monitor rooms, separated by at least 8/200/12 glazed windows, should be designed to 0.2–0.35 s. The introduction of stereo recording and broadcast imposed further discipline to studio techniques. Conditions should not be 'dead'/semi-anechoic, as this would be unpleasant to work in. The ear is better able to locate an image in a stereo field with some reflections and there is the 'single pass' concept of monitoring: part of the room is left 'hard' behind the loudspeakers, with the area behind the engineers treated with absorption lining. Control room ambient noise should match the main studio conditions.

The need to have reliably acceptable conditions at all times follows from 'live' broadcast, where a retake cannot be relied upon if some intrusion occurs. British Broadcasting Corporation advice may be taken from Rose's *Guide to Acoustic Practice* [26]. The technique of mix of office and technical facilities by installing factory-built rooms-in-rooms also works well for fitting out 'shell' interiors and consists of modular studios made of standard wall, roof and floor panels. These are made of double metal skins with mineral quilt filling to the cavity between skins. The room-side metal skin can be perforated: this reduces the sound insulation properties but contributes most of the general absorption within rooms. Low frequency absorber boxes may also be required. Small announcer booths can be installed on a subfloor, each with a standard dedicated room air conditioning unit, either recirculation or connected to a header supply and extract ducts via attenuators (Figure 2.40). The system is one-third the weight of the equivalent masonry structure, is quick to erect, and can be reconfigured relatively easily.

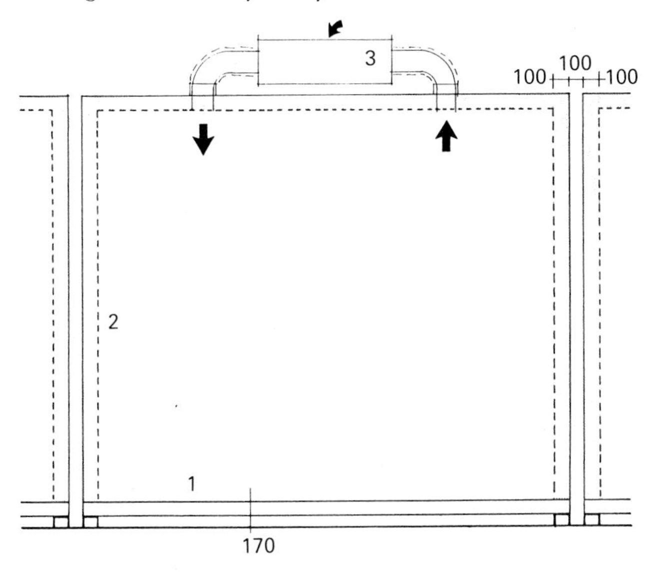

1 100-mm modular floor units on isolating rails with cable voids
2 100-mm modular wall units: metal skins, inner face perforated absorbent quilt fill
3 Dedicated room ventilation unit, recirculated air with some fresh air intake

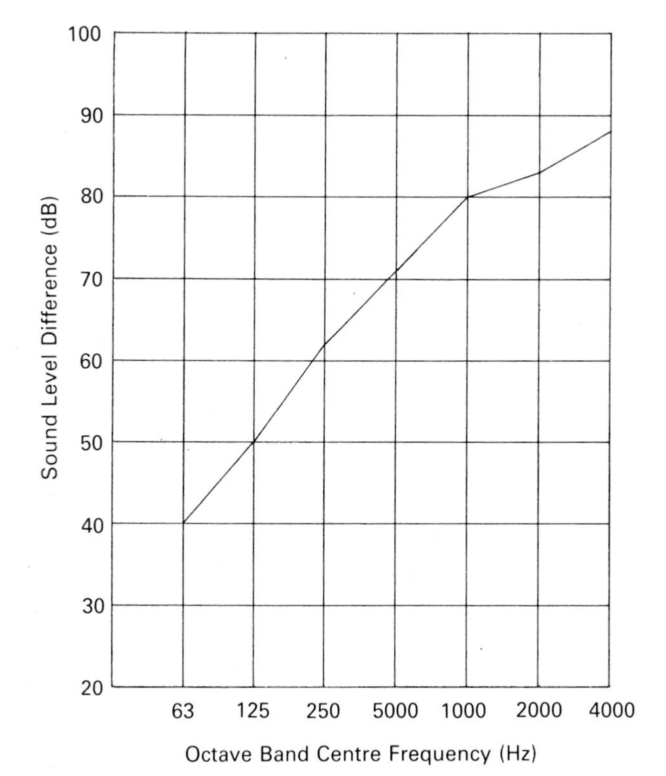

sound level difference between studios: twin modular IAC panels

Figure 2.40 *Modular studios*

BBC checks of background noise are against three reference curves, which very approximately equate to NR5, NR10, and NR15:

	⅓ octave band values/octave 63	125	250	500	1k	2k	4k (Hz)
1. Radio light entertainment studios	41	31	24	18	13	9	6
2. Other radio studios, control rooms. Television studios	36	26	19	13	8	4	1
3. Radio drama studios	31	21	14	8	3	−1	−4

Close matching of curves has been pursued to ensure there is some masking noise without rumble or hiss. This is difficult in practice as attenuator selection to suit low frequency performance results in noise levels well below the curve at upper frequencies. Flow noise at secondary attenuators or diffusers can be a means of selectively adding back higher frequency noise.

Theatres

The term 'theatre' covers a wide range of auditoria from community halls (naturally lit and ventilated, multi-use and small) to large national theatres for resident companies. Traditional theatres and opera houses have a 'hammerhead' plan shape, a proscenium opening between the stage and audience, an orchestra pit, and acoustically comprise a high, bare space (alternatively filled with 'flats') coupled to a 'dead' auditorium. For this reason it is useful to provide built-in absorption linings to the flytower which will balance sound decay in the two parts of the coupled volume, and to some extent soak up residual intrusive noise down from the fire-shuttered fly tower lantern.

Small theatres may seat less than 500, larger theatres seating 1000 to 1200, and the largest exceeding 1500 seats (Association of British Theatre Technicians definition). Compact seating arrangements ('lining the walls with people') ensure that everyone is within 15–20 m of the stage so nuances of acting and speech can be followed. Reference can be made to case study collections [27, 28].

Overhead ceiling panels can be sound-reflecting and angled to give useful early reflections to 'carry' sound to rear seats (Figure 2.41). Finishes should be sound-absorbing except near the stage, and perimeter surfaces should be non-parallel.

Balcony fronts may need to be modelled or have sound absorption applied, in order to avoid distracting back reflection effects for performers on stage. Angled proscenium walls can be modelled to reflect sound to seating by facetting, which can improve audibility. As with concert halls, acoustic modelling can be useful. In providing a setting for other events, like opera or music concerts, a means of adapting the acoustics to these uses can be incorporated. Banners or similar devices do not help, as the space is already 'dead' acoustically. There are a number of electroacoustic systems which are able to increase the RT

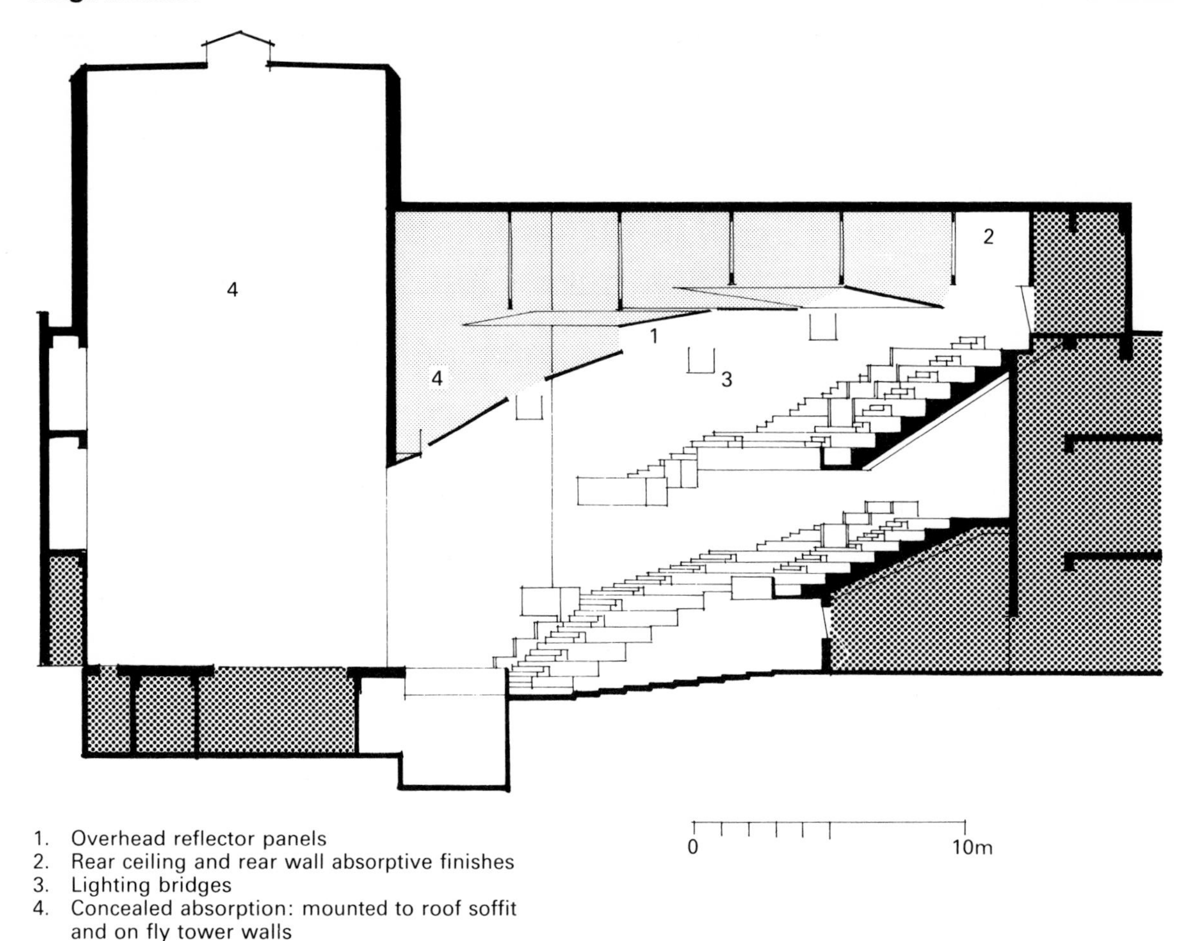

Figure 2.41 *Overhead reflector panels in a 1000-seat theatre. (Courtesy of Bucks CC/BDP)*

artificially and compensate on stage for the lack of local reflecting surfaces. An example that is installed in an existing UK theatre is illustrated in Figure 2.42 and Table 2.11. The basis of the system is the sampling of the direct sound field at the stage, by an array of microphones. Signals are relayed to a single processing rack which, by equalization and signal delays, simulates early reflections and modified reverberation. Broad-band loudspeakers around the hall transmit the modified signals to supplement the natural sound decay in the auditorium. The system claims to 'reshape' a hall as well as just increase reverberation, giving early reflections to an orchestra to make up for the absence of an orchestral shell, and ability to balance early and late sound as well as increasing lateral sound. There may be some overlap with the house sound system.

Such systems are in a fast state of developoment and care in commissioning is required so that any artificial acoustics are not unrealistically 'special effect' and hence unconvincing to professional musicians.

For good speech intelligibility, the seating should be grouped as closely as possible to the stage and good viewlines and reasonable rake provided. Ventilation should be low noise, i.e. less than NR25.

An excellent briefing checklist for theatre and other performing arts spaces is the *Arts Council Guide to Building for the Arts* [29].

Table 2.11 *ACS electroacoustics system as installed in the Gordon Craig Theatre, Stevenage*

Setting[a]	RT[b] (s)	Use
0	1.0	System off: natural hall acoustics for speech and theatrical performances
1	1.3	Piano, jazz, ballet, musicals
2	1.5	Chamber music, recitals, opera
3	1.8	Chamber and Baroque music
4	2.0	Symphonies
5	2.1	Symphonies
6	2.6	Choral music
7	3.0	Organ (dedicated setting)

[a]Commissioned use of different settings.
[b]Measured reverberation time at 500 Hz; plotted on Figure 2.42.

Trading rooms

The acoustic environment on trading floors is often regarded as horrific. When interviewed in a survey, traders stated a preference for a balance of some degree of trader privacy and the ability to overhear messages across the

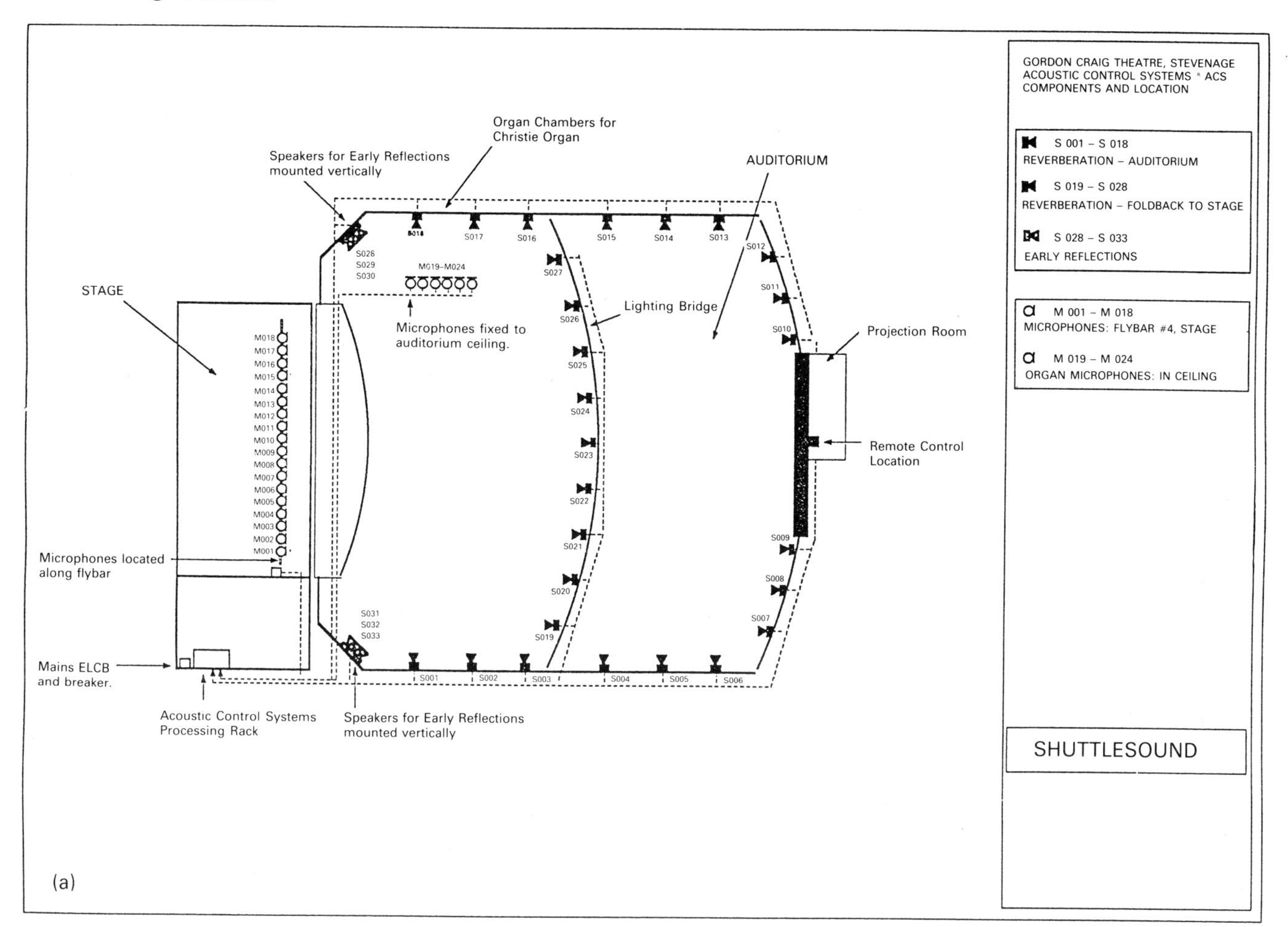

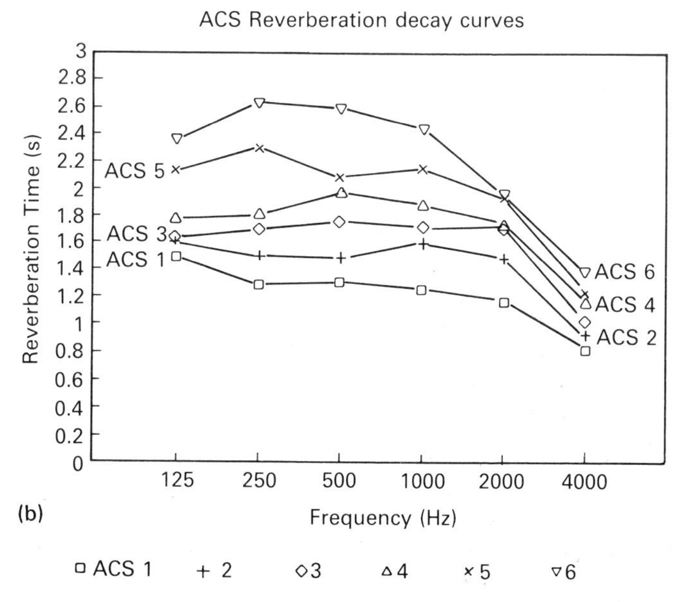

Figure 2.42(a) & (b) *ACS: Adaptable reverberation by electroacoustics*

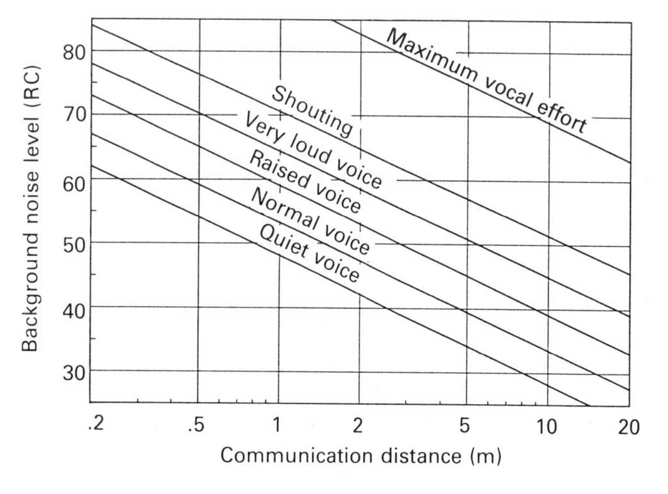

Figure 2.43 *Effect of background noise on speech intelligibility*

room, but this has proved elusive. What firms want is good audibility within a working group, and acceptable use of telephone and intercoms. The whole room is expected to have some 'buzz' to generate excitement. Trading floors ambient levels can be +10 to +15 dBA above office interior levels, which causes a poor signal-to-noise ratio so occupants shout into their phones and make the din worse; trader work-stations are more tightly spaced, 5 m^2/person rather than the 12 m^2/person of open-plan offices. In the same survey, floor vibration (e.g. from footfall vibration) had not been thought to adversely affect VDUs, even though dealing rooms tend to be planned in long-span, column-free structures.

The aim should be to create good communication in local groups but diffuse and damp longer sound paths across the room which contribute to the build-up of reverberant noise (Figure 2.43). In reasonable conditions, speech communication could be assumed to embrace 12 to 20 colleagues, across about 5 m; this is a slight contrast to the Mohave Desert where experiments earlier this century shows a maximum speech propagation distance of 42 m. Communication beyond the working group is best done by intercom – new headset designs claim to be much more efficient at cutting out intrusive noise, by improved microphone technology.

The tall exchange rooms of the past and double height/mezzanine rooms in some new trading floors provide few surfaces providing useful local reflections for talk betwene close colleagues. The typical flat low ceiling of acoustic tiles is not a good answer: the surface is more sound reflective at glancing angles of incidence, not less. A better ceiling design is modelled ceiling planes, coffers, or suspended baffles with absorption on vertical surfaces to absorb long sound paths, and patches of reflection over groups to boost local audibility.

The high heat loads make for significant ventilation load and hence high duct velocities: NR 40 or even NR 45 is likely to be acceptable.

The USA practice is to stack screens and local storage on flat tables which look untidy but allow fast change. The UK 'upright piano' desks as racking to monitors give better local enclosure but limits viewing to other dealers. The ideal is a compromise between local aural field and longer distance awareness of proceedings. A strong determinant on future design will be how quickly voice-activated computer terminals come in (at present these are at the design stage). Perhaps deals will eschew telephone handsets, wear headsets and microphones, and rely on clearer mimic display systems; dealing rooms may come to resemble airport control rooms and become quieter and more ordered.

References

1. BS 5821: 1984 Methods for rating the sound insulation in buildings and of building elements, British Standards Institution, Milton Keynes
2. Building Regulations *Part E – Sound*, 1985, amended June 1992
3. BS 648: 1964 Schedule of weights of building materials, British Standards Institution, Milton Keynes
4. BS 2750: Part 9: 1980 Method for laboratory measurement of room-to-room airborne sound insulation of a suspended ceiling with a plenum above it, British Standards Institution, Milton Keynes
5. Gade, A. C. *Acoustical Survey of Eleven European Concert Halls*, Report 44, Technical University of Denmark, 1989
6. Talaske, R. H. (ed.) *Halls for Music Performance 1962–1982*, Acoustical Society of America, New York, 1982
7. Bickerdyke, J. and Gregory, A. Code of Practice: *An Evaluation of Hearing Damage Risk to Attenders at Discotheques*, Department of the Environment, DGR/481/99, London, 1980
8. *Noise at Work Regulations*, Health and Safety Executive Guidance, 1990
9. *Acoustics in Educational Buildings*, Building Bulletin 51, HMSO, London, 1966
10. *Secondary Schools Design: Drama and Music*, Building Bulletin 30, HMSO, London, 1981
11. *Guidelines for Environmental Design and Fuel Conservation in Edcuational Buildings*, DES Design Note 17, HMSO, London, 1981
12. *Lighting and Acoustic Criteria for the Visually Handicapped and Hearing Impaired in Schools*, DES Design Note 25, HMSO, London, 1987
13. BS 8233: 1987 Code of practice for sound insulation and noise reduction for buildings, British Standards Institution, Milton Keynes
14. Miller, John, *Building Regulations and Health*, Building Research Establishment Report, 1986, London
15. Miller, John, *Sound Control for Homes*, Report no. 114, CIRIA, 1986
16. *Thermal, Visual, and Acoustic Requirements in Buildings*, BRE Digest 266, HMSO, London, 1979
17. BS 4142: 1990 Method of rating industrial noise affecting mixed residential and industrial areas, British Standards Institution, Milton Keynes
18. *100 Practical Applications of Noise Reduction Methods*, HMSO/HSE, London, 1987
19. *Noise Control in Industry*, 3rd edn, Sound Research Laboratories, Spon, London, 1991
20. de Ruiter, E. Ph. J., *Atria in Shopping Centres, Office Buildings and Hospitals*, IOA proceedings, **10**(8), 1988
21. Morland, J. B. Articulation index. *Noise Control Engineering*
22. BS 3638: 1987 Method for measurement of sound absorption in a reverberant room, British Standards Institution, Milton Keynes
23. Specification for Studio Centres, *Engineering Code of practice for Independent Local Radio* – Issue 2, Independent Broadcasting Authority, London, February 1988
24. *Acoustical Properties of Control Rooms and Listening Rooms for the Assessment of Broadcast Programmes*, Report R22, European Broadcasting Union, London, 1985
25. Borwick, J. *Sound Recording Practice*, Oxford University Press, Oxford, 1987
26. Rose, K. *Guide to Acoustic Practice*, 2nd edn, BBC Engineering, Oxford, 1990
27. Talaske, R. H. (ed.) *Theatres for Drama Performance*, Acoustical Society of America, London, 1985
28. Forsyth, M. *Auditoria: Designing for the Performing Arts*, American Institute of Physics, New York, 1987
29. Strong, J. *Arts Council Guide to Building for the Arts*, Arts Council, 1990

Chapter 3 Services noise and vibration

Peter Sacre and Duncan Templeton

Background

The control of noise from mechanical and electrical plant can be a vital area of design, as failure to meet criteria is more readily perceived than, say, room acoustics criteria. The designer needs to be aware of the need to limit noise inside and outside the building:

- in occupied internal areas, where noise can be irritating or distracting, or can affect working efficiency;
- in industrial premises, where processes rather than occupied rooms are serviced;
- in the areas immediately surrounding the building, which may be used for circulation or leisure, where excessive noise can be intrusive and may present an environmentally unsatisfactory character;
- beyond the site boundary – excessive noise from plant may cause nuisance, leading to complaints and legal proceedings, especially in residential areas.

The importance of noise depends mainly on two factors:

- the type of building – whether its use depends on low noise levels, and
- the location – particularly the proximity of other noise-sensitive areas beyond the boundary.

Noise control should be an integral part of the design procedure. Too often noise aspects are introduced into a design too late, and in an *ad hoc* way. The role of any advice from an acoustics consultant or silencing specialist should be proactive rather than reactive. The cost is many times greater for a retrofit compared to the original inclusion of adequate noise control measures.

The key participants who can influence services noise are:

- the client: including noise criteria in the brief, from site layout to room data sheets
- the architect: provision of adequate structure, sensible location and area allocation of plant rooms, planning in distribution routes
- the mechanical engineer: total design of HVAC systems, setting criteria
- the electrical engineer: design of substations, emergency generators, lifts design
- the mechanical and electrical engineering subcontractors: detailed selections of components and installation
- the acoustics consultant or engineer: design advice from briefing to commissioning
- the specialist supplier of noise control hardware: providing data and goods to match initial selections.

Setting design objectives

The areas of design for which noise criteria need to be set are as follows:

- *Central plant.* 'Plant rooms' are often split (air-handling units in roof-level housing; boilers and pumps in ground level or basement rooms). Chiller condenser units may be in a louvre-screened compound open to atmosphere.
- *Local plant.* Decentralized ventilation systems can have advantages of lower cost (no long duct runs) and flexibility (zoned units). However, plant noise sources are taken out of the sound-insulating plant room and into user spaces and so care is needed in their siting.

There are already a number of guides on the noise control of building services. For a number of years, the noise control products trade has concentrated on noise from fans in central ventilation or air-conditioning plant transmitted through air distribution ductwork. Specialist acoustic suppliers via technical sales personnel can select and supply the appropriate package ductwork and associated attenuators. As a result, there should be relatively few noise problems in buildings due to inadequate fan noise silencing through ductwork. However, many problems exist due to poor ductwork layout or high velocities causing regenerated noise.

For the typical commercial or public sector building, the following potential noise sources and transmission paths may need to be considered:

- *Internal noise*
  - Central air-handling plant
    - Fan noise to ducts
    - Airflow-generated noise in ductwork, at duct fittings or dampers
    - Noise break-out through duct walls
    - Noise generated at grilles and diffusers
  - Local air-conditioning plant and room units
    - Fan-coil units
    - Volume-control terminal units
    - Heat pumps
    - Local extract fans
    - Fan convectors
    - Warm air curtains
  - Piped services
    - Pump noise or flow noise radiated from pipework or from building surfaces to which pipework is fixed
    - Water hammer
    - Flow noise from drains, particularly WC, soil pipes and rainwater pipes from roofs
  - Electrical equipment
    - Emergency generators
    - Uninterruptible power supplies, generator sets
    - Transformers
    - Thyristor speed controllers and light dimmers, fluorescent lamp ballasts
  - Airborne or structureborne noise transfer through plant room envelope to adjacent areas
  - Other sources
    - Kitchen equipment
    - Laundry equipment
    - Workshop machines

Waste compactors
Lifts
Escalators
Document transfer systems
Computers (integral fans)

- *External noise*
  - Atmosphere terminations
    - Flues
    - Louvres opening into plant rooms
    - Louvres ducted to fan inlets or exhausts
    - Boiler flues
    - Generator flues
  - External equipment
    - Airhandling plant
    - Roof extract fans
    - Cooling towers
    - Air-cooled condensers
    - Packaged chiller plant

Design approach

The building services are most frequently designed by a Mechanical Services Consulting Engineer, with the building envelope and support structure being the responsibility of other consultants. Other specialists may be involved; silencing products designers, acoustics and energy consultants for example. The role of such specialists may be limited to provide 'design intent' sketches, draft specifications, and guidance, which 'main profession' consultants will incorporate in contract documentation for construction purposes. It is important to realize that structure, layout, furnishings, fittings and furniture will have a fundamental effect on the noise from building services installations.

The design output typically consists of drawings and specifications which form the basis of a legal agreement between client and contractors. It is essential that the design intent and the necessary design details for noise control are clearly and exhaustively covered in the drawings and specifications. Many noise problems in finished buildings are due to inadequate information, or to poor communication with the contractors, rather than to a fundamentally deficient design.

Information at the out-to-tender stage varies from line diagram schematics, with minimal sizing of ducts and no identification of silencers, to full documentation: ductwork to scale, duct velocities, sized attenuators and schedules. There is a danger that an engineer will take a specific specialist supplier's quotation very literally and in its entirety, including octave band values for particular silencers. These are in fact specific to the fan being used and cannot be checked at the commissioning stage, so a schedule of sizes, types and location will suffice, room criteria being the commissioning target.

Buildings designed by integrated practices, or by 'design and build' contractors, should present fewer problems caused by inadequate flow of information between professions. However, the need remains for a formalized procedure for noise control design.

The noise input required at the various stages of design are identified in Table 3.1. In some cases, noise from services may be a critical issue which fundamentally affects the layout or structure of the building, as well as the approach to building services. The design team can address each services noise issue as it arises, from 'is the plant room big enough and far enough away?' to 'is there enough space in the ceiling void we've assumed?' Noise control is a basic requirement because higher standards (lower-value criteria) mean lower duct velocities, and hence larger ducts, for the same duty. Room criteria are a big clue to the type of system suitable in a ventilation system, for example NR 25 implies two-stage attenuation and fully-ducted systems to air-handling units in a separate plant room, NR 35 or 40 implies single-stage attenuation, relatively high velocities, and common use of ceiling voids as plenums or for volume-control terminals.

Table 3.1 *Services noise advice timing*

Activity/decision	Noise input
Feasibility, preliminary design:	Provisional design criteria
	Identify critical spaces
Site location	Suitable plant type
Site use	Suitable plant locations
Space layout	External noise survey
Building structure	Provisional external criteria
Location of central plant	Adequate structure
Method of servicing rooms	
Location of external plant	
Detailed design:	Develop design criteria
Size plant, confirm location	Planning conditions
	Internal plant selection
Size ductwork	External plant selection
Duct runs, damper positions	Louvre locations
	Duct sizing, routing
Plant-room structures	Duct attenuator selection
Select terminals, room units	Terminal units
	Grille/diffuser selection
	Vibration isolation
	Plant room construction
Drawings and specification	Check drawings/notes
	Design criteria stated
	Plant maximum noise levels
	Attenuator schedules
	Vibration isolation schedules
	Specifications for noise control equipment
Equipment order	Works noise tests
	Check equipment against specification
	Approve changes and waivers
Installation	Site checks
Commissioning	Measure noise levels
	Highlight/diagnose/solve problems

Design criteria

There is a general need to control noise and set criteria both within and external to buildings. There are several levels of noise requirements within buildings. Examples are:

- technical areas where particularly high standards of noise control are required. These may be small (music practice rooms, audiometry suites, continuity voice-over studios) or large (concert halls, conference rooms);
- working areas where noise from services can be the dominant noise source, for example offices, laboratories, and hospital wards. Noise control should not only be adequate to not intrude on conversations or telephone calls, but for ventilation to be unobtrusive;
- industrial buildings and the interiors of large plant rooms where high noise levels can be produced and occupants may need to be protected. An example is in newspaper printing centres (Figure 3.1). Only in these refuges looking onto the presses, with sound reduced to below 65 dBA, can controllers take off ear defenders and make telephone calls or conversation. In very noisy environments an additional difficulty is the audibility of alarms and announcements.

The acoustic consultant and silencing specialist are essential in the first category but also have a lot to offer in the design of the other categories.

To determine acceptable criteria, the first point of reference should be the project client who, on a large scheme, may have his own specialist advisers, for example hospital resident engineers or property managers and maintenance engineers. Ideally, a brief would be enlarged by room data sheets including environmental standards as baseline data. Where existing premises are to be altered or extended, surveys of existing noise levels, noise climate and airborne and impact separation, may be useful.

The room activity and use combined with the existing ambient noise from other internal and external sources will suggest a criterion for services noise. This criterion is often established by reference to a Noise Rating (NR) curve. This allows a check against the sound character which is limited by the use of a single-figure measurement, i.e. dBA. Single-figure units fail to pick up the annoying effect of narrow-band tonal noise and discriminate against low-frequency noise. Usually NR curves rather than Noise Criteria (NC) are used in the UK for considering services noise. Generally, NR and NC values can be interchangeable but there are variations:

- at low frequencies, NR values exceed the equivalent NC values;
- at high frequencies, NC values exceed the equivalent NR values;
- NR values run between 31.5 Hz and 8 kHz, NC between 63 Hz and 8 kHz although in practice 63 Hz to 4 kHz is typically used in design checks for either;
- NR values are on true curves determined by formula; a spectrum can be stated as any NR, e.g. NR 31. NC curves are defined in steps of 5, e.g. NC 30, NC 35. The limitations of this can be avoided by stating the excess over the lower value, e.g. NC 30 + 1.

Air-conditioning noise can also be checked by Room Criteria (RC) curves as recommended by ASHRAE (the American Society of Heating, Refrigerating and Air Conditioning Engineers), but neither RC or PNC (Preferred Noise Criterion) curves are in general use in the UK (Tables 3.2 and 3.3 and Figures 5.6–5.9).

Attitudes vary on the relative values of ventilation noise levels and background noise levels, although values are generally based on CIBSE banding. If ventilation noise levels are set very low, it will be difficult to commission the system because ambient noise will tend to dominate, and the client will not be getting value for money in an over-silenced installation. If the levels are too high, the system itself is obtrusive. Around or slightly below (0 to −5 dBA) averaged activity noise is usually found to be

Figure 3.1 *Noise havens: printing hall*

Table 3.2 *Design criteria*[a]

Environment	NC or NR
Radio drama	10
Radio talks, continuity studios, live television studios	15
Recording studios, audiometric rooms, concert halls, opera halls	20
Theatres, cathedrals and large churches, commercial television studios, large conference and lecture theatres, music practice rooms, hotel bedrooms, courtrooms	25
Senior management offices, small conference and lecture rooms, multipurpose venues, libraries	30
Cellular offices, multiplex cinemas, restaurants	35
Circulation in public buildings, open-plan offices, ice rinks, swimming pools, cafeterias	40
Shops, bars, WCs, supermarkets	45
Warehouses, industrial premises, laundries kitchens	50

[a]See also Table 2.9 with regard to health care facilities criteria.

Table 3.3 *Criteria values (in dB)*[a]

		OBCF (Hz)								
Criterion value	Criterion	31.5	63	125	250	500	1 k	2 k	4 k	8 k
15	NR	66	47	35	26	19	15	12	9	7
	NC	—	47	36	29	22	17	14	12	11
	PNC	58	43	35	28	21	15	10	8	8
	RC	—	—	35	30	25	20	15	10	—
20	NR	69	51	39	31	24	20	17	14	13
	NC	—	51	40	33	26	22	19	17	16
	PNC	59	46	39	32	26	20	15	13	13
	RC	—	—	40	35	30	25	20	15	—
25	NR	72	55	44	35	29	25	22	20	18
	NC	—	54	44	37	31	27	24	22	21
	PNC	60	49	43	37	31	25	20	18	18
	RC	—	—	45	40	35	30	25	20	—
30	NR	76	59	48	40	34	30	27	25	23
	NC	—	57	48	41	35	31	29	28	27
	PNC	61	52	46	41	35	30	25	23	23
	RC	—	55	50	45	40	35	30	25	—
35	NR	79	63	52	45	39	35	32	30	28
	NC	—	60	52	45	40	36	34	33	32
	PNC	62	55	50	45	40	35	30	28	28
	RC	—	60	55	50	45	40	35	30	—
40	NR	83	67	57	49	44	40	37	35	33
	NC	—	64	56	50	45	41	39	38	37
	PNC	64	59	54	50	45	40	36	33	33
	RC	—	65	60	55	50	45	40	35	—
45	NR	86	71	61	54	49	45	42	40	38
	NC	—	67	60	54	49	46	44	43	42
	PNC	67	63	58	54	50	45	41	38	38
	RC	—	70	65	60	55	50	45	40	—

[a]For NR, NC and PNC curves, see Chapter 5, Figures 5.6–5.9.

acceptable; ventilation noise can have a useful masking effect in room-to-room speech privacy (as indicated in Table 2.8 in Chapter 2). In order not to add noise from different services sources, 'local' ventilation noise from room supply and extract grilles should exceed by at least 5 dBA noise break-out, or re-radiated noise, from primary plant. In considering a criterion, the 'steady state' ventilation noise to be introduced has to be considered relative to both background noise from traffic noise break-in and equipment noise like computer fans, and varying ambient noise from occupants' activity.

Broadcasting authorities

The BBC, IBA and EBU (European Broadcasting Union) all have recommended criteria. Reference should be made to *Guide to Acoustic Practice* [1], EBU Report No. R22, *Acoustical Properties of Control Rooms and Listening Rooms for the Assessment of Broadcast Programmes* [2], and the IBA's *Specification for Studio Centres* [3].

External noise

Environmental noise control is generally covered in Chapter 1, under Industrial Noise. Emission limits from fixed plant and processes may be written in as a planning condition, or should be established early in the design process to avoid the local environmental health officer agreeing with complainants that the plant noise is a public nuisance to neighbouring properties and is therefore actionable under statutes.

Different local authorities take different attitudes. Some, not wanting prevailing neighbourhood noise levels to 'creep' up due to the addition to new to existing sound, will ask for existing levels not to be raised. This is very onerous, as to reliably enable this, the new sources will have to be 10 dBA less than existing levels. It is normally acceptable to control daytime noise emission at the nearest noise-sensitive properties to not exceed prevailing levels, and by this means limit any possible increase to 3 dBA maximum. Night-time noise control is likely to be stricter, say −5 dBA on existing levels. There is then the task of the consultant as client's representative, to agree with the local authority an interpretation of 'existing noise climate'. This is the generator of many sound levels surveys.

Assessments of noise complaints will often be based on BS 4142: 1990, Method of rating industrial noise affecting mixed residential and industrial areas [4], seen as an important guide by local authorities. Problems can arise in derelict areas, previously bustling with industrial or commercial activity but now awaiting redevelopment, because the reference existing background noise levels are 'temporarily' particularly low.

There is dependence on the operating hours of plant. Office ventilation systems may cut off in the evening but computers and refrigeration plant need to stay on for the full 24-h cycle. Hotel public rooms and kitchens have systems which can go on into the early hours, to the possible consternation of overnight guests. Ice rinks and swimming pools have large-duty plant which has to keep the ice frozen, or water conditioned, continuously. The designer needs to be sure of the cut-off times before designing all plant to daytime background noise levels.

Intermittent noise may be the subject of negotiation with the local authority, for a relaxation of noise control (say +5 to +10 dBA or NR on night-time criterion values) of standby generators, knowing that they will be 'run up' regularly but for relatively short periods during the daytime. Any continuous operation will be exceptional, for example power failure, and an increase in the night-time noise level will be temporarily acceptable if the units have to be run continuously. Noise break-out for other emergency plant, for example fire pumps, powered smoke extract systems, would normally be exempt.

Vibration

Perception of vibration

People can be quite sensitive to vibration, particularly where the source of the vibration cannot be seen. Satisfactory levels of vibration for people in different building types (Figure 1.14) and in the three different axes, i.e. vertical and the two horizontal directions, are given in BS 6472: 1984, *Guide to evaluation of human exposure to vibration in buildings* (1 Hz to 80 Hz) [5]. The velocity curves from BS 6472 are given in Figure 3.2. However, these 'satisfactory' levels will be greater than those which can be

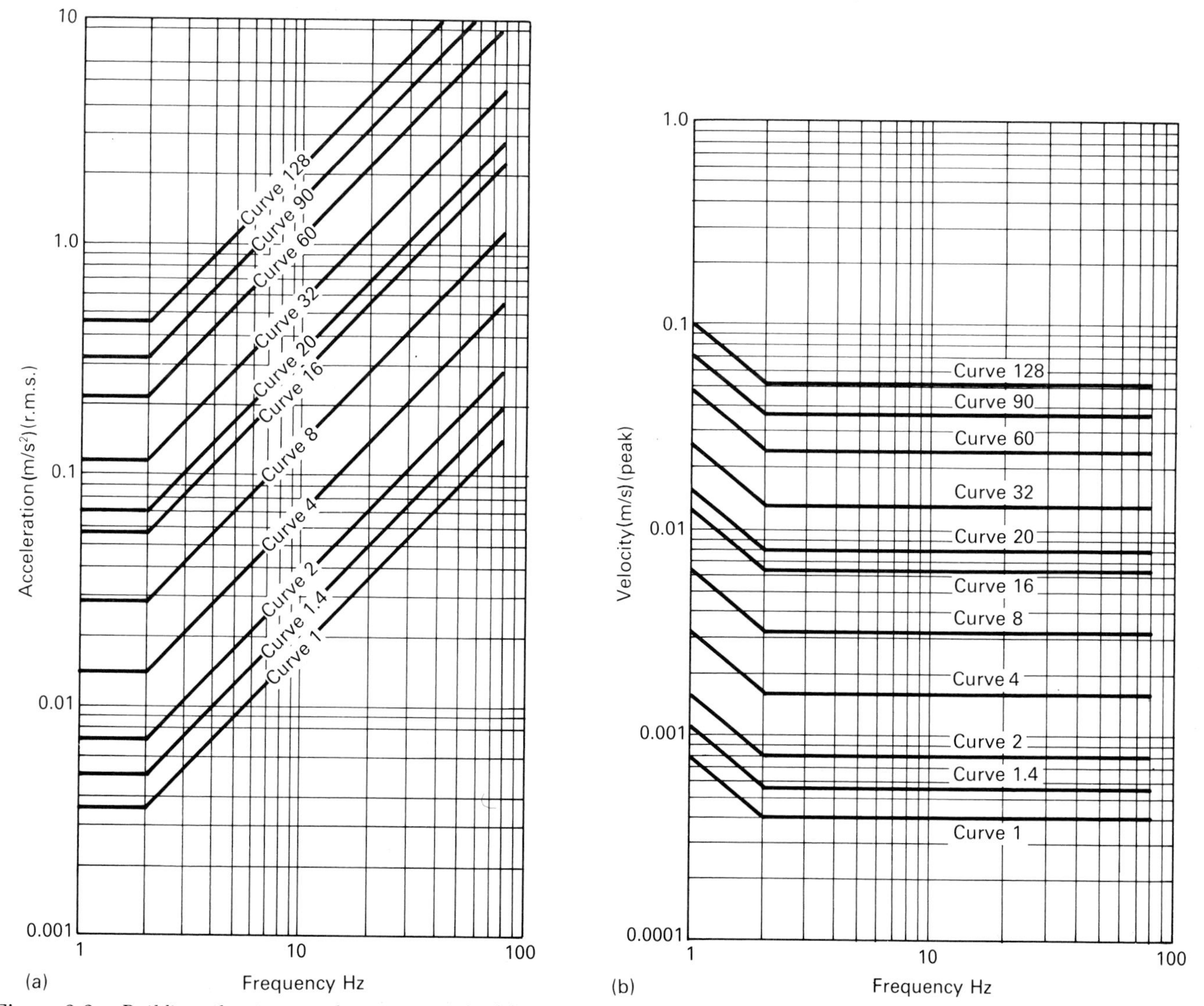

Figure 3.2 *Building vibration x- and y-axis curves for (a) peak acceleration and (b) peak velocity. Extracts from BS 6472: 1992 are reproduced with the permission of BSI. Complete copies of the standard can be obtained by post from BSI Publications, Linford Wood, Milton Keynes, MK14 6LE*

perceived or felt. Vibration effects in buildings are typically in the 2–50 Hz frequency range.

In designing a building, it is necessary not only to take account of people, where different standards could occur, e.g. industrial environment compared to office, but also the possible effect of vibration on sensitive equipment. This could be projectors in multiplex cinemas (projectors are effectively optical instruments) or more exacting still, microchip manufacturing processes or delicate balances in laboratories. Offices may not seem sensitive areas, but structurally adequate long-span floor structures can exhibit substantial movement even due to people walking across them and the effect can be very disturbing to staff.

Structureborne noise

A by-product of vibration is structureborne noise and although the vibration levels in a building may be low and satisfactory, the noise levels radiated from a structure due to vibration may exceed required ambient noise levels. This is normally only a problem where low noise levels are required in areas such as auditoria, conference rooms and bedrooms, and only where these areas do not have windows that would allow low frequency break-in noise to mask the structureborne noise. Although reduction of these audio frequencies relating to structureborne noise can be achieved by isolating the building, reductions of 5–10 dB only are likely to result. The use of resilient pads to isolate the building will not significantly reduce low frequency vibration levels.

Building damage

Consideration also needs to be given to the protection of a building or other structure in order to prevent damage from such activities as piling or press operation. Guidance on vibration levels for different building types to avoid damage is given in DIN 4150: 1986, Part 3, Structural vibration in buildings; effects on structures [6]. However, it is worth noting that the levels of perception by people and the levels that are considered to be satisfactory by BS 6472 are well below those vibration levels that could cause damage to buildings. Further details relating to groundborne vibration are given in Chapter 1.

Sources

There are various potential sources of vibration:

- industrial activities such as presses or generators,
- building services plant either associated with general air-conditioning/ventilation systems or installations such as lifts,
- footfall due to the movement of people.

Natural or resonant frequency is discussed in Chapter 5. Floors, walls and indeed entire buildings have their own resonant frequency characteristic. This resonant frequency can be excited by a single blow such as footfall, as described earlier for long-span floors.

Mechanical plant as used in building services, in addition to having its own natural frequency due to its mounted condition, also has forcing frequencies which are a function of its own operating conditions, e.g. running speed.

Vibration can occur in any combination of six modes: vertical, longitudinal, horizontal/traverse (linear motions), rolling, pitching and yawing (rotational motions). Services designers' concern will be primarily with the vertical mode.

The effect of vibration transmission into a building structure from mechanical plant will be determined by the relationship between the forcing and natural frequencies. This is illustrated in Figure 5.10 for a single one-degree-of-freedom system. When the forcing and natural frequencies are close then the vibration from the plant will be easily transmitted to the supporting structure, therefore to control vibration transmission the ratio of the frequencies must be changed. The natural frequency of the supported plant is dependent not only on its own mounting but also on the supporting structure. Long-span lightweight construction can inherently be easily 'driven' and Steffens's *Structural Vibration and Damage* [7] quotes other BRE guidance to stay clear of low (approaching 5 Hz) natural frequency characteristics. Higher values of 10 Hz or more should be sought for floor structures.

Having obtained a rigid structure from the structural engineer, it is necessary to introduce resilience into the support of the plant using vibration isolators. The correct selection of vibration isolator has to take into account the natural frequency of the supporting structure. Disappointing performances will be obtained for isolators selected to achieve 95% isolation if the machine is mounted on the mid span of lightweight steelwork.

Other parameters associated with vibration isolation include static deflection which is a function of natural frequency and, since it is a more readily identifiable unit, is very useful. Damping will also need to be considered in assessing vibration isolation, since it will reduce the anticipated isolation performance of a system.

Design considerations

Types of equipment

During the design phase, it will be necessary to consider a wide range of potential noise and vibration sources. Some of the major items are discussed below. General advice is given but it will be necessary to obtain measured data from manufacturers and follow detailed prediction routines.

Fans

Noise in fans is generated by:

- blade action
- airstream effects at fan surfaces
- resonant fan casing vibration
- fan drive/motor drive and vibration

For the usual constant fan speed, the least noise occurs when the fan is on or around its maximum efficiency. 'Stall' speed or overspeed should be avoided. Derating pulley changes resulting from 'over-engineered' systems may not improve the noise character.

Typically, turbulent flow is random and causes broadband noise acros the audible frequency range. There are pure tones at the blade rotation frequency and its higher harmonics, as an overlay to this broadband noise.

Fan noise data of concern are the octave band sound power levels of noise via the intake, exhaust terminations, and as radiated via the fan casing and external motor. Reference bodies which may be quoted in a performance specification include CIBSE [8] in the UK and ASHRAE in

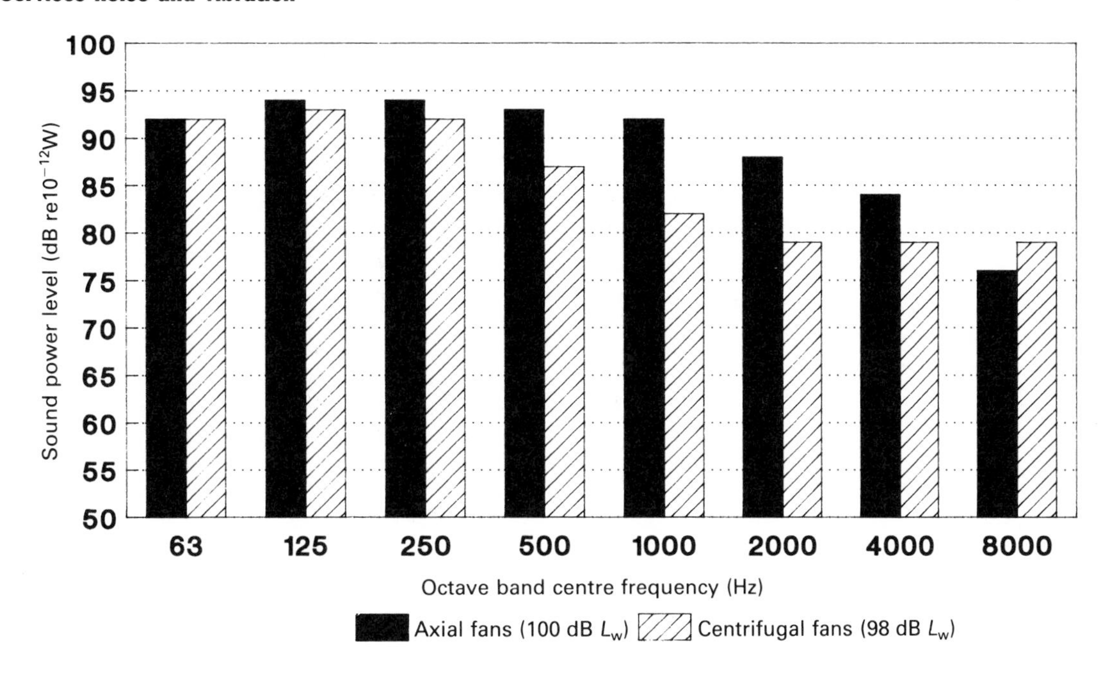

Figure 3.3 *Typical fan noise frequency spectra*

the US. Pending selection or as a cross check of manufacturers' claims, an estimate can be made from empirical formulae and typical spectra (Figure 3.3) based on the duty of the fan.

An empirical formula for sound power level at inlet or outlet is:

$$L_w = 40 + 10 \log V + 20 \log h$$

where V is delivered volume (m^3/s), h is fan static pressure (N/m^2). SRL's book, *Noise Control in Building Services* [9], has a useful initial guide (Table 3.4). Axial and centrifugal fans produce similar sound power, with axial fans having higher high-frequency values (Figures 3.4 and 3.5).

Air-handling plant

Air-handling plant of modern design consists of the fan unit itself (which may be centrifugal or axial), flexible connections, casing and chassis, filter, coils, mixing boxes, and possibly integral attenuators and dampers.

Cooling tower/condenser units

Noise arises from the fan, fan motor assembly, and water turbulence down to sumps. Regenerated noise may occur via water circulation pipes. An indicative sound power level is given by:

$$L_w = 11.5 + 10 \log P$$

where P is the total rated fan power output (in watts).

Directivity factors will be important as condensers are typically screened in a compound rather than fully enclosed. A typical spectrum is shown in Figure 3.6.

Refrigeration units

Fridge plant compressors may be annoying by virtue of intermittency of operation. Typical spectra are shown in Figure 3.7.

Boilers

Sound pressure levels are fairly similar for different types of fuel and representative octave band SPLs are given in

Table 3.4 *Typical fan noise spectra*

Equipment	OBCF (Hz) 63	125	250	500	1 k	2 k	4 k	8 k
Fans (up to 75 mm static pressure)								
25 hp	95	94	91	84	79	74	69	64
40 hp	98	97	94	87	82	77	72	67
100 hp	101	100	87	90	85	80	85	80
250 hp	104	103	100	93	88	83	78	83
Fans (150 mm static pressure or over)								
50 hp	107	106	103	96	91	86	81	76
100 hp	111	110	107	99	94	89	84	79
250 hp	113	112	109	102	97	92	87	82

Figure 3.4 *Axial fan*

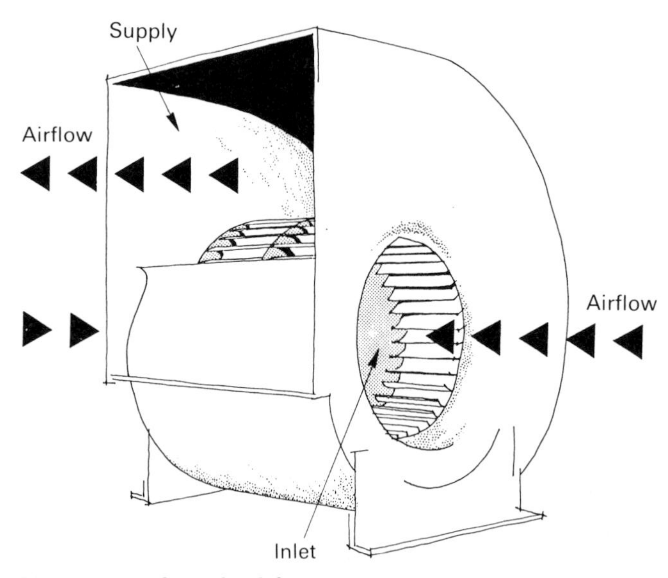

Figure 3.5 *Centrifugal fan*

Figure 3.8. The principal noise sources are the fuel burner units and combustion air fans. Noise will also be discharged up the flues and is predominantly of a low frequency character. Prediction formulae are given in CIBSE/ ASHRAE but caution is necessary in view of the large number of variables (flue height, directivity, cross-sectional area, linings, etc.) inherent in empirical formulae.

Generators

In many public buildings, emergency power for light and safety procedures is provided by battery sets. Some industrial premises, hospitals, broadcasting centres, and offices with vital constant power needs for computers, etc., will include emergency power generators, frequently in the form of diesel engines. Noise comes from:

- the engine itself
- exhaust,
- air intake,
- cooling fan,
- ventilation openings to engine enclosure.

It is misleading to measure engine sets when run-up in routine tests because full load cannot be applied. Noise control measures should be applied as a kit:

- enclosure, with controlled ventilation supply and extract openings by attenuators.
- exhaust silencing, possibly two-stage,
- inlet silencing,
- vibration isolation, 10-mm static deflection rubber engine mounts are usually adequate for basement floor slab placement.

Gas turbines

Gas turbines are industrial engines used for power generation, pumping and compression. The sound power level will depend on the engine rating but at mid frequencies can be around 120 dB. As with standby generators, noise emanates from the turbine itself, the exhaust, and noise from the discharge of turbulent hot

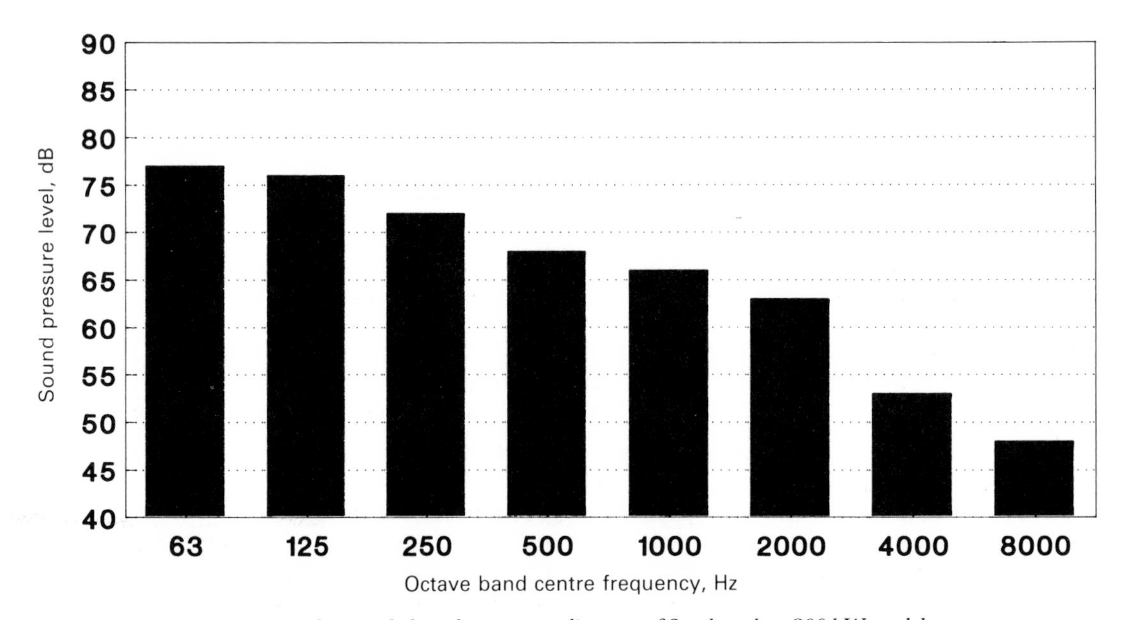

Figure 3.6 *Typical frequency spectrum of air-cooled condensers at a distance of 3 m based on 300 kW model*

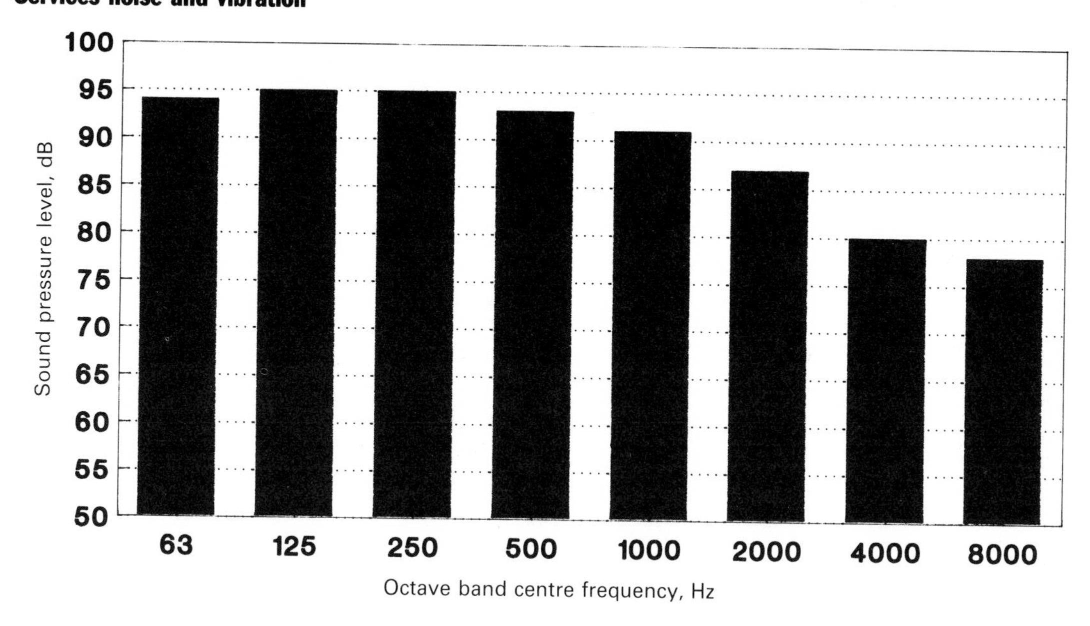

Figure 3.7 *Typical frequency spectrum of reciprocating chillers at a distance of 1 m based on 600 kW model*

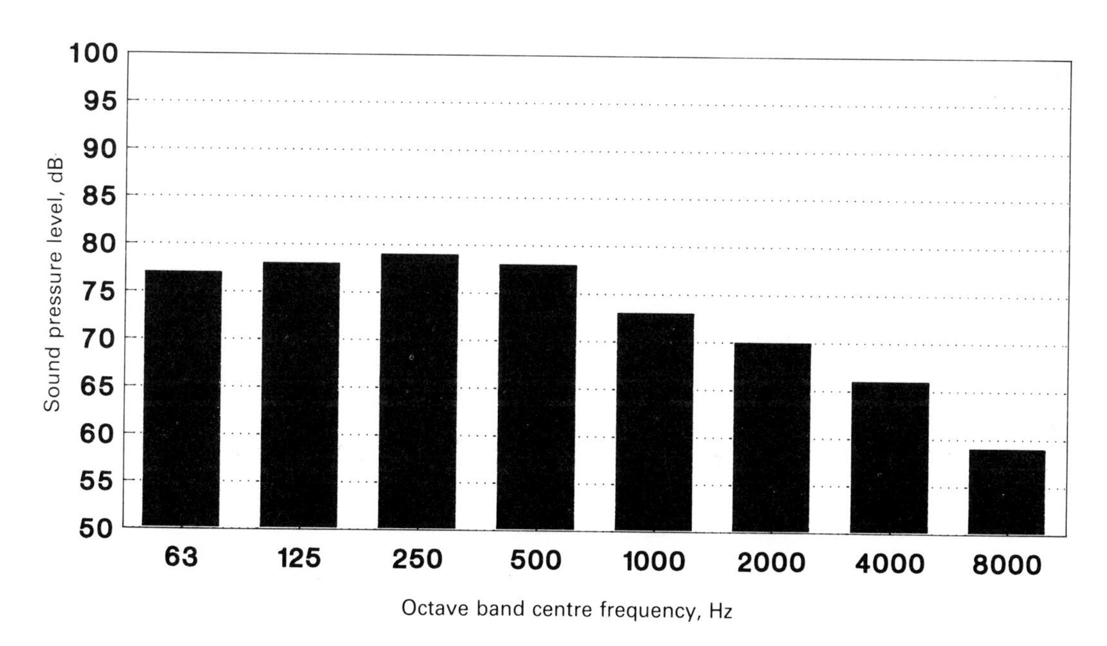

Figure 3.8 *Typical boiler room noise levels*

gases. Treatment can be by silencers, specially designed to withstand the hot gases, and some form of acoustic enclosure.

Noise design concerns for the above items of plant are addressed in the following summary:

- plan location of plant, or plant rooms, away from critical areas,
- provide substantial structures around the plant including acoustic doorsets where necessary,
- ensure any service penetrations of the structure do not downgrade its sound insulation performance,
- efficiently isolate the plant to control vibration transmission,
- provide any noise control necessary to reduce noise external to the building,
- fans or air-handling units will typically require attenuation on both the intake and exhaust sides of the fan,
- give a limiting sound power level for all major items of plant.

Valves

Valves are frequently the source of peak noise events at industrial plant complexes. The valves may be either for emergency only or to control flow in a system. By their nature, valves produce noise as a by-product of high pressure air, or other gas, relief. The sound power level of a valve may be in the order of 150 dB. 'Streamlined' valves help to a degree but the type and location can improve the situation: reducing jet size can shift the sound energy to higher frequencies which are more easily attenuated by screening and distance. 'Blow off attenuators' are available for mounting on 'pepper pot' arrays of small outlets. Such installations are the province of specialist process engineering designers.

Lifts

Lifts can present an annoying intermittent noise source in blocks of flats, hospitals, or even office buildings. Low-rise hydraulic lifts have fewer moving mechanical parts but are best specified with submersible pump and motor, casings lined with sound deadening material, and suitable vibration isolation not only of the pump and motor but also of the pipework.

The lift motor should be mounted on vibration isolators and the doors should be selected for quiet operation. High-speed lifts should not give wind whistle and air pressure rattles at doors panels. The intermittency of motor and doors operation can draw attention to the noise, in an otherwise quiet environment. Diverter sheaves can cause vibration in older models; good maintenance will help lifts to keep running smoothly and quietly. For both fire and acoustic separation, any builder's work holes in lift shaft walls should be made good.

Modern commercial buildings may have lift shafts in dry construction (multi-layer plasterboard for example), rather than masonry; the isolation of lift noise can be as effective with care and attention to the sealing of the multiboard edges at junctions.

A large property managing group has the following standard criteria as a performance specification in office schemes:

- Door noise 1.5 m from floor and 1 m inside door shall not exceed 65 dB (L_{Amax}), precision SLM set on 'fast' response.
- Noise levels at maximum car velocity, measured as above, should not exceed 55 dB L_{Amax} for lifts of velocities 0.5–2 m/s or 60 dB L_{Amax} for lifts of velocities 2–7 m/s.
- Lift noise within lift lobbies measured as above to be within 55 dB L_{Amax}.

A checklist of lift noise design concerns is as follows:

- plan lifts next to non-critical areas, e.g. stairs, stores;
- provide a substantial shaft, preferably with structural breaks to the main building structure;
- use large-diameter resilient wheels to counterweights and close tolerances at guide rails;
- isolate motor room, and form attenuating lined tube penetrations for suspension cables;
- allow controlled ventilation openings to shaft to avoid 'air pump' effects of pressure build-up during lift movements;
- use low-noise high-quality doors and signal bells.

Escalators

Escalators can give rise to noise levels around 50–55 dBA locally, with strong dependence on tread speed (+12 dBA/doubling of speed); squeal and clanks can arise from treads and handrails in worn examples.

A checklist of noise concerns is as follows:

- use sound-insulating casing to drive, gearing, and chains,
- apply damping to resonant panels,
- ensure maintenance to treads and handrail,
- review electrical noise sources: armature design, mountings.

Vibration can occur from the escalator operation and also from personnel movements on and off it. Vibration on the escalator steps should be less than 1 mm/s rms vertical vibration velocity (above 5 Hz). Operation should be imperceptible on adjacent floor slabs beyond 2 m from the end combs.

Lighting

This is only likely to be of concern in low-noise (NR 25 or less) rooms. Fluorescent lights have noisy ballasts (chokes), so should be avoided in such areas, unless remote starter chokes are planned in. Low-voltage lighting with local transformers can produce noise. While an individual light fitting may not seem a noisy item, an array of 50 in a small lecture room may allow distinct tones, buzzes or harmonics to build up.

There may also be noise from light fittings incorporating return air slots. Sound power levels are 35–40 dB at 0.04 m^3/s through a fitting. Such fittings also allow a route for room-to-room noise, so attenuation may have to be incorporated if light fittings are within 1.5 m either side of partitions or if a high performance separation is required.

High-level mercury discharge lighting in sports halls may produce around NR 30 which will be acceptable for sports use but may need reviewing if other events are held there.

Ventilation systems

Fan noise

Noise is basically due to the fan, and air flow causing regenerated noise at dampers, control branches and bends, and terminals. CIBSE B12 gives a step-wise procedure for calculating the noise control requirements for fan noise and a design example is shown in Table 3.5. Guidance is also given in ESDU 82002 (Reduction of sound in ventilation and similar air distribution systems) [10], ESDU 81043 (Sound in low velocity ventilation ducts) [11] and ESDU 82003 (Example to illustrate the use of data items on noise from ducted ventilation and airconditioning systems) [12].

Initial design is assisted by the availability of microcomputing routines where parameters are fed in for the calculation of resulting room noise levels. Software packages are included in most mechanical engineers' microcomputer menus.

The main sound components in assessing the need to control fan noise are:

- fan sound power entering the system,
- attenuation at branches and straight runs of ductwork,
- insertion losses at attenuators,
- diffuser and end reflection effects,
- room effect.

Table 3.5 *Example of determination of fan noise through ductwork system at 125 Hz frequency*

Stage		Level, dB at 125 Hz frequency
1. Determine fan sound power level, L_{wf}		
L_{wf}		90
2. Determine ductwork system losses		
Duct attenuation	10 m of 700 × 500 ignore remainder	−4
Bend attenuation	2 × vaned bends	0
Branch attenuation	$\simeq 10 \log \left[\frac{\text{volume at grille}}{\text{fan volume}}\right]$	−11
End reflection at termination	grille area = 0.2 m^2	−5
3. Resultant sound power level at each grille, L_{wg}		70
4. Determine room losses		
Reverberant correction to obtain $L_{prev} = +10 \log n - 10 \log V + 10 \log RT + 14$ dB		
n = no. of grilles	= 4	
V = volume of office	= 360 m^3	−7
RT = reverberation time in office	= 0.7 s	
5. Reverberant sound level, L_{prev}		63
Direct correction to obtain $L_{p\,dir} = -20 \log r - 11 + D$ dB		
r = distance to grille	= 1.5 m	−10
D = directivity factor		
6. Direct sound level, $L_{p\,dir}$		60
7. Determine total sound level in room		
$L_{p\,total} = L_{p\,rev} + L_{p\,dir}$		65
8. Design criterion	NR 35	52
9. Attenuation required		13

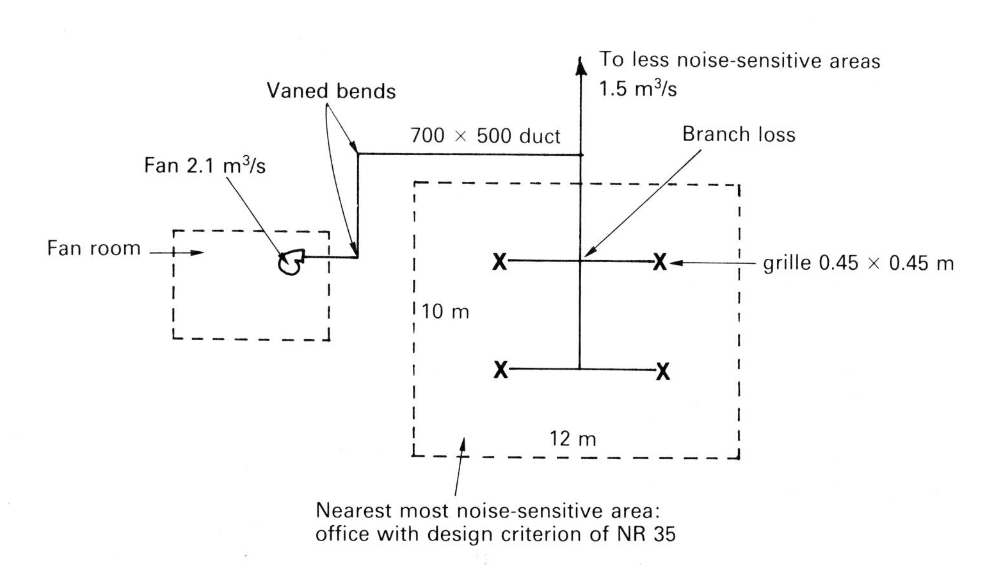

Schematic of ductwork system

Table 3.6 *Typical attenuator insertion loss (in dB)*[a]

Length (mm)	OBCF (Hz)							
	63	125	250	500	1 k	2 k	4 k	8 k
500	5	7	10	15	23	17	13	11
1 000	8	11	19	31	48	37	28	21
1 500	10	16	27	45	50	50	39	31

[a]The attenuator performance depends not only on its length but also on the ratio of airway-to-splitter width (refer Figure 3.11). These insertion loss figures are for an attenuator unit with an approximate ratio of 1:2.5 (airway:splitter).

Attenuation
Attenuation along ducted systems can be achieved by:

- length of duct run,
- lining of internal surfaces, a practice favoured in the USA over attenuators. Linings ducts is more effective for small ducts and higher duct velocities,
- bends,
- plenum chambers.

Attenuators, or silencers, are purpose-made sections of lined ductwork with splitters to incorporate a large surface area of absorption along the attenuator length. It is usual for the attenuator to have greater cross-sectional area than the duct it is in, to avoid undue pressure drop. The location is important and the performance is assessed by:

- insertion loss (typical insertion losses for particular lengths of attenuator are shown in Table 3.6),
- pressure loss,
- airflow noise.

In low velocity, low static pressure systems, the fan may be the only significant noise source, i.e. there are no great regeneration problems. High pressure, high velocity systems need more detailed calculation.

Regenerated noise
The basic layout of the ductwork and the air velocity within it influences noise levels most. The optimal placement of attenuators and other in-line duct items is critical. Recommended maximum duct velocities are as shown in Table 3.7, for low velocity systems.

The flow rate of air in a duct can be checked by a calibrated inlet device or by static suction in the early part of the system.

Regenerated duct noise can be created by:

- transition pieces,
- bends (turning vanes alleviate noise),
- dampers, grilles,
- branches inducing turbulence.

Cross-talk can be designed out by layout or allowed for by lining, attenuation or adjustment of grille size. Typical requirements are set out in Table 3.8. The method of assessing cross-talk requirements is given in Diagram 3.1.

In office buildings, ceiling voids or underfloor voids are often used as supply or extract plenums, with air passage via grilles and possibly light fittings. Such systems in offices are more economic, saving on ductwork, but limit the scope for attenuation and can lead to room-to-room cross-talk problems.

Duct noise break-out
To reduce fan noise break-out from ductwork, ducts can be lagged by a barrier mat (quilt with lead foil interlayer), Keene's Cement (not now favoured) or studwork panel casing. A layer of 12 kg/m^2 lead on mineral wool will increase the sound reduction of the ductwork by 5 dB/octave.

Duct shape influences in-duct noise and duct noise break-out characteristics. Circular ducts are more rigid and of minimum perimeter for a particular cross-sectional area thus reducing noise transmitted into rooms or ceiling void. Hence circular ductwork is often preferred in exposed system installations within spaces. Rectangular ducts have less rigid walls and the flat metal is more easily excited, and although it may provide low-frequency in-duct attenuation, it allows more noise break out at low frequencies. This can lead to 'drumming' heard within the room through which the duct passes.

Table 3.7 *Maximum recommended duct velocities*

NR or NC design requirement	In-duct air velocity (m/s)		
	Main	Branch	Final run-outs
20	4.5	3.5	2.0
25	5.0	4.5	2.5
30	6.5	5.5	3.25
35	7.5	6.0	4.0
40	9.0	7.0	5.0

Table 3.8 Cross-talk attenuation

Requirement in receiver room	Attenuator length (mm)	Noise reduction at 500 Hz (dB)
NR 40	750	25
NR 35	1000	30
NR 30	1250	35
NR 25	1500	40

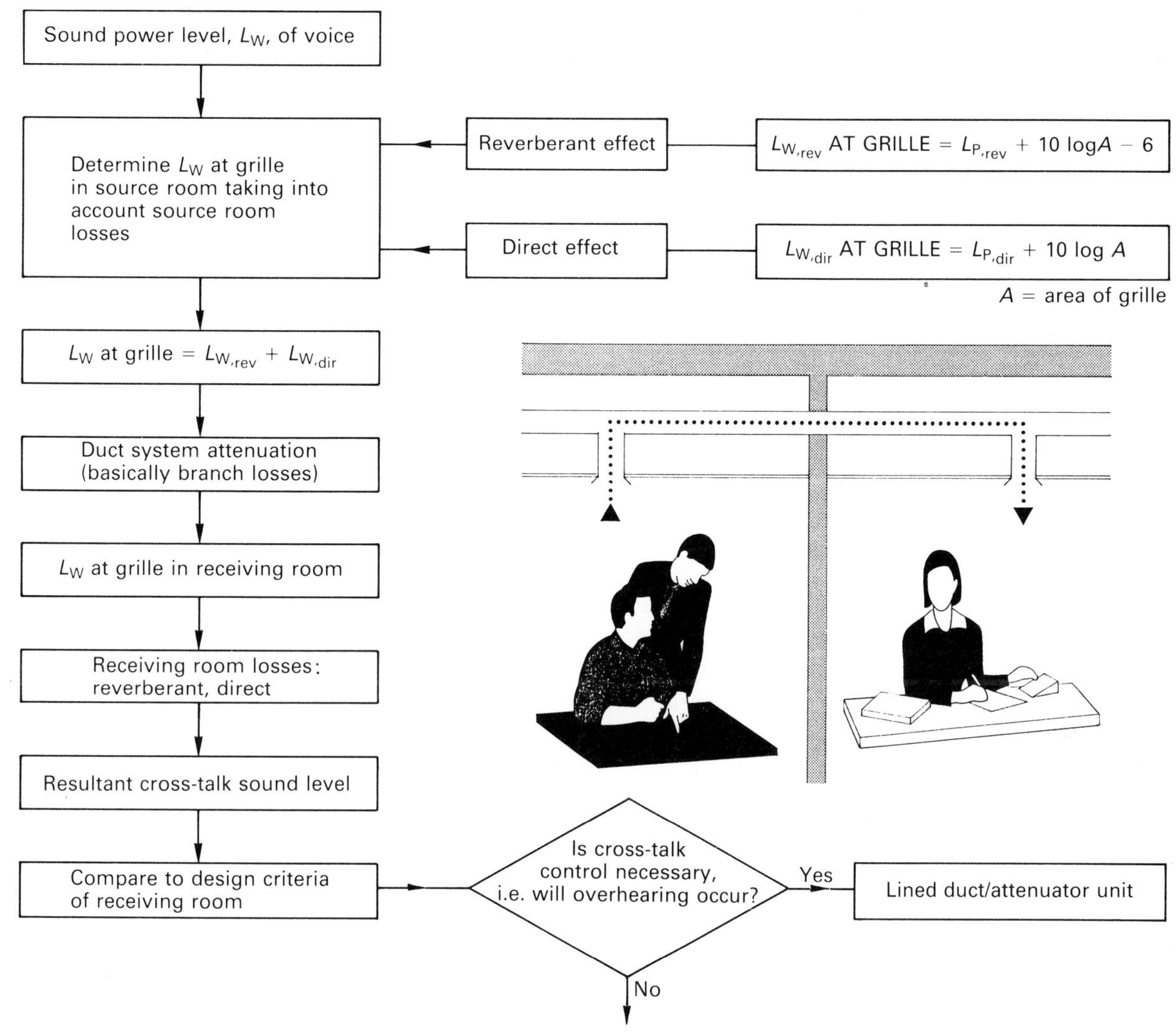

Diagram 3.1 *Method of determining cross talk*

Builder's work ducts can be used for low velocity systems in, for example, auditoria. These are long plenum chambers formed in airtight masonry or plasterboard (Figure 3.9). The advantages are:

- lower cost than equivalent very large-scale metal ducts,
- easier installation,
- efficient use of building's space.

The disadvantages are:

- mixed responsibilities of main contractor and ductwork supplier,
- difficulties in avoiding pressure drop-off from a take off at one end to one at the other,
- good workmanship is required to ensure an airtight chamber.

The builder's work details at ductwork penetrations of wall need careful attenuation (Figure 3.10).

Riser ducts are a feature of the distribution from plant rooms in multi-storey buildings. They can be either masonry or dry construction. Access doors to high velocity riser ductwork should be acoustically rated.

Transfer grilles are frequently used to save ductwork runs in ventilating adjoining small rooms, but negate acoustic separation. They may be used only for acoustically non-critical partitions or doors.

Large public spaces

Large public spaces present a conflict between large-scale air distribution and noise generation, particularly in large noise-sensitive volumes like auditoria. High occupancy demands good ventilation but high ceilings mean long 'throws' for ceiling-mounted supply systems against the 'natural' convection currents. An alternative claimed to save half the cooling load on a recent theatre project is the European common practice of low-level, underseat supply and overhead extraction, all at low velocity.

Another issue in large low-velocity systems is the point at which it becomes more economic to go from complete metal duct systems (which are the responsibility of the engineering subcontractor), to a mix of ducted systems and 'builder's work' air-sealed plenum chambers for supply or extract (Figure 3.7).

In office buildings, ceiling voids or underfloor voids may be used as supply or extract plenums, with air passage via

Figure 3.9 *Air plenum above recital hall*

grilles and possibly light fittings. Such systems in offices are more economic, saving on ductwork, but limit the scope for attenuation (single-stage only) and lead to room-to-room cross-talk problems.

Central and room units

Preference for centralized systems rather than room or sector ventilation units is based on ease of maintenance and separation of plant rooms from served spaces. However, there are now many good package units available and a sensible compromise which may be considered is to have centralized plant providing basic air-handling to user areas, with an overlay of room units for specific high-load areas. This is often a solution for 'tenant's plant' installed as a fit-out contract to supplement the landlord's serviced shell.

Room units can be of different types:

- air-handling only, by a packaged unit recirculating air through a chamber where it is tempered by chilled or heated pipework, and then mixed with a proportion of fresh air make up;
- air conditioning by wall-mounted unit, with a fan coil on the room side and a condenser unit outside;
- fan coil units which recirculate and temper air, with control by varying fan speed;
- induction units which by jet action move the air within the space many times the supply air velocity;
- ducted air systems, as a scaled down version of a centralized ventilation system;
- terminal units, which alter 'mains' supply of ducted air locally.

Each system has a characteristic noise and, unlike centralized plant, the ventilation units are within the user space rather than segregated in a plant room. Fan coil units have a typical sound power spectrum of around NR 50–55 profile, terminal units around NR 40–45. Some optional extra improvement can be gained by selection of a quieter standard model, damping casing radiation, fitting attenuators (the back pressure implication needs checking), and acoustic lagging.

Manufacturers of fan coil units and also grilles invariably quote an achieved NR or NC level by their units based on an estimated room loss. This room loss is often 8 dB and assumes only one unit serving it, whereas the actual value could be 3–5 dB thus underestimating the as-installed noise levels situation.

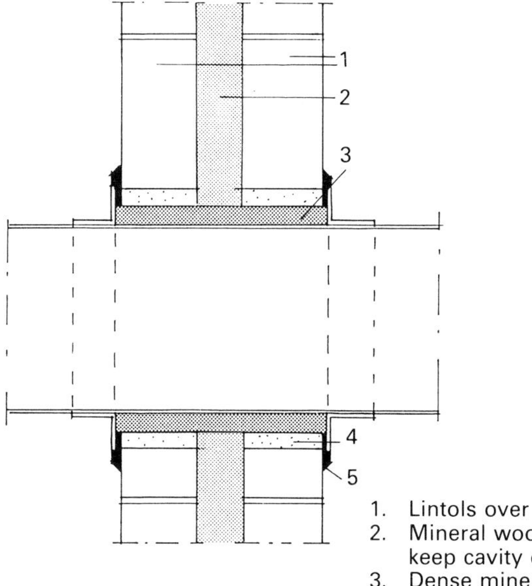

Figure 3.10(a) *Builder's work penetrations: duct through wall*

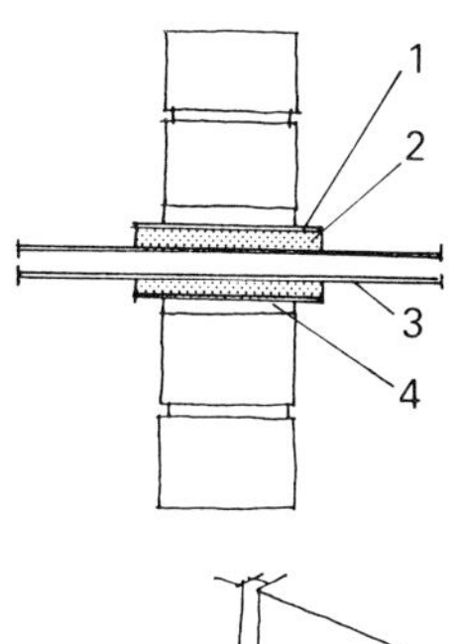

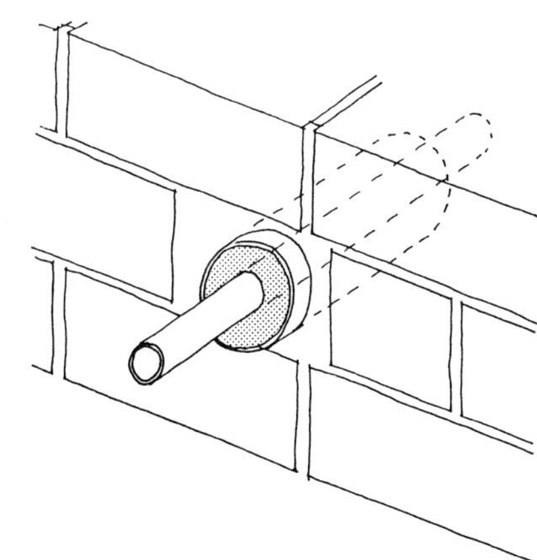

Figure 3.10(b) *Builder's work penetrations: pipe through wall*

Plant rooms

Plant in older buildings was placed on the basis of 'boilers in the basement, tanks on the roof'. Modern plant rooms house not only boilers and pumps but also air-handling units, lift motors, compressors, and open-to-atmosphere chiller plant. Plant-room noise levels are typically in the range NR 70–85 for boiler rooms, NR 60–75 for air-handling plant rooms. Masonry structures – concrete floors, concrete, brick or blockwork walls – are essential, with metal acoustically-rated access doors. An SRI of 50 dB (100–3150 Hz) is a minimum requirement for walls and floors. Additional airborne attenuation through the floor can be achieved by the introduction of a floating floor.

The roof structure sound insulation will need considering in the case of roof-mounted freestanding air-handling units often used on commercial, retail or multiplex cinemas projects.

Absorption in plant rooms may reduce reverberant sound pressure levels by about 5 dB but it is usually more cost effective to have noise control at source or increase the sound insulation of the plant room structure. If possible, expensive shrouds to units should be avoided as after initial maintenance there is a tendency to leave enclosure panels loose or detached altogether.

Plant room structure

Once the location of plant or plant rooms is fixed, consideration needs to be given to providing an adequate plant room structure. In addition to determining the appropriate main construction which is typically masonry, any openings and penetrations by ductwork or pipework have to be carefully designed.

Acoustic doorsets may have to be specified. Metal doorsets are capable of achieving a higher sound insulation than a timber type. In specifying acoustic doorsets, care must be taken in selecting the appropriate performance from manufacturers' data.

Ventilation openings will normally need to be acoustically controlled by attenuator units or acoustic louvres.

Service penetrations will need to be effectively sealed. Suitable details are given in Figure 3.10.

External plant

Noise to the outside can be from plant which by its nature needs to be open to the atmosphere, screened areas or freestanding plant rooms holding generators or chillers, or openings into fully enclosed plant rooms – air inlets or exhausts, and plant room natural ventilation.

Condensers are often roof mounted and therefore do not benefit from ground attenuation or natural screenings by proximity to walls. Such units often have to run at night and so have to be considered relative to low background noise levels. Screening will be effective if the line of sight to receiver is blocked. Following the concern over *Legionella*, cooling towers are advisedly placed further from any occupied areas in any case.

Noise control measures may entail 'top hat' units on chiller units. This could raise planning concerns because the plant is visually more obtrusive.

At the preliminary stage, the location and building configuration may be affected by noise break-out considerations.

Noise control methods

The main requirements for controlling noise from building services are identifed in Diagram 3.2. Internally this is normally achieved by an adequate plant room structure, the use of attenuator units, and careful selection of

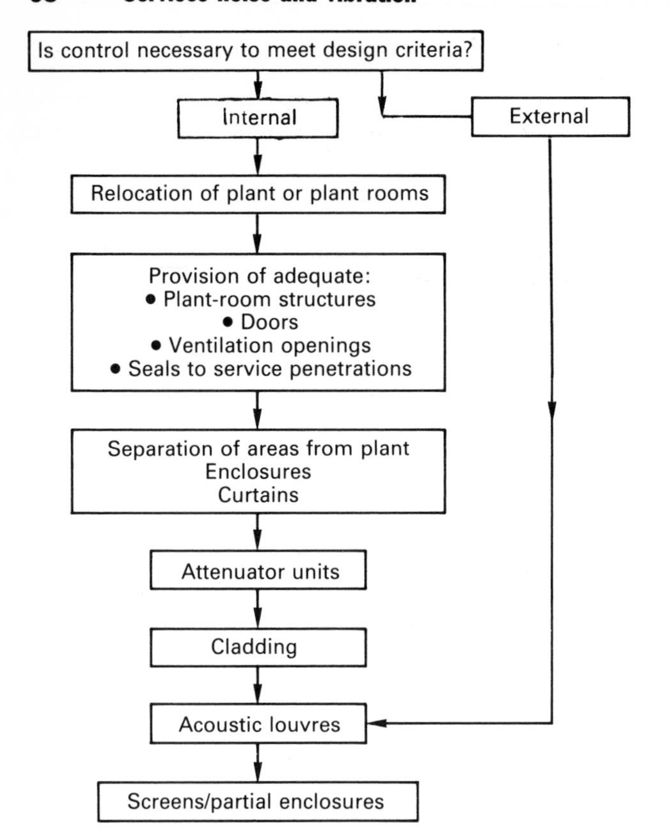

Diagram 3.2 *Requirements – noise control from building services*

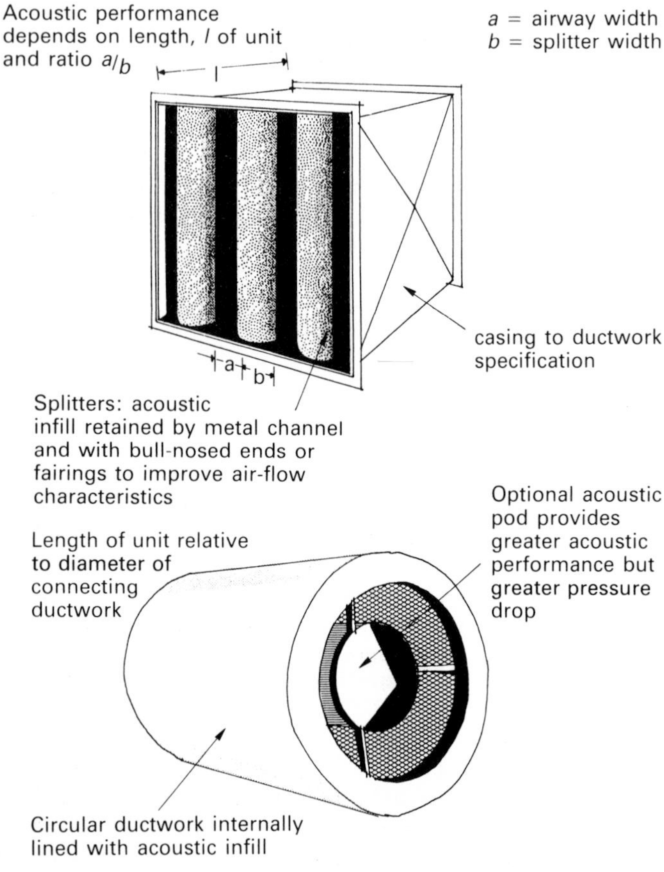

Figure 3.11 *Acoustic attenuator construction*

equipment. In addition, external noise control can be achieved by acoustic louvres, screening and enclosures.

Attenuator units

As discussed briefly earlier, attenuator units are components in a ductwork system used primarily to control fan noise, and would normally be positioned as close to the fan as possible. When the ductwork is serving sensitive areas, e.g. auditoria, secondary attenuators are used, which are normally located at or near the termination into the area, not only to further control fan noise, but also to control any noise regenerated close to the terminations. Attenuators can be used to control cross-talk.

Attenuator units can be circular or rectangular, as shown in Figure 3.11, and basically comprise a casing containing acoustically-absorbent material held in splitters, side linings or pods.

In specifying suitable attenuators, the parameters of maximum dimensions, i.e. length, width and height, dynamic insertion loss required, and maximum resistance to airflow (typically 50 Pa), should be given to meet the design airflow and noise criteria required.

The construction of the attenuator units needs to meet the following:

- casing should be constructed to the relevant ductwork specification;
- acoustically-absorbent infill should be inert, fire proof, inorganic, vermin proof, non-hygroscopic, and preferably retained by perforated metal or equivalent;
- splitters should ideally be constructed with bullnose fairings to both entry and exit, thus reducing the resistance to airflow;
- duct connections need to match the duct flanges or spigots.

In some instances there are requirements to use plastic ductwork or provide special lining materials.

Acoustic louvres

Acoustic louvres are slatted blades angled to keep out rain and block line of sight through, but have an additional feature over conventional louvres: the undersides of the blades are lined with absorptive material behind perforated mesh. The arrangement allows attenuation from one side of the louvre bank to the other, whilst maintaining a relatively high free area value. The deeper the louvre bank, the more effective it is, but it would provide a high pressure loss as a penalty. Compared to in-duct silencers, the insertion loss performance is poor at about 12 dB at 500 Hz for a 300-mm-deep louvre (Figure 3.12).

Besides controlling air intakes or exhausts direct into plant rooms, acoustic louvres can also be found as atmosphere terminations to plenum chambers and as open screens around chiller plant on roofs or around transformer plant in the open. As screens, the limited through-the-vanes performance is not as important, as it only has to prevent more sound going through the screen than over the top by diffraction.

Cladding

Cladding, or lagging, is a measure to be considered to either damp pipes or ducts which are excited by air or gas

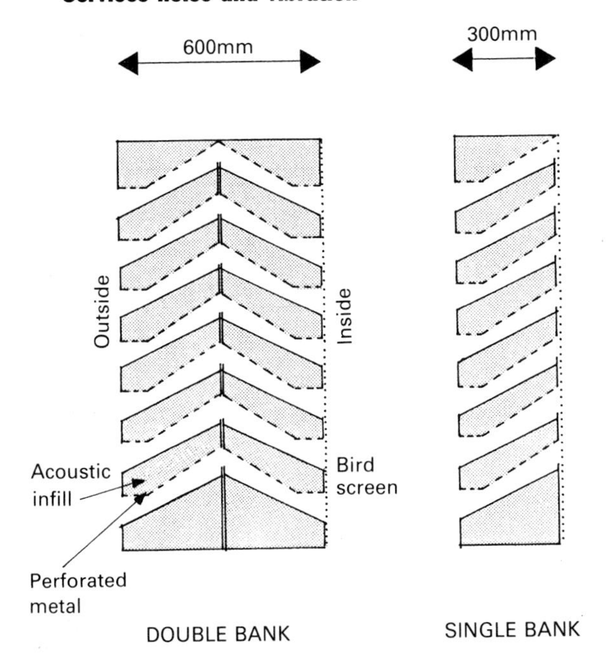

Typical Insertion Loss of Acoustic Louvres

Frequency	63	125	250	500	1k	2k	4k	8k	Hz
Single bank	5	5	7	12	18	21	16	16	dB
Double bank	8	9	12	21	32	34	32	32	dB

Double-bank louvre causes a greater pressure drop than single-bank louvre

Figure 3.12 *Acoustic louvres construction*

turbulence within, or to uprate the thin duct or pipe wall to prevent noise break-out from high airborne noise levels within. The former can re-radiate noise to user areas that they run through either direct via the duct or pipe walls, or via vibration effects arising from transmitted excitation to casings or suspension to local structure. A light cladding, for example thermal-grade duct lagging, will not reliably damp the movement or have adequate surface mass to contribute sound insulation. The solution is to use a form of cladding, such as:

- builder's work enclosure, for example framing plus plasterboard panels,
- double skinning ductwork, i.e. a substitute for conventional ductwork along selected sections of premade composite construction lengths of ductwork. The composite consists of an outer metal skin, glass fibre or mineral wool core, and an inner metal skin,
- barrier quilting of mineral wool with lead foil interlayer. The weight of such treatments varies from 5 to 15 kg/m^2.

Enclosures

Enclosures in industrial premises may either be used to protect workers, in the form of 'noise havens', or control rooms off-line from the noisy processes, or to localize high noise levels to one area within a complex. While a screen's performance is an unpredictable 5–10 dBA noise reduction, an enclosure may provide 20–40 dBA reduction by careful design, depending on construction. There are a number of proprietary modular systems. The most suitable construction is for sound absorption to be present inside the enclosures' faces, without compromising the sound insulation of the solid sheets behind. In some instances partial enclosures can provide useful noise reductions, as indicated in Figure 3.13.

Acoustic curtains

Acoustic curtains, in the form of limp blankets hung from and draped over a framework, can restrict noise break-out from noisy machine areas within an industrial interior, whilst still allowing some airflow. Such a treatment could reduce reverberant levels in an area alongside by about 5–10 dBA. The curtains have the advantage of simplicity of framing/suspension and hence can be relocated easily.

Screens/partial enclosures

To control noise from externally located plant, it is sometimes necessary to shield the plant. This is achieved by a solid barrier of relatively lightweight material, it only having to prevent more sound going through the barrier than over the top by diffraction, or in some instances it may be an acoustic louvre as described earlier.

Provision of a screen should allow adequate ventilation of mechanical plant.

Vibration

Design control

Vibration isolation of machinery is not a recent innovation; Cassell's Magazine of 1880 notes as a new idea 'the use of india-rubber cushions under workbench legs, or kegs of sand or sawdust used for the same purpose. An ordinary anvil so mounted may be used in a dwellinghouse without annoying the other inmates'. This would seem optimistic, even given current isolator technology.

All mechanical plant associated with building services is a potential source of vibration. The major concerns are typically with larger items but even small pumps and motors could cause problems. The main sources of vibration are as follows:

- fans
- refrigeration equipment
- cooling towers
- air-cooled condensers
- pumps
- generators

However, lifts, escalators, boilers, waste compactors, delivery bays, and kitchen equipment also need to be considered.

Before a vibration isolation scheme is formulated, endeavours should be made to relocate plant as far away from sensitive areas as possible and to provide a rigid structure for support. This would also apply conversely to sensitive equipment.

Once the most appropriate location for plant or sensitive equipment has been agreed, vibration will be controlled by reducing the transmissibility either from the plant into the structure or from the structure to the sensitive equipment. This vibration isolation will be achieved by introducing resilient supports.

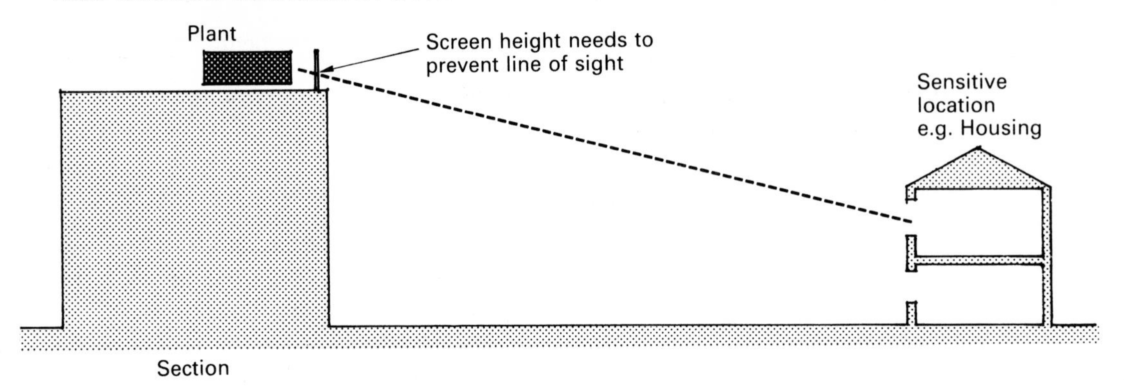

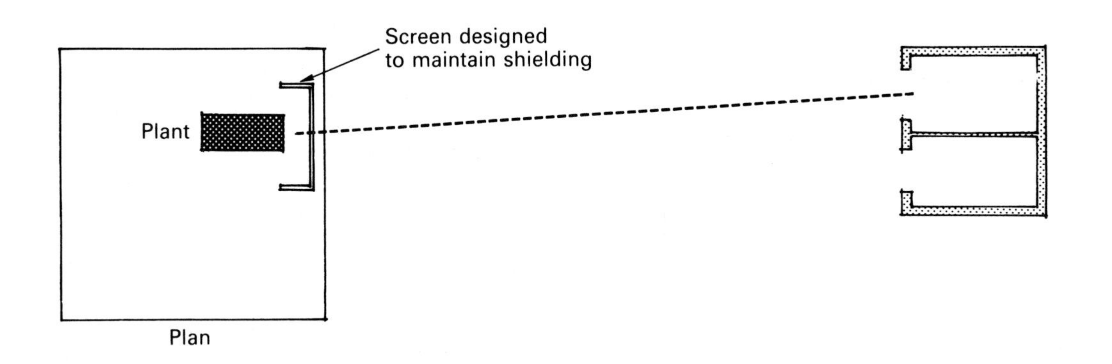

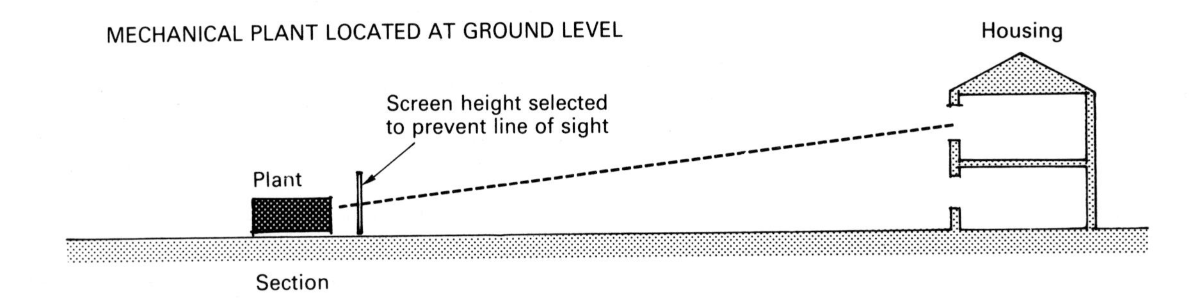

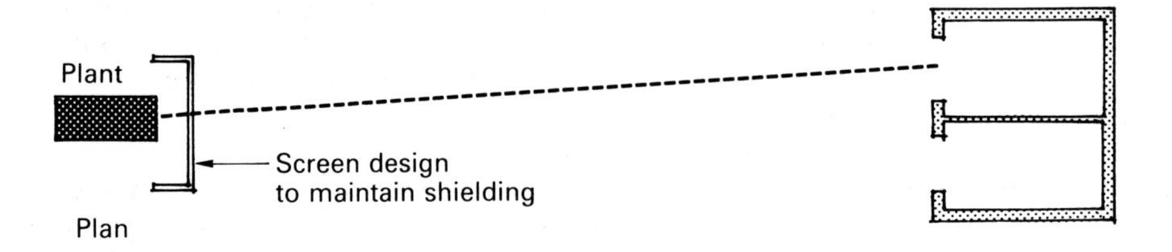

Figure 3.13 *Examples of partial noise enclosures*

Table 3.9 *Static deflection: guidance for resilience to provide vibration isolation and avoid resonances. (Source: A. T. Fry)*

Equipment	Minimum static deflection (mm)				
	Basement	Floor span[a]			
		6 m	9 m	12 m	15 m
Refrigeration machines					
Absorption	6	12	25	50	50
Packaged hermetic	6	12	50	60	90
Open centrifugal	6	12	50	60	90
Reciprocating chillers:					
500–750 rev/min	25	50	50	60	90
751 rev/min and over	25	25	50	60	60
Reciprocating air or refrigeration compressors					
500–750 rev/min	25	40	60	70	90
751 rev/min and over	25	25	40	60	70
Boilers or steam generators	6	6	25	40	70
Pumps (water)					
Close coupled					
up to 5 hp	6	12	25	25	25
7½ hp and over	20	25	50	60	90
Base mounted					
up to 5 hp	9	12	40	50	60
7½ hp and over	25	25	50	60	90
Packaged unitary air handling units (low pressure up to 750 Pa)					
Suspended up to 5 hp	20	25	25	25	25
7½ hp and over					
up to 500 rev/min	30	40	40	50	60
above 501 rev/min	25	25	25	40	50
Floor mounted up to 5 hp	6	25	25	25	25
7½ hp and over					
up to 500 rev/min	12	40	50	50	60
above 501 rev/min	12	25	25	40	50
Axial fans (floor mounted)					
Up to 5 hp	6	25	25	25	25
6–20 hp up to 500 rev/min	12	40	50	50	60
above 501 rev/min	12	25	25	40	50
25 hp and over					
up to 500 rev/min	20	50	60	70	90
above 501 rev/min	12	25	30	40	50
Centrifugal fans (floor mounted)					
Low pressure (up to 750 Pa)					
Up to 5 hp	6	25	25	25	25
7½ hp and over					
up to 500 rev/min	12	40	50	50	60
above 501 rev/min	12	25	25	40	50
High pressure above 750 Pa					
Up to 20 hp					
175–300 rev/min	9	60	60	90	120
301–500 rev/min	12	50	50	60	90
above 501 rev/min	9	30	30	50	60
25 hp and over					
175–300 rev/min	40	60	90	120	140
301–500 rev/min	25	50	60	90	120
above 501 rev/min	12	30	50	60	90
Cooling towers					
Up to 500 rev/min	12	12	50	60	90
Above 501 rev/min	9	9	25	40	60
Internal combustion engines (standby power generation)					
Up to 25 hp	9	12	50	60	60
30–100 hp	12	50	60	90	90
Above 125 hp	25	60	90	120	120
Gas turbines (standby power generation)					
Up to 5 MW	6	6	6	9	9

[a]The floor span refers to the largest dimension between supporting columns. The equipment is assumed to be at mid-span.

In determining the most appropriate resilient supports, static deflection is the most important parameter. It is a function of the resonant frequency of a system and is a design parameter that can easily be checked on site.

In selecting vibration isolation for plant, account must be taken not only of the lowest forcing frequency of the machine (typically its running speed) but also the resonant frequency of the loaded supporting structure, for example floor slab.

An example of the typical selection charts used to determine the preferred static deflection is given in Table 3.9 for different types of plant and supporting floor spans.

Once the static deflection has been determined, the method of mounting the equipment has to be assessed. There are normally three options, simply shown in Figure 3.14:

- provide antivibration mounts directly between the equipment and the supporting structure. This includes isolators as an integral part of a packaged unit, e.g. isolation of the fan from the casing of an airhandling unit, (a);

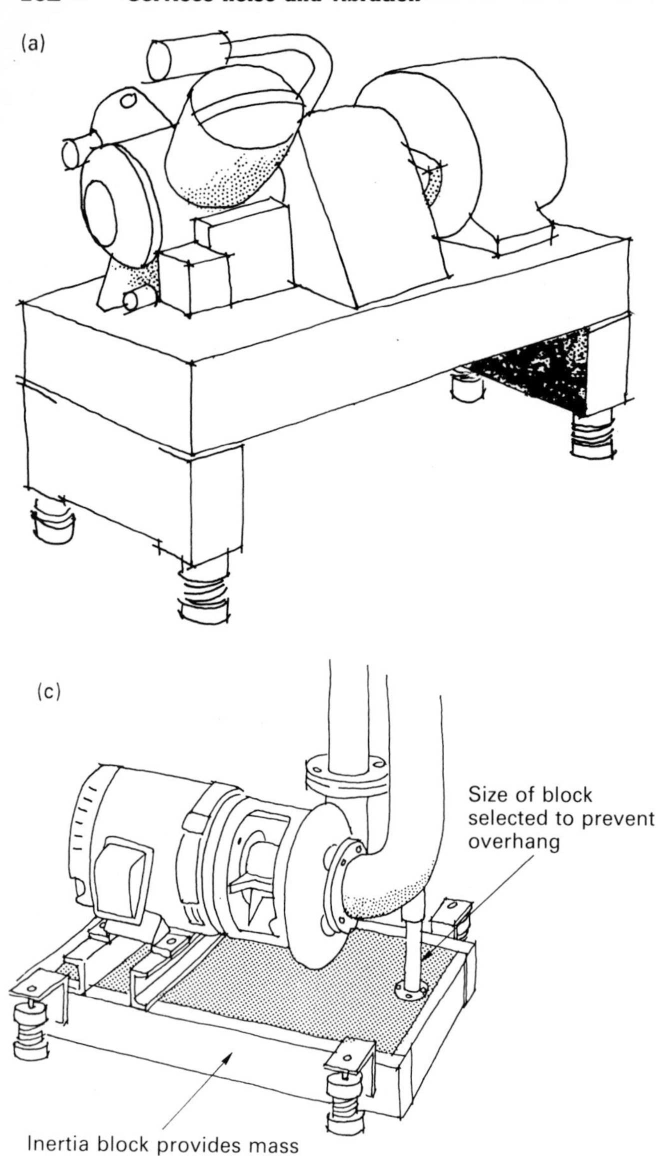

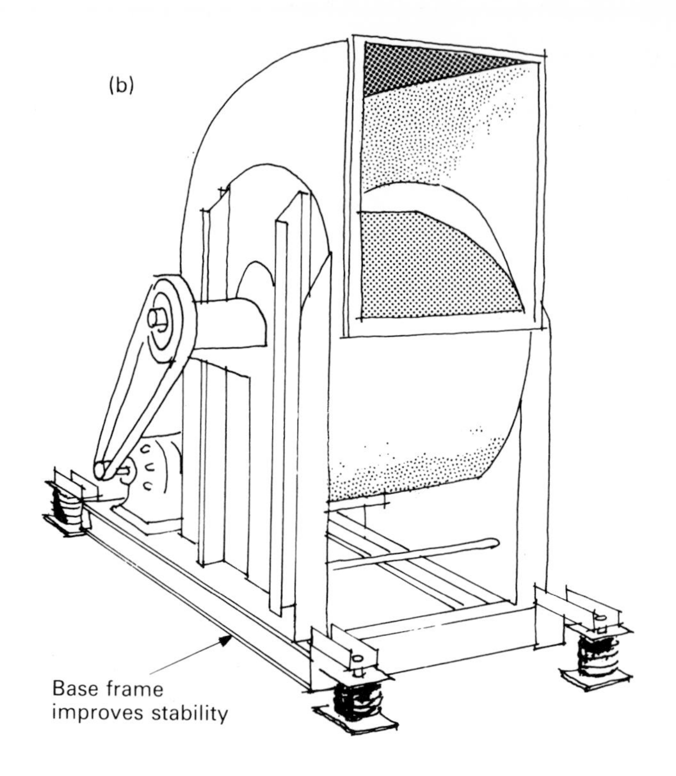

Figure 3.14 *Vibration isolation mounting options: (a) machine mounted directly on vibration isolators; (b) equipment mounted on a specially designed base frame with integral vibration isolators; (c) equipment mounted on an isolated inertia block*

- support the equipment on a steel base frame and then provide antivibration mounts between the frame and the main supporting structure, (b);
- mount the equipment on a concrete inertia block which is isolated from the main supporting structures, (c).

Unless the equipment is rigid and not liable to distortion if supported by its existing feet individually on isolators, then either a steel base frame or concrete inertia block should be used. Both the frame and the block add rigidity to the equipment and can be designed to a size, greater than the base size of the equipment, to improve the stability and lower the centre of gravity of a system. The provision of the concrete inertia block will increase the mass of the system, giving a more even weight distribution (where equipment alone could be significantly heavier at one end than the

Table 3.10 *Types of commonly available vibration isolator*

Type	Material	Range of static deflection (mm)
Mats or pads	Rubber Neoprene-coated Glass fibre	0.5– 5 0.5–15
Turret mounts	Neoprene	5–10
Springs	Steel	5–75

Table 3.11 *Vibration isolation for services plant*

Item	Type
Pumps, compressors	– rubber/neoprene in shear – damped metal springs – inertia slabs + mounts
Air-handling units	– (direct fix if integral AVs) – rubber/neoprene in shear – undamped metal springs
Extract fans	– resilient hangers + flexible connection
Standby diesels	– solid layer pads – rubber/neoprene in shear – damped metal springs
Transformers	– rubber/neoprene composites in compression pads

other), will minimize the vibration effects of changing equipment speeds or loads, will reduce problems likely to occur due to coupled vibration modes, and can act as a local acoustic barrier.

Once the preferred mounting method has been selected, the type and location of vibration isolators or antivibration mounts can be determined. The types of vibration isolation commonly available are as shown in Tables 3.10 and 3.11.

In areas where high isolation, i.e. high static deflection, is required, such as in the semiconductor manufacturing industry, air mounts are likely to be required. These consist of an air-filled bag which can be coupled to a leveling control device enabling equipment to be maintained at the correct level regardless of load.

The selection of isolators should be undertaken by an acoustics engineer or the vibration isolator supplier. They will need to determine the preferred location of the isolators, taking into account the loads imposed at support positions.

In describing vibration isolation, a simple system is normally used taking into account the vertical direction only. To minimize the risk of the selected isolation being reduced by the effect of other modes of vibration, it should be ensured that the horizontal stiffness of the isolation is similar to the vertical stiffness, and the centre of gravity of the system is as low as possible.

In addition to careful selection of the vibration isolation for equipment, it will be necessary to ensure that there will be no flanking/short circuiting/bridging of the isolation. Flanking could occur due to service connections, poor installation, or the impact of ancillary operations, for example infill 'builder's work'.

Flexibility for pipe, duct and other connections is essential to avoid unacceptable stresses in stiff pipe and other connections. This can be provided by the use of flexible connectors in the pipe or duct system close to the equipment.

In high-pressure pipe systems, allowances must be made for the force tending to stretch or compress the flexible hose, and restrained connectors would normally be used. In this case, the connection will be stiffened and suitable pipe layouts must be used to improve the situation. In addition, reliance is placed on the natural flexibility of the pipework itself and this will be increased by changes of direction. However, until the vibration has been reduced effectively, the pipework needs to be isolated from the supporting structure, typically for a distance of 50 pipe diameters from the equipment in normal circumstances. This isolation is achieved by hangers or clamps, having a similar performance to the isolators under the equipment. Any other rigid connections should be flexibly joined to the isolated system, e.g. electrical conduits. Care will also need to be taken during installation to ensure that bridging of the isolator does not occur, e.g. via hangers or by fixing bolts of isolator touching the equipment base.

On site, other trades may influence the performance of an apparently isolated system. Examples of problems are concrete being spread such that it covers isolators or debris building up under isolated equipment bases, particularly inertia blocks.

In ensuring that the most efficient form of vibration isolation is selected, the mechanical engineer in formulating his specification should include the following:

- advice from an acoustics engineer;
- an appropriate schedule detailing the equipment to be isolated and its method of isolation;
- all isolators to provide the required minimum static deflection given in the schedule under the imposed load of the equipment, the selection of isolators taking account of eccentric load distribution.

To enable the correct selection of vibration isolation, the following details are required:

- type of equipment,
- weight of equipment,
- centre of gravity location,
- number and position of mounting points,
- operating speed and nature of the operating mechanism,
- details of the supporting structural floor, particularly spans.

Isolator types

All isolators offered to the contract must meet the specifications and must be suitable for the loads, and operating and environmental conditions which prevail.

Mats and pads should be manufactured from synthetic rubber, neoprene, or glass-fibre coated with neoprene. The materials and design should render them impervious to contamination from oils and attacking chemicals and be rot and vermin proof.

Turret compression mounts should be fabricated from synthetic rubber or neoprene between two steel plates. The materials must be oil and corrosion resistant with the metal protected from corrosion by painting. Friction surfaces must be provided to the bottom and the top. Bolt holes must be provided to allow fixing.

Spring compression mounts should comprise a high-strength, low-stress, laterally-stable, open-steel spring located by a steel pressure plate on the top and bottom. The bottom plate should include a bonded ribbed neoprene pad to the underside of minimum thickness 6 mm, and be pre-drilled for bolting down. Each isolator should be identifiable by a colour-code mark and provided with a levelling facility with final lock nuts. The spring element should have an overload capacity of 50% with an outside diameter of at least 80% of the operating height. The horizontal stiffness should be not less than the vertical spring stiffness.

Captive spring mounts are as above but the steel spring should be encased in a neoprene-covered body to achieve horizontal and vertical snubbing.

Vertically-restrained spring mounts are as for the spring compression mounts but the steel spring should be mounted within a hanger box constructed from steel of minimum thickness 1.6 mm. The box should be vertically restrained by noise-isolated bolts. In addition, any horizontal buffers or snubbers should be manufactured from synthetic rubber or neoprene. During normal operation, the snubbers should be out of contact.

Hangers are turret or spring compression mounts (or a combination of the two) to the specifications above incorporated within a hanger box. The hanger box should be constructed from steel of 1.6 mm thickness complete with a hole for the suspension rod and an enlarged lower hole for the drop-rod to equipment. The lower hanger rod should be allowed to move laterally at least 15° without any contact with the hanger box.

Steel base frames should be purpose-built using a welded steel framed with attachment points for suitable vibration isolation. The frame should provide adequate support of the equipment without flexing or significant deflection.

Concrete inertia bases should be purpose-built, using welded steel frame formwork containing reinforced concrete 35 mm above the bottom of the base. The base should be designed with sufficient strength and rigidity to support the equipment and compensate for dynamic reactions due to operation of the equipment. The size of the base should be sufficient to give support for all integral parts of the equipment including inlet and discharge manifolds. The design of the base should provide a minimum clearance of 18 mm between the underside of the base and the structural floor below, with the installed mountings operating at their design static deflection and under full plant rating.

Pipework flexible couplings provide flexible hose couplings for connecting pipework, comprising nylon fabric or a steel mesh carcass with a waterproof cover. The design of the coupling should take into account the conveyed fluid temperatures and pressures imposed.

Ductwork flexible connections provide flexible coupling of fans/air-handling units and connecting ductwork designed to provide a minimum operating length of 100 mm.

In addition to vibration isolation of the main items of plant by these methods, there are likely to be situations where small plant items need to be isolated from the structure to prevent structureborne noise. This can normally be achieved by neoprene or synthetic rubber inserts.

Installation

Laboratory tests

An awareness of laboratory tests in services noise control is important in order to understand the limitations of the data provided by manufacturers. The concerns are:

- adequacy of the facilities and the reputation of the testing house,
- relevance of the data quoted to design application,
- traceability to British, US or International standards.

The 'laboratory' should be an approved testing house rather than a corner of the works, because the facilities must provide for testing between 40 Hz and 16 kHz with effective isolation to background noise and vibration. There should be an airflow testing rig and test rooms for both reflection and absorption available, by means of reverberation room and anechoic chamber. Linked rooms enable the testing of sound insulation in a transmission suite.

Airflow testing facilities need a silenced air supply to ensure that tests reflect the device performance – the device may be an attenuator, diffuser, grille, louvre terminal unit, damper or ventilator.

Reverberation rooms have bare walls, floor and ceiling so that there is a diffuse and even distribution of sound energy around the room.

Anechoic chambers (Figure 2.17) have deep wedges of absorption on all surfaces to create a 'dead' acoustic. They give 'free-field' conditions so the direction of sound energy from a test item can be investigated.

Transmission suites are a pair of reverberant rooms to test the sound reduction index to BS 2750 [13], of doors, windows or equipment like cross-talk attenuators or acoustic louvres. An area of 10 m^2 is a minimum surface area of aperture. Floor and ceiling transmission can be tested in a vertical suite of rooms.

Test methods need to be standardized to reference bodies' test codes, either:

- British Standards Institute, BSI
- International Organisation for Standardisation, ISO
- European Heating and Ventilation Society, Eurovent
- American Society of Heating, Refrigeration and Air Conditioning Engineers, ASHRAE

A schedule of relevant codes is given in Chapter 5. A summary for services noise components is as follows:

- attenuators – BS 4718 [14]
- impact sound – BS 2750, BS 5821 [15]
- sound absorption coefficient – BS 3638 [16]
- airborne sound reduction index – BS 2750
- power-generated SWL measurements – BS 4718, BS 4773 [17], BS 4856 [18], BS 4857 [19], BS 4979 [20]
- in-duct plates or transformations – BS 1042 [21]
- fan testing – BS 848 [22]

Installation: works tests

Reasons for tests

On a large project, there may be many room units of one type, so it is important to know the characteristics of a sample unit. The unit may be an air-conditioning package, fan coil unit, generator, or other item of plant. Frequently, the manufacturer's data is inadequate, or not specific to the model ordered.

Timing

The timing of tests are critical – prototype testing and the obligation for the supplier to cooperate must be built into the programme and contract documentation. The responsibility and cost of any retests needs to be set out, as a disincentive for too early 'time wasting' testing or too hasty requests for retest when the supplier knows full well that the unit under retest will still not comply.

Checklist of considerations

Typically, the test will take place in a less than ideal situation, far removed from a laboratory. There are a number of considerations:

- Mounting: is it on a ground floor slab rather than, say, a raised computer floor? Are there integral AV mounts set correctly?
- Access: can measurements be taken at different sides and heights to check directivity?
- Operation: can the system be run in its normal modes of operation, i.e. under load? There may be a number of settings, each of which should be checked.
- Completeness of system: it may not give a realistic picture to works test one component which is being incorporated in a package arrangement, for example a

room unit in a computer room may need a mock-up of cladding, sample areas of ceiling and raised floor incorporating intake and return air attenuators, even though separate suppliers may be involved.

- Background noise: correct for the difference, identified by measurements with and without the system operating, between noise from the system under test and other noise prevalent.
- Reverberant character of works: correct for the factory compared to the final application.
- Status of tests: is the consultant testing on behalf of the client or is compliance to be demonstrated by the supplier's or subcontractor's representative? If the tests identify improvements necessary, who bears the cost and programme delay implications?

Commissioning tests

These advisedly take the form of a two-stage inspection. The first is a pre-commissioning exercise which typically involves a walk-around as a visual check that antivibration mounts are properly adjusted (i.e. not either fixed down or bolted solid), builder's work around penetrations is complete, and inertia bases are clear of debris (it is a very convenient place to sweep material out of sight). At this stage, it may also be possible to undertaken initial testing of systems to troubleshoot any major problems prior to the main commissioning exercise. The second stage is the official commissioning survey.

The commissioning process usually starts immediately before handover on a building contract, handover meaning the client can have beneficial occupation of the building and its systems rather than when everything is totally completed. Systems handover may need to be phased, and adjustment and testing extend beyond handover to when the building has been in use for a while.

For tests to be meaningful, correct design duties should be achieved by the system. Levels can then be taken to check room design criteria have been achieved. In larger rooms, levels may be taken at standing and/or seated head height, for a variety of positions. Any 'hot spots' directly under supply or extract grilles, or discernible rattles (it is amazing what people leave in ducts) need to be noted.

There may be a requirement to include measured levels in engineering manuals at handover. Care should be taken to take measurements for equipment in normal use, or at its various settings to provide complete information. This may be difficult for automatically controlled chiller plant, measured in winter. Diagrams 3.2 and 3.3 show requirements and a basic problem-solving path.

References considering commissioning and testing of systems are as follows: CIBSE Commissioning Codes: Series A Air Distribution, Series B Boiler Plant, Series R Refrigerating Systems [23]; BSRIA Applications Guide 1/87, 1: Operating and Maintenance Manuals for Building Services Installations [24]; RIBA Architect's Job Book Vol 1: K8 Commissioning Services Installation [25].

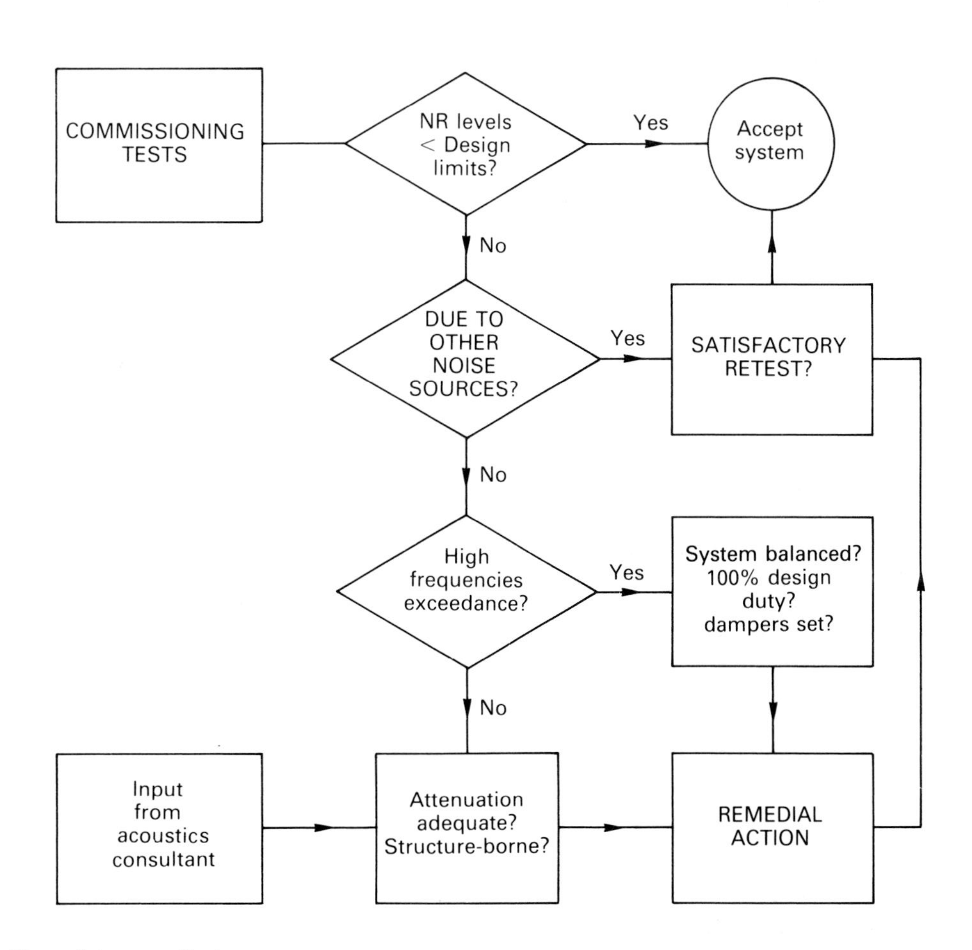

Diagram 3.3 *Problem solving: ventilation system*

References

1. Rose, K. *Guide to Acoustic Practice*, 2nd edn, BBC Engineering, London, 1990
2. European Broadcasting Union, *Acoustical Properties of Control Rooms and Listening Rooms for the Assessment of Broadcast Programmes*, Report No. R22, EBU, 2nd edn, 1985
3. *Specification for Studio Centres*, Engineering Code of Practice for Independent Local Radio – 1984, Issue 2, Independent Broadcasting Authority, London, February 1988
4. BS 4142: 1990 *Method of rating industrial noise affecting mixed residential and industrial areas*, British Standards Institution, Milton Keynes
5. BS 6472: 1992 *Guide to evaluation of human exposure to vibration in buildings (1 Hz to 80 Hz)*, British Standards Institution, Milton Keynes
6. DIN 4150: 1986: Part 3 *Structural vibration in buildings: effects on structures*
7. Steffens, R. J. *Structural Vibration and Damage*, BRE Report, HMSO, London, 1974
8. CIBSE, *Sound Control*, Section B12, *Guide Volume B: Installation and equipment data*, Chartered Institute of Building Services Engineers, London, 1986
9. Fry, Alan (ed) *Noise Control in Building Services*, Sound Research Laboratories Ltd, Pergamon, Oxford, 1980
10. ESDU 82002 'Reduction of Sound in Ventilation and Similar Air Distribution Systems', *Noise in Air Conditioning Systems*, Fluid Mechanics in Internal Flow, **9,** Engineering Services Data Unit, London, July 1982
11. ESDU 81043 *Sound in Low-Velocity Ventilation Ducts*, 'Noise in Air Conditioning Systems', Fluid Mechanics in Internal Flow, **9**, Engineering Services, Data Unit, London, December 1981
12. ESDU 82003 *Example to Illustrate the Use of Data Items on Noise from Ducted Ventilation and Air Conditioning Systems*, 'Noise in Air Conditioning Systems', Fluid Mechanics in Internal Flow, **9**, Engineering Services Data Unit, London, December 1982
13. BS 2750: 1980 *Measurement of sound insulation in buildings and of building elements*, British Standards Institution, Milton Keynes
14. BS 4718: 1971 *Methods of test for silencers for air distribution systems*, British Standards Institution, Milton Keynes
15. BS 5821: 1984 *Methods for rating the sound insulation in buildings and of building elements*, British Standards Institution, Milton Keynes
16. BS 3638: 1987 *Method for measurement of sound absorption in a reverberant room*, British Standards Institution, Milton Keynes
17. BS 4773: 1989 *Methods for testing and rating air terminal devices for air distribution systems*, British Standards Institution, Milton Keynes
18. BS 4856: 1978 *Methods for testing and rating fan coil units, unit heaters, and unit coolers*, British Standards Institution, Milton Keynes
19. BS 4857: 1983 *Acoustic testing and rating*, British Standards Institution, Milton Keynes
20. BS 4979: 1986 *Aerodynamic testing of constant and variable dual or single duct boxes, single duct units and induction boxes for air distribution systems*, British Standards Institution, Milton Keynes
21. BS 1042: various dates *Pressure differential devices*, British Standards Institution, Milton Keynes
22. BS 848: Part 2: 1985 *Fans for general purposes: methods of noise testing*, British Standards Institution, Milton Keynes
23. CIBSE Commissioning Code Series A: *Air distribution*, CIBSE, London, 1971
    CIBSE Commissioning Code Series B: *Boiler plant*, CIBSE, London, 1975
    CIBSE Commissioning Code Series R: *Refrigeration plant*, CIBSE, London, 1991
24. BSRIA *Operating and Maintenance Manuals for Building Services Installations*, Building Services Research and Information Association Applications Guide 1/87, Bracknell, 1987
25. RIBA *Commissioning Services Installations*, Royal Institute of British Architects, Architect's Job Book, **1**: K8

Chapter 4 Sound systems

Peter Mapp

Introduction

Sound and communications systems are becoming increasingly integrated into building design, forming part of an emergency warning/fire alarm and evacuation system, a general announcement (paging) system, or as a special design feature enabling background music to be played, or live entertainment to be played or relayed. An even greater need for early consideration and appropriate integration of the sound system occurs in public buildings housing auditoria, lecture theatres and conference facilities, where high-quality sound reinforcement systems will be required. The system chosen must complement the natural acoustics of such halls.

In buildings with hostile acoustic environments, such as noisy and/or reverberant leisure centres, ice rinks and swimming pools, special care needs to be taken with the loudspeaker type and locations and the overall system design, in order to ensure that adequate speech intelligibility and clarity will be achieved.

In sports stadia and similar venues where crowd noise levels can be high, the system must be capable of comfortably overcoming the anticipated noise level and providing a clear and distortion-free signal. Similar considerations also apply to noisy industrial process areas and plant rooms.

In addition to the more usual speech and music signals, sound systems may be used to transmit warning tones and signals, or information tones/signals (for example, factory processes, swimming pool sessions). Sound systems are frequently required to interface with or form an integral part of other systems such as fire alarm and emergency evacuation control systems, or systems for the hard of hearing, audiovisual (A/V) systems, video and film projection systems, languages translation/interpretation systems and broadcast and relay systems.

In order to achieve an appropriate sound quality, clarity and intelligibility, a sound system must meet certain performance criteria, which will be dependent upon the type and application of the system concerned. The basic parameters to be considered include:

- sound level,
- overall frequency response,
- signal-to-noise ratio,
- direct-to-reverberant sound level ratio,
- freedom from echoes and distortion.

Other secondary requirements include system stability and reliability, freedom from interference both to and from other systems operating within the vicinity. Even humidity and climatic conditions influence how a system performs on individual occasions.

Systems used for fire alarm and emergency announcement purposes must be appropriately fire protected and monitored, and incorporate secondary power supply back-up and switch-over facilities.

Local acoustic and environmental conditions must also be fully taken into account at the system planning and design stage, together with zoning and emergency override and priority requirements.

Table 4.1 gives a brief summary of typical building types and sound installation requirements. In many buildings such as hotels, conference centres and theatres, more than one type of system may be required, e.g. stage music, speech reinforcement, and perhaps even electroacoustics, systems for the main auditorium, public paging/announcement for the front-of-house, circulation and ancillary areas, and staff or technical paging/announcements behind the scenes. These self-contained systems may, however, need to be linked or interface with other systems or subsystems for emergency announcements, evacuation or fire alarm purposes.

The design and installation of a sound system may involve a number of different disciplines including:

- the architect, who will have interest in the loudspeaker types, appearance and placement, and the overall integration of the system into the project design;
- the M & E consultant, who will be involved with the provision of electrical power supplies and cable routes, etc., and if a voice alarm system is included, possible interfacing to the fire alarm detection and control system;
- specialist consultants, including theatre consultants, audiovisual consultants and acoustic consultants, who will all be interested in the way the sound system interfaces with their particular areas and installations;
- sound systems design consultants: although many of the smaller systems are designed either by the M & E consultant often with the aid of a manufacturer, specialist independent sound-systems design consultants are available, who are able to design a system without a vested interest in a particular range of equipment. Equally, a specialist consultant will have a wider view of the market, equipment performance, current techniques, and experience of systems design.

A specialist consultant will take responsibility for his design and because he is not trying to sell anything can be acceptably integrated into the design team. He is therefore in a much better position to argue the case for critical loudspeaker locations or the requirement for specific acoustic treatments, and realistic budgets, then can a manufacturer's representative, or a non-specialist consultant.

Most of the larger manufacturers offer a design service, but schemes will be based on their own equipment ranges, which may not necessarily be suitable for the project in hand. Furthermore, few manufactuers' designers have a background in acoustics and loudspeaker technology – an area which currently is one of the less well understood areas of system design.

A number of computer-aided design programs are now available to assist with sound systems design, and are primarily aimed at improving the loudspeaker coverage/room acoustics problem. However, the majority of the programs have been developed by loudspeaker manufacturers with an underlying aim, not surprisingly, of selling more loudspeakers. At the present time, the accuracy of the programs is still being evaluated and improved, as the

Table 4.1 *Sound systems for different buildings*

Building type	Live reinforcement	Paging and general announcements	Emergency announcements	Audible signal information	Off air	BGM	High quality recorded programme	AFILS/IR	Interface requirements	Notes
Airports	–	✓	✓	–	–	✓	–	✓	FA/E	Multiple inputs and zones required noise levels vary.
Auditoria (general)	SM (ES)	Rehearsal only	✓	–	–	–	✓	✓	FA/E, ES A/V	Paging only during rehearsals. Relay to other areas.
Canteens	(SM)	✓	✓	(✓)	✓	✓	(✓) (See note)	(✓) (See note)	FA/E	May be used for live entertainment. Use separate system as auditoria.
Clubs	SM, ES	As auditoria Theatres	✓	–	(✓)	✓	✓	✓	FA/E A/V, ES	Zone areas. See 'Theatre'.
Concert halls (see theatres)	SM, ES	Rehearsal only (see theatres)	✓	–	(✓)	–	✓	✓	FA/E A/V, ES	Relay to other areas. Zone peripheral areas
Council chambers	S	–	✓	–	(✓)	–	–	–	FA/E A/V	Relay to other areas. Zone separately.
Adjacent areas	–	–	✓	–	–	–	–	–	–	May act as overflow area. Zone accordingly. See Offices.
Court buildings (public areas)	–	(✓)	✓	–	–	–	–	–	FA/E	Announcements to public areas
Court rooms	S	–	(✓)	–	–	–	(✓)	–	FA/E A/V	High quality recording and playback usually required.
Conference rooms/ centres Public areas	S (ES) (M) (SI)	(See note)	✓ ✓	(✓) (✓)	(✓) (✓)	Public areas only	✓ –	✓ (✓)	FA/E, ES, A/V, SI	Paging generally restricted to areas outside conference rooms. Zone systems.
Department stores	–	(✓)	✓	–	(✓)	–	–	–	FA/E	Zone different floors/ areas. Spot announce m/c?
Discos	S (ES) (M)	✓	✓	–	(✓)	Only in areas isolated from dance floors	✓	–	FA/E, ES, A/V, M/S	Noise limiters FA/E to override. Zone areas.
Factories	–	✓	✓	✓	(✓)	(✓)	–	In noisy areas for communi-cation	FA/E	Zone system.
Halls/assembly rooms	SM ES	(See note)	✓	–	(✓)	(✓)	✓	✓	FA/E, A?V, MS, ES	Treat as theatre/ conference centre.
Hospitals	–	(See note)	✓	?	✓	(✓)	–	–	FA/E HR	Treat patients, public and staff areas separately.

Table 4.1 (*Continued*)

Building type	Live reinforcement	Paging and general announcements	Emergency announcements	Audible signal information	Off air	BGM	High quality recorded programme	AFILS/IR	Interface requirements	Notes
Hotels	✓ S+M (See note)	✓	✓	–	✓	✓	(✓)	? (See note)	FA/E A/V	Incorporate many types of system (See other categories).
Lecture theatres	✓ S (M) (SI)	– (See note)	✓	–	(✓)	–	✓	✓	FA/E (SI) A/V (ES)	Not usual to page but external areas should be; SI facilities sometimes needed
Offices	–	✓	✓	(✓)	(✓)	(✓)	–	–	FA/E	Zone as required
Passenger termini	–	✓	✓	–	(✓)	(✓)	–	✓	FA/E	Acoustics and noise levels, zone areas; local mic/ control facilities.
Places of worship	✓ S (M)	(✓)	(✓)	–	–	–	(✓)	✓	FA/E	Acoustics; some require high quality music.
Railway stations	–	✓	✓	–	–	✓	–	✓	FA/E	Acoustics and noise. Zone platforms?
Restaurants	✓ (M)	(✓)	(✓)	–	(✓)	✓	–	–	FA/E	Many need to zone different areas; some have live entertainment.
Pubs (see clubs)	(✓) SM	(✓)	(✓)	–	(✓)	✓	(✓)	–	FA/E M/S A/V	Zone bars, performance area.
Discos (see clubs)	✓ SM	✓	✓	–	(✓)	✓	✓	–	FA/E M/S A/V	Noise level limiters FA/ overrides.
Schools/ educational establishments	✓ SM	✓	✓	✓	✓	–	✓	(✓) (See note)	FA/E M/S A/V	Many have separate auditorium system; sometimes require special loop systems.
Shopping precincts	–	✓	✓	–	–	✓	–	(✓)	FA/E	Acoustics zone areas cater for local systems and O/R.
Sports Stadia										
Indoor	SM (✓)	✓	✓	(✓)	(✓)	✓	✓	(✓)	FA/E (ES)	Crowd noise acoustics
Outdoor	SM (✓)	✓	✓	(✓)	(✓)	✓	✓	(✓)	FA/E ES	Zone stands separately + crowd noise.
Swimming pools	–	✓	✓	✓	(✓)	✓	(✓)	–	FA/E	Pool hall must be separately zoned; consider acoustics.
Leisure centres	(✓)	✓	✓	–	(✓)	✓	(✓)	(✓)	FA/E	Careful zoning of different areas required.
Theatres										
Auditorium	(ES) SM	(✓) (See note)	✓	–	✓	–	✓	✓	FA/E A/V ES	Paging to auditorium only under rehearsal conditions.
Public areas	(✓) (See note) SM	✓	✓	✓	✓	✓	✓	–	FA/E	Possible connection to other systems and local inputs; foyer entertainment.

Table 4.1 (*Continued*)

Building type	Live reinforcement	Paging and general announcements	Emergency announcements	Audible signal information	Off air	BGM	High quality recorded programme	AFILS/IR	Interface requirements	Notes
Back stage	–	(See note)	✓	–	(✓)	–	–	–	FA/E Other theatre systems	Specialist show relay and technical intercoms.
Exhibition halls	(✓) S	✓	✓	–	(✓)	✓	(✓)	(✓)	FA/E	Consider acoustics empty and full; local microphone inputs and zoning.
Ice rinks	(✓)	✓	✓	✓	(✓)	✓	✓	(✓)	FA/E Music Systems	Consider acoustics zone rink separately.
Plant rooms, engine rooms and machinery spaces	–	✓	✓	(✓)	–	–	–	(✓)	FA/E Equipment warning zone	Acoustics and noise considerations. Also use of visual indicators to alert staff.
Museums, art galleries	(✓) (See note)	✓	✓	–	(✓)	–	✓	✓	FA/E AFILS/IR	Sometimes live reinforcement of events and lectures are given; a portable system may be appropriate; separate systems required for some exhibits. F/R 'Tour' systems may be employed.

Key: –, not usually required; ✓, usually or often required; (✓), may be required/should be considered; S, speech; M, music; ES, external system (brought in), may operate and need to interface with permanent systems, e.g. rock band PA systems or broadcast systems; SI, simultaneous interpretation systems; FA, fire alarm system; E, emergency evacuation system; A/V, audiovisual (video projection) systems; M/S, music and speech; HR, hospital radio majority of them rely on fairly simple algorithms and design techniques but this is an area which is continually developing and soon more sophisticated and complex design studies will be possible, offering greater accuracy. The programs can offer a design discussion aid, quotation substantiation, and allow non-specialists to gain a visual impression of the design and the effects of any constraints involved.

A number of acoustical consultants also offer a sound system design service, but care should be taken to ensure the consultant involved has a thorough understanding of the complete design and installation process, and will be able to take on the electronic and electrical aspects of the systems design as well as the basic acoustical ones. Many will provide a basic performance specification which is an alternative approach but caution is needed regarding who takes responsibility for the installed system leading from the specification.

System planning

Site survey

The success of any sound system will depend on its initial planning and detailed consideration of the system's primary function and operational requirements. Adequate forethought should be given to how the system is to be operated and who will operate it.

A detailed site survey should be carried out at an early stage, in order that the sound pressure levels required from the system may be more accurately assessed. The need for group or automatic noise sensing controls may also be established.

Areas with long reverberation times should be noted. The measurement and assessment of room acoustic data requires specialist equipment and consultancy should be sought accordingly.

If an induction loop system for the hard of hearing is to form part of the installation, a background magnetic noise survey should also be carried out.

The site survey should include an assessment of potential cable routes, equipment locations and fixing methods. Any areas requiring special fixings or access should be noted.

Discussions should be held with the client in order to establish microphone/control points and system input priorities. Wherever possible, early discussions should be held with the architect in order to establish suitable loudspeaker and equipment locations that can be integrated in the general design.

Design check-list

Uses

A check-list may be used as a discipline in design, for example it must be decided which of the following tasks, in priority order, is the system to cater for:

- general paging and announcements
- emergency announcements
- speech reinforcement
- music playback (background music)
- music reinforcement (of jazz, cabaret, rock bands)
- relay of broadcasts (from other areas/services including radio)
- use in conjunction with other A/V systems including film and video media
- altering the room acoustics, i.e. electroacoustics (see Chapter 2)

In addition, the system will have to conform to British or International/European standards and perhaps also local regulations. A selection from the above list of the range of features will decide the overall quality of system.

The system operator – will he be one of the following:

- full-time, trained sound operator,
- part-time operator and/or semi-trained staff, e.g. receptionist or caretaker,
- personnel with some limited operational training, e.g. security staff, local management,
- anybody – actors, musicians?

Answers to these questions will help decide on the type and complexity of controls and the security of the system against inadvertent misuse.

Quality of system

'Quality' in this context means suitability for purpose and environment and required speech intelligibility. This may be best defined by considering division by building type, i.e. what area does the installation serve:

- underground station,
- plant room,
- hotel foyer,
- hotel function room,
- conference room,
- theatre auditorium,
- sports stadium,
- airport terminal,
- church or cathedral,
- swimming pool or ice rink,
- shopping mall?

From the above considerations, judgements can be made of the system fidelity and quality of sound.

Acoustic considerations

Acoustic considerations relate to the natural acoustics of the spaces served, which will affect the perceived performance of the system:

(a) Ambient noise levels
- Are they low or high?
- Do they vary?
- Are they broadband or tonal in character?
- Are they primarily low or high frequency in character?
- Are they intermittently high?

(b) Reverberation
- Is the space highly reverberant, $>$2.2 s?
- Or is it moderately live, 1.5–2.2 s?
- Or is it acoustically fairly dead, $<$1.5 s?

(*NB* The above values need to be considered in conjunction with the volume of the space.)

(c) Surfaces
- Are there highly reflective or curved surfaces within the space/building?

Environmental considerations

(a) Climatic conditions – Normal humidity, and maximum likely?
– Temperature range?
– Wind forces?
– Direct exposure of equipment (particularly loudspeakers) to moisture, rain, and snow?

(b) Atmospheric – Likely air pressure range?
– Any acidity?
– Any other pollution aspects?

(c) Security – Potential for vandalism?
– Access for tampering by unauthorized personnel?

Loudspeakers

(a) Choice – Appearance important?
– Loudspeakers capable of providing adequate sound levels?

(b) Locations – Will locations allow appropriate maintenance access?
– Are loudspeakers secure?
– Adequate field of coverage?
– Loudspeakers selected and distance from listener adequate for clarity and intelligibility?

(c) Fixings – Loudspeakers to be suspended above the audience of normal building users/occupants?
– Need to remove from view on occasions?

(d) Connections – Cable fixings/connections to the loudspeakers via plug and sockets or permanent connectors?
– Signal distribution to be via 100-V line (high impedance) or low impedance (e.g. 8 Ω) network?

Equipment locations and housings

(a) Does main equipment need to be in a secure or fire-resistant/rated room?
(m) What environment does the equipment need to be in, e.g. temperature, humidity?
(c) Should equipment be centralized or decentralized to satellite racks/units?
(d) Does equipment need to be rack-mounted and how much space is required?
(e) Can servicing be from the front or is side or rear access required?
(f) What are the main power requirements for the equipment?
(g) Will the equipment produce a significant heat load and require ventilation/air conditioning?
(h) What is the floor loading requirement?
(i) Are there maximum cable run distance restrictions on any items of equipment?
(j) Do equipment racks require locking doors, or are tamper-proof security covers required for certain items?
(k) Will cable access be adequate, considering future extensions to system?

Cable routes and cable types

(a) Are there any special requirements for the cable route, or cables, between main equipment and input lines or output cables?
(b) Does cable route pass through restricted access areas?
(c) Are special new conduits, trunkings or ducts required?
(d) Do cable routes cause a maintenance problem?
(e) Are cable route lengths permissible for the equipment selected, e.g. will excessive voltage drops occur or are repeater stages needed for digital signals?
(f) Do cable routes pass close by to likely sources of interference or interference-susceptible areas, equipment or other cables?
(g) Are selected cable routes secure and vandal-proof?
(h) Is the cable type suitable for the task, e.g. screened, fire-proofed, mechanically-strong and protected?
(i) Is the cable type suitable for the environment?
(j) Is the cable adequately sized to cater for capacity, anticipating voltage losses, etc.?
(k) Does the cable require any special identification or markers?
(l) Are cable terminations suitable for the load, environment and signal type?
(m) What cable termination identification system is to be adopted?
(n) Are the cables to be surface-clipped or run in trunking, conduits, or cable trays?

Input and control requirements

(a) How many inputs to the system are required?
(b) Where are they to be located?
(c) Are inputs to be at line or microphone level?
(d) Are inputs to be balanced?
(e) Are there any electrical isolation requirements between inputs, outputs, or other system equipment?
(f) Who is to control the system or have access to the system?
(g) Are individual inputs to be separately controllable or fixed in level?
(h) Will input or system configurations change?
(i) Are any remote control facilities required?
(j) Are inputs to be prioritized; if so, what are system priorities?
(k) Is the system to be zoned?
(l) If so: – How many zones are required?
– Are zones to follow the building's designated fire zones?
– Are different but simultaneous inputs required to be routed to different zones?
– Is message stacking required?
(m) Do different inputs need to address different zones or groups of zones?
(n) Is a mimic display or control panel required to help inputs/zones/signals routing, etc.?
(o) Does the system need to interface with other audio or control systems?
(p) Is the system to provide part of a fire alarm or emergency announcement/evacuation system?

(q) If so, is the system to include the control logic or will it receive logic control signals from the FA/VA/ evacuation system?
(r) Does the system require monitoring?
(s) What type of microphones are to be used on the system?
(t) What facilities do any remote microphone or paging stations require?

Statutory requirements
(a) Does the system have to conform and comply with any BS, IEC or ISO performance criteria, design or safety standards, EMC standards, or Codes of Practice?
(b) Does the system have to comply with any local authority requirements, Building Regulations, fire, police or licensing authority requirements?
(c) Does the system or any part of the system require a licence, and what are the licence conditions/ requirements?

Budget
(a) What is the system budget?
(b) Is it adequate, including both supply and full installation?
(c) Has due allowance been made for associated costs, such as Builder's Work, conduits, and site attendancies, commissions, discounts to Main Contractor, etc.?
(d) Is there overlap with electrical engineering subcontract works and costs?

Time scale
(a) What is the time scale for the project?
(b) Is it adequate and practicable?
(c) What factors and restraints could affect satisfactory completion on time?

System commissioning
(a) Is there adequate time within the installation programme for the appropriate system commissioning?
(b) Who will carry out the commissioning?
(c) Is a formal series of acceptance tests to be carried out?
(d) What commissioning procedures are to be adopted?
(e) What documentation is required?

Maintenance and operation
(a) How detailed should the Operator's Manual and Maintenance Manual be?
(b) Is a simple user guide required?
(c) How many copies of the Manuals are required?
(d) Is a Maintenance Contract required?
(e) What spares should be provided with the system installation?

Design principles

When designing a sound system, the most important factors to consider are the intelligibility and clarity of the required speech, warning tone or music signals, and the overall frequency response. The following discussion is primarily concerned with the factors and methods of obtaining good speech intelligibility for, if this can be achieved, then the clarity of other warning signals or music is almost automatically assured.

The three primary factors which must be considered are:

- loudness,
- frequency,
- echoes and reverberation.

Loudness

A prerequisite of any sound system is that it is loud enough to be clearly heard. Research has shown that under normal, quiet listening conditions, optimum speech intelligibility is achieved at sound levels of 65–75 dBA with the range 70–75 dBA preferred (see Chapter 5 for explanation of dBA). This level should therefore be designed for wherever possible.

However, in conditions of higher background noise, the overall level of the sound system will need to be increased to compensate for the effect of the background noise masking out the speech.

Ideally, a speech-to-background noise or signal-to-noise ratio of 10 dBA should be aimed for. The maximum signal level should not normally exceed 90–95 dBA, except under exceptional conditions or in areas where hearing protection is worn.

In extremely quiet areas, where paging signals for example are not to be intrusive, levels towards the lower end of the 65–75 dBA range should be used but the final setting should always be achieved/agreed subjectively.

The target sound levels given above refer to the long-term average level of the speech signal. However, the level of speech fluctuates considerably, with the short-term peaks being some 10–12 dB higher than the mean. When planning a sound system, adequate provision must be made to cater for this significant peak-to-mean ratio.

Ideally, a 10 dB headroom margin should be allowed for, if the power amplifiers are not to clip. 'Clipping' can result in excessive distortion and hence loss of intelligibility, and possible damage to loudspeakers if sustained signal clipping occurs.

In terms of the required amplifier power output, a 10 dB margin is a significant factor, begin equivalent to a multiplication factor of 10. However, techniques such as compression and limiting of speech input signals are available which enable smaller power margins to be employed. The amount of 'headroom' required will vary a great deal upon the type of system being considered. For speech paging/announcement systems, a signal headroom of 6 dB may be adequate, particularly if compression or limiting is incorporated.

Where ambient noise levels vary considerably, the signal-to-noise ratio of the sound system can be maintained by incorporating either local group or zone volume controls, which allow the level of the sound system signal to be adjusted as required. Alternatively, automatic noise sensing and level adjustment circuitry may be incorporated. This automatically and continuously monitors the noise level at a point and adjusts the output of the sound system accordingly in order to maintain an adequate signal-to-noise (S/N) ratio.

Frequency response

When selecting equipment for a sound system, due consideration should be given to the frequency response characteristics of the equipment and the nature of the task it is required to perform.

The minimum frequency range over which a sound system must operate for intelligible speech is 400 Hz to 4 kHz. However, the quality of such a limited range is extremely poor. For reasonable quality reproduction a sound system should be capable of reproducing the range of at least 150 Hz to 6 kHz. For high quality sound reinforcement, the range should extend to 8 or 10 kHz. For high quality reproduction of music, a range extending from at least 100 Hz to 10 kHz is required.

The coverage (dispersion) angles of the loudspeakers must be taken fully into account when designing the layout and potential coverage/sound level variations. The coverage angles are defined as the angles at which the output from the loudspeakers is 6 dB less than that produced on the main axis of the loudspeaker. Usually, the total included angle is quoted. This angle will change significantly with frequency, decreasing as the frequency increases with most loudspeaker types. When designing sound systems, the 4 kHz coverage angle should be used.

When laying out and positioning loudspeakers, consideration should be given to the difference in Sound Pressure Level (SPL) which will be produced between a position on the main axis of the loudspeaker, and an 'off-axis' position; at the coverage angle for example where, by definition, the SPL will be 6 dB lower than on the main axis. It may in fact be slightly more than 6 dB lower when account is also taken of the additional angular path distance involved. A 6 dB difference is quite noticeable, and is sufficient to affect significantly the potential intelligibility of the system in a noisy or reverberant environment by decreasing the apparent S/N (D/R) ratio as calculated for the on-axis case. In such circumstances, loudspeakers should be positioned so that their coverage angles either meet or preferably overlap so there are no gaps in the coverage.

In highly reverberant areas, it is often desirable to limit the low frequency response of the system in order to reduce the sound masking effect of the low frequency reverberation and hence improve the potential intelligibility of the reproduced speech. Although the basic tone control filters found on most Public Address (PA) and sound reinforcement equipment can be extremely useful, their control range is limited. The use of a multi-band filter or graphic equalizer to contour and adjust the overall frequency response of a sound system is an extremely useful and powerful aid to system intelligibility and should be considered for all but the smallest system.

Echoes and reverberation

Secondary sources of sound, for example reflections from room surfaces, or the output from supplementary or repeater loudspeakers, which arrive at a listening position within a period of up to 35–50 ms after the original sound, will integrate or merge with the original direct sound (whether it is a person speaking or a loudspeaker) and combine to produce one overall louder sound (i.e. intelligibility is enhanced).

Secondary sounds arriving after 50 ms do not fully integrate and, depending on their level relative to the initial sound, may be heard as echoes which will have the effect of reducing the overall intelligibility of the speech signal.

Sound systems should be designed to ensure that the generation of long-delayed secondary sound signals, caused either by structural reflections or by secondary loudspeakers, do not occur or are well controlled.

Loudspeakers should be placed no more than 12–15 m apart when facing in the same direction or 25 m apart when directed to cover an area as shown in Figure 4.1. The loudspeaker should be angled downwards to cover the appropriate area and to limit the overspill to adjacent areas. A row of loudspeakers even though individually spaced at 12–15 m apart can still give rise to an appreciable echo effect due to the difference in distance between the nearest and subsequent sets of loudspeakers whose output, due to the effects of the Inverse Square Law, may still be significant.

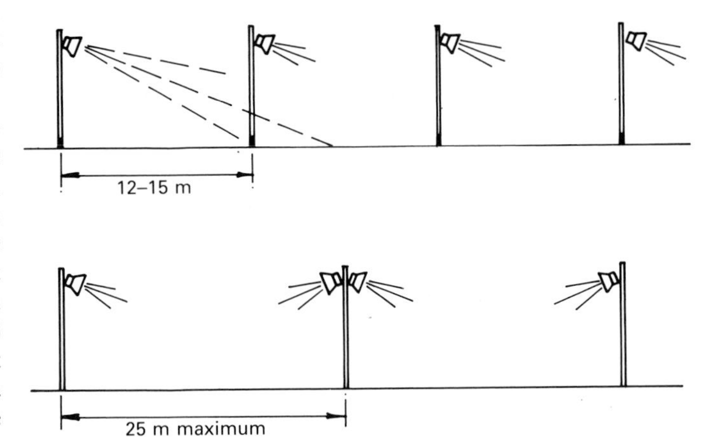

Figure 4.1 *Typical loudspeaker arrangement to reduce echoes*

Electroacoustic systems

A particularly special area requiring a detailed understanding of natural room acoustics and electronic systems is that of electroacoustic enhancement of auditorium acoustics. As explained in Chapter 2, the optimal acoustic conditions for speech and music are quite different. An auditorium designed primarily for music will not be ideal for speech and vice versa. Where an auditorium is to be used for both speech and music, a way round the problem is to design the acoustics so that they are relatively dry, for example in the range 1.0–1.5 s RT, and then increase the reverberation time electroacoustically up to 2 s plus. Other acoustical parameters such as early reflection sequence, spaciousness and envelopment can also be adjusted electronically. A number of commercial systems are now available and although differing in detail, adopt the same basic concept. Essentially a series of microphones sample the sound field either close to or on the stage and then via a series of electronic digital delays and reverberators, and frequency equalizers, the signal is replayed via carefully positioned loudspeakers located throughout the auditorium. The electronic delays, etc., are set to simulate natural acoustic reflections and reverberation. Modern systems allow comprehensive control of the reflection sequences and their intensity, together with the overall reverberation effect. Although some of the early systems were criticized as being rather 'coloured' or artificial, the latest developments and faster digital electronic processors have virtually eliminated these problems.

Distance and power considerations

In general, it can be assumed that the Sound Pressure Level (SPL) decreases linearly with distance from loudspeaker, reducing by 6 dB every time that the distance is doubled. For example, the SPL at 2 m from a loudspeaker is 6 dB less than at 1 m, whilst at 4 m the SPL will be 6 dB less than at 2 m, and 12 dB less than at 1 m. It can be seen therefore that the SPL initially decreases rapidly with distance but as the absolute distance increases, the relative change becomes less. For example, the SPL 100 m away from a loudspeaker is only 6 dB less than at 50 m, yet it is 40 dB less than at 1 m. Furthermore, the SPL at 52 m is not measurably different than at 50 m.

Each 3 dB increase in SPL output required from a loudspeaker needs a doubling of the audio frequency power delivered to the loudspeaker, i.e. the output from a loudspeaker follows a 10 log Power Law, whilst the fall-off with distance follows a 20 log Distance Law.

An example calculation applying these principles may be given as follows. If the sensitivity of a particular loudspeaker is 90 dB for an input of 1 W when measured at a 1 m distance (the standard rating conditions), what will be the SPL at 8 m, and how much additional power will be required to achieve an SPL of 85 dB?

At 8 m from the LS, the SPL will ahve decreased by: $20 \log 8 = 18$ dB, $90 - 18 = 72$ dB SPL.

To achieve 85 dB requires a power increase equivalent to: $85 - 72 = 13$ dB $= 19.95$ W, i.e. 20 W.

The above approach is sufficiently accurate for both outdoor and indoor calculations for planning purposes. In indoor situations, the contribution from the reflected or reverberant sound components should be considered also as this will affect not only the perceived loudness (this will be increased) but also the potential intelligibility if the reverberant component is too high with respect to the direct sound. For large reverberant areas, expert advice from a specialist sound consultant should be sought.

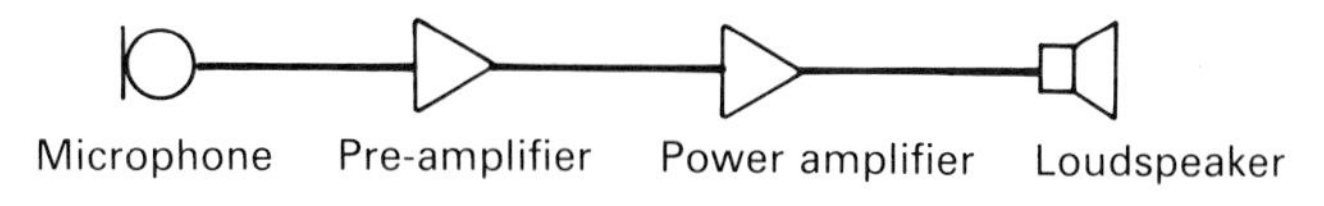

Figure 4.2(a) *Basic sound system components*

System design and components

A sound system consists essentially of four basic components, as shown in Figure 4.2a:

- microphone, or signal input,
- pre-amplifier (control unit),
- power amplifier,
- loudspeaker.

Additionally, loudspeaker zoning or group switching may be required. In order to obtain the maximum performance from a system, e.g. to maintain an adequate S/N ratio or provide an appropriate frequency response, a number of other additional stages may be added such as compressor/limiters, equalizers, and volume controls, as shown in Figure 4.2b.

Microphones

A wide variety of microphone types and characteristics are available. They may be classed by their sound pick-up pattern and form of generating element. The majority of microphones encountered will either be of the omnidirectional or unidirectional type.

Omnidirectional microphones pick up sound equally from all directions, and should therefore not be used when discrimination is required against either reverberant sound pick-up or unwanted sound or noise pick-up.

Unidirectional microphones may be in the form of either a cardioid or hyper-cardioid microphone, so called because of the shape of the corresponding sound pick-up patterns which receive sound from the front but reject sound pick-up from the rear. They may as a result be used in situations where it is desirable to discriminate against sounds arriving from a particular direction or to reduce reverberant sound pick-up. Unidirectional microphones are also useful in sound reinforcement situations where they may be used to increase the potential gain of a sound system before acoustic feedback occurs. This is particularly useful where the microphone has to operate in the close proximity of the loudspeakers.

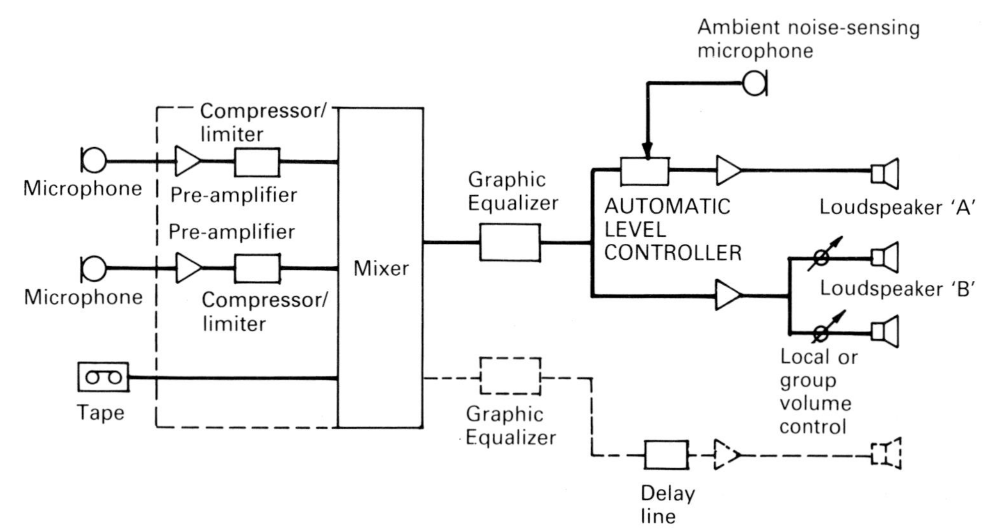

Figure 4.2(b) *Simplified sound system schematic diagram of signal processing equipment*

The two most widely used forms of microphone are the dynamic (moving coil) and the electret. Special microphone types include the lavalier, tie-clip, PZM, noise-cancelling, gun or rifle, and wireless models.

Dynamic microphones are generally extremely robust and offer a wide range of performance characteristics which are little influenced by temperature and humidity, making them potentially suitable for both indoor and outdoor use. Good-quality moving coil microphones are able to offer a wide frequency response but this may be deliberately tailored to a particular characteristic to aid intelligibility, e.g. by applying low frequency cut-off.

Electret microphones are capable of providing a very smooth and extended frequency response at a higher output level than the dynamic. They do, however, require their own external power supply to operate and are generally not as robust as the dynamic microphone, but their higher sensitivity enables very small electret capsules to be made which can be employed in a number of special types of microphone. Some models have an internal battery, but are best avoided in professional installations due to their continual maintenance requirements.

Lavalier microphones are small, usually dynamic, with an omnidirectional response, designed for speech pick-up from lecturers, etc.; they are fitted with a neck cord and worn by the person speaking.

Tie-clip microphones are small and lightweight, usually electret, which may be clipped to the clothing of the person speaking by clip or lapel badge. These are generally omnidirectional but unidirectional types are available.

PZM (pressure zone microphones) are a newer form based on a small electret capsule fitted onto a special mounting plate. They are most useful for recording purposes or sound pick-up of more than one person; the form and mounting arrangement can substantially reduce the 'reverberant' sounding pick-up of conventional microphones.

Noise-cancelling microphones are used in high noise environments for PA purposes. The microphone is highly insensitive to indirect sound pick-up and requires a close speaking distance; generally, the response is limited and optimized for speech.

Wireless microphones may either be in the form of a small tie clip and pocket transmitter or an all-in-one hand-held or stand-mounted unit. They require a separate receiver unit tuned to the microphone's transmission frequency, but allow freedom of movement. They should not be used where a secure transmission is required. Lecturers should remember to remove them, to avoid candid transmittal from a visit to the washroom.

The output from a microphone is generally of the order of a few millivolts. A pre-amplifier with a considerable degree of gain is therefore required, typically 60–70 dB. Particular care should be taken to ensure that microphone circuits are not subjected to either RF or other forms of electrical interference.

Microphone circuits should be low impedance and balanced (in terms of run out), cables must be segregated from other services, by at least 300 mm or by the use of dedicated and compartmentalized trunkings or conduits.

When mounting microphones on lecterns or table tops, consideration should be given to the use of shock/vibration isolating mounts to reduce unwanted noise or vibration pick-up. Dynamic (cardioid) microphones are the most commonly used. The microphone should exhibit a smooth frequency response. The distance between the person speaking and the microphone should be kept to a minimum, preferably within 300 mm in difficult situations. Microphones should be of low impedance with a balanced output. As few a number of microphones as possible should be 'live' at any given time (each time the number of open microphone channels is increased, the gain before feedback margin reduces by 3 dB). Microphones should be kept out of the direct field of high power or centralized loudspeakers if optimum gain before feedback is to be attained.

Pre-amplifiers

The pre-amplifier (control unit or mixer) amplifies low-level signals and combines or selects them as required before power amplification takes place. Pre-amplifiers may have inputs for either microphone level or line level (typically 0.5–1.0 V) or 0 dBm signals. Tone control or basic equalization facilities may be incorporated, either as user-operated or as pre-set controls.

The pre-amplifier may also incorporate a compressor/limiter stage and/or automatic gain control to help maintain a more constant signal level, compensating for different voice levels.

Compression reduces the peak-to-mean ratio of speech and so can be used to reduce the 'headroom' requirements of the power amplifier loading. For example, a compression ratio of 2 reduces the peak-to-mean ratio by 6 dB, i.e. from 10–12 dB to 4–6 dB, whereas a compression ratio of 3 is equivalent to a 9-dB reduction and should be considered as the normally permissible ratio (1.2–2 is optimal). Compressor/limiters may either be incorporated into each pre-amplifier input or alternatively a single unit may be connected between the output of the pre-amplifier and the input to the power amplifier.

Each microphone or system input should be capable of being individually controlled. Microphone inputs should ideally be provided with tone controls or equalization facilities which are separate from music or other auxiliary inputs. In live sound reinforcement applications, it is generally not possible to pre-set the input level controls as the volume required will depend on the voice level of the person speaking and the additional acoustic damping provided by the audience. In reverberant rooms with little natural sound absorption, e.g. by carpet or acoustically-absorbent finishes, the audience can significantly affect the degree of gain available before feedback. Outputs for recording or feeds to other areas' services, e.g. simultaneous interpretation systems or audio induction loops, may also be required. Wherever possible, a Sound Reinforcement system should be controlled by a trained operator. Paging and announcement systems where the microphone is not in the same acoustic space as the loudspeakers can be pre-set, though some compensation for different voice levels may be required – the use of either an automatic gain control (AGC) or compressor is preferable.

Signal processing equipment

Apart from compression and limiting, a number of other additional items of electronic signal processing equipment are commonly incorporated into the sound system design.

Two processors that are particularly important for optimizing speech intelligibility are automatic noise sensing/level control, and equalization. Other processors include phase/frequency shifters for improving the gain of a sound system before feedback, and audio delay lines which enable the signal arrivals from distant loudspeakers to be synchronized in order to overcome potential echo problems or to enhance apparent localization of a sound source.

Automatic noise-sensing and level controllers consist essentially of two basic elements, the ambient noise-sensing microphone, and the PA signal and control circuit.

Where the ambient noise level in an area regularly fluctuates considerably, say by more than 6–8 dBA, an ambient noise sensing and automatic level adjustment system can help to maintain an appropriate S/N ratio. Examples of such areas are industrial or process areas, workshops or locations, and spaces affected by traffic or occupational noise, including airports, rail termini, sports stadia and swimming pools. A sensing microphone is located at a suitable point where it is able to monitor the ambient noise level within the designated area. The sensing microphone must be carefully positioned to ensure that it is not affected by very local events. It is therefore usually located at a fairly high position, away from local noise sources such as machinery, air-conditioning equipment, or isolated groups of occupants.

The signal from the microphone is used to drive, after suitable processing, a voltage-controlled amplifier/attenuator which adjusts the level of the signal routed to the power amplifier(s) feeding the particular area or zone in question. The signal from the microphone is time-averaged so that occasional transient events do not affect the overall level. When an announcement is made, the level is automatically set so that the announcement does not fluctuate, in order to maintain a nominally consistent S/N ratio.

Correct equalization of a sound system is becoming increasingly recognized as an essential aid to optimizing both speech intelligibility and sound quality/clarity. An equalizer is essentially a multi-frequency tone control, typically consisting of either 10–12 octave band filters or 27–30 one-third octave band filters.

The one-third octave band filter provides extensive control over the entire audio band with usually ±12 dB of cut or boost available at each of the one-third octave frequencies. Although the unit is particularly suitable for sound reinforcement applications, it is becoming widely used in more general PA systems. The more restricted control of the octave band equalizer tends to limit it to only basic PA applications. Other forms of tunable equalizer are also available.

The equalizer is used to help compensate both for deficiencies or irregularities in the response of the system loudspeakers and/or the acoustic characteristics of the space that they are serving. The response of most horn loudspeakers for example begins to roll off above 3–4 kHz whilst the response of many cone loudspeakers begins to fall off above 5 or 6 kHz – the equalizer may be used to extend and flatten the response of the loudspeakers.

In difficult acoustic environments, e.g. those with long reverberation times and/or high ambient noise levels, the equalizer is used to shape the response of the broadcast sound to either help reduce reverberant excitation or accentuate the frequencies most important to intelligibility and clarity, i.e. 1–5 kHz.

Where different types of loudspeakers are employed, or where a number of acoustically different areas are served by the PA system, a number of equalizer channels may be required. In addition, care should be taken with the routing of such signals, as the equalization filter curve set for one type of loudspeaker may not be suitable for another. Furthermore, the curve set for a reverberant hangar or noisy workshop is unlikely to be suitable for office areas or other 'deader', i.e. less reverberant, spaces. Figure 4.2b illustrates how the above signal processing connects into the signal chain. In large PA systems, several equalizers and level controllers may be required, as each difficult area should ideally be individually treated.

Power amplifiers

The signal presented to the power amplifier is normally at the standard audio signal line level of 0 dBm. The input should normally be balanced where a substantial distance separates the pre-amplifier (or control unit/mixer) from its power amplifier. Unbalanced inputs may be used where the distance/cable run is short.

Amplifier outputs may be either of low impedance, e.g. to suit 4–8 Ω loudspeakers, or high impedance for 100 V line working. The majority of PA systems employ 100 V line working to reduce line losses and for ease of connection, each loudspeaker being simply connected to the transmission line in a parallel arrangement via its individual tapped matching transformer. Foreground music or high-power/high-quality sound reinforcement systems will normally be low impedance types.

Amplifier output powers are generally standardized, e.g. 30, 50, 100 and 200 W up to 500 to 1000 W. All power amplifiers should be protected against open- or short-circuiting of the amplifier output and thermal runaway. The loudspeaker circuits must not be loaded to more than the rated output power of the amplifier.

When calculating the amplifier power requirements, consideration must be given to the peak-to-mean (headroom) requirements of the input signals, together with an allowance for cable losses.

Loudspeaker cable losses are generally assumed to be resistive for calculation purposes, but particular care should be taken when using long runs of MICS cable due to its greater capacitive reactance adversely affecting the impedance of a loudspeaker circuit.

Where a number of power amplifiers are located together, care needs to be taken to ensure that the mains or stand-by power supply can adequately cope with the initial switch on surge. It is common practice in large installations to employ an automatic sequential switch on, thereby limiting the inrush current requirements. The heat load generated by the amplifiers must also be taken into account.

Loudspeakers

The loudspeaker converts the electrical power from the amplifier into an acoustic signal via the vibration of a cone or diaphragm which sets up sound pressure waves. The important parameters to note are:

- rated or maximum power, capacity,
- sensitivity (SPL at 1 W, 1 m),

- frequency response,
- directional information, i.e. dispersion or coverage angles.

The choice of loudspeaker depends on the intended use, power/SPL output capability, directional characteristics, quality required, location (e.g. indoor or outdoor use) and local acoustic environment. The aesthetic requirements of the unit may also need to be considered. The most common forms of loudspeakers are:

- the cabinet,
- recessed ceiling loudspeaker,
- the bidirectional and wedge cabinet loudspeaker,
- the column loudspeaker,
- the re-entrant horn loudspeaker,
- the Constant Directivity horn loudspeaker,
- the full-range, integrated, high power loudspeaker.

Cabinet loudspeakers are suitable for music, speech and paging systems in areas where the floor-to-ceiling height is typically 4 m or less. Good reproduction units typically provide a coverage of a 60–90° cone from wall mountings.

Recessed ceiling loudspeakers may be used in areas with ceiling heights up to 5–6 m, as well as lower ceiling areas. The coverage from a ceiling-mounted loudspeaker is generally less than from a wall-mounted cabinet; approximately 25 m^2 for a ceiling loudspeaker at a mounting height of between 2.8 and 4.8 m, as opposed to 30–50 m^2 for a wall cabinet paging/PA system, based on a 60–90° cone. When calculating the area covered by a ceiling loudspeaker, the height of the listeners' ears above floor level must be taken into account, as shown in Figure 4.3. A high-density ceiling loudspeaker installation can form the basis of a very high quality sound reinforcement system with very uniform sound distribution and coverage. When designing a high quality sound reinforcement system, the coverage angle at 4 kHz should be used; typically, this will be 60° or less for a 200-mm cone loudspeaker.

Bidirectional and wedge loudspeakers may either be directly mounted on a wall or ceiling, or suspended below a high ceiling by a chain or wire. The loudspeaker produces two cones of sound (60–90°). Directly under the loudspeaker, a high-frequencies dead spot can occur.

Column or line source loudspeakers provide wide coverage in the horizontal plane but deliberately restrict the sound output in the vertical plane (e.g. 90–120° horizontal, 15–20° vertical). This characteristic enables the column loudspeakers to have the potential of providing greater intelligibility in reverberant areas or greater coverage than a conventional cone loudspeaker in less hostile environments. The column loudspeakers may be used for both PA and sound reinforcement applications. It is essential however to correctly aim the column loudspeaker so that the beam of sound it produces is directed at the listeners. Normally, column loudspeakers should not be mounted flat against a wall but should be provided with suitable angle brackets allowing correct alignment. By aiming the column loudspeaker towards the centre of the area to be covered, a fairly uniform coverage is obtained as the nearer listeners will be located out of the main beam of sound which can therefore be increased to reach the rear of the area. When using repeater columns, e.g. in long churches, the repeater units should be located within 12–15 m of the primary or any other supplementary loudspeakers, unless signal delay lines are employed (see Figure 4.4).

Re-entrant horn loudspeakers have a restricted frequency response and are used mainly for speech. They are more efficient than either the cabinet or column loudspeaker and are therefore suitable for use in areas with high background noise levels, large enclosed areas, and outdoors. The coverage angles of horn loudspeakers vary considerably but typically are between 40 and 80°.

Loudspeakers should be selected and located to provide as uniform a coverage of an area as practicable. The inverse

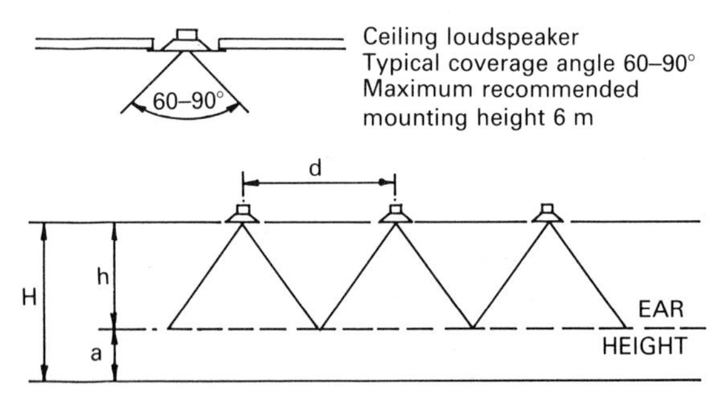

The area covered by each loudspeaker should be calculated using the distance $h = H - a$, a = Height of seated or standing listener as appropriate. For a seated listener take a = 1.2 m.

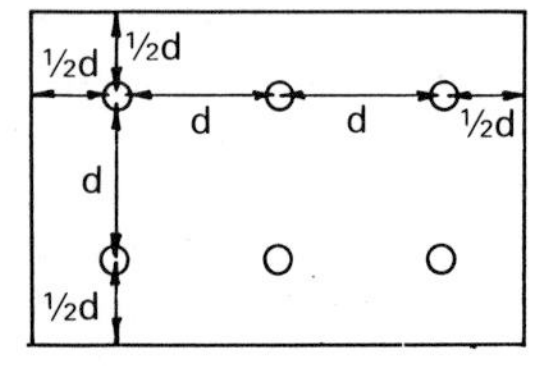

Loudspeaker spacing

For very good coverage, e.g. conference rooms - Spacing *d* should be 1h to 1.2h (typically 3 m)

For good coverage, e.g. canteens, office areas, etc. – Spacing *d* should be 1h to 2h (typically 3 – 5 m)

For variable coverage, e.g. corridors, general areas – Spacing *d* should be 2 to 3h or 3 to 4h in corridors (typically 5 – 9 m)

Figure 4.3 *Loudspeaker coverage*

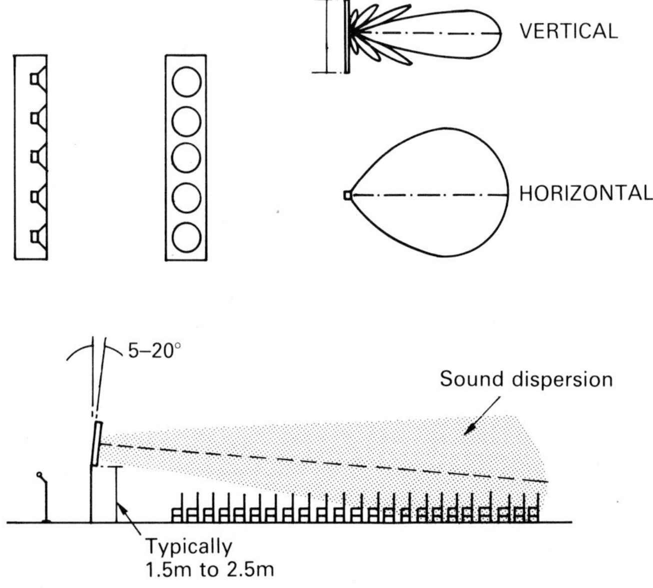

Figure 4.4 *Column/line source loudspeakers*

square law and coverage angle data should be used initially to predict the coverage and plan an installation. For PA/paging systems, the maximum variation should be less than 10 dB at 2 kHz. In areas of high noise or reverberation, a smaller variation will be required if intelligibility is not to be downgraded. In sound reinforcement systems, the variation should be within 6 dB and preferably within 4 dB at 2 and 4 kHz.

The majority of loudspeakers used for PA and SR systems are operated on a 100-V line distribution system, with each loudspeaker fitted with an individual tapped transformer. Tappings provide a useful range of output adjustment and enable the loudspeakers to be set to compensate for different local conditions such as noise level, room volume, and length of throw, as described earlier in this chapter.

Group or zone volume controls can be employed to adjust the overall sound level in a given area.

Loudspeakers are affected by the acoustic environment and peaks in their response coupled with the room acoustic characteristics can cause feedback at high gain settings in sound reinforcement systems. Correct selection, placement and equalization of the loudspeakers is essential. Points that should be considered are:

- type of loudspeaker,
- position of loudspeaker,
- dispersion (coverage) angle,
- power rating/SPL capacity,
- frequency range and response.

Loudspeaker coverage of a room can essentially be achieved in either of two ways:

- by use of a centralized loudspeaker system cluster,
- by use of a low-level grid of ceiling-mounted loudspeakers.

Centralized loudspeaker systems typically comprise column, CD horn, or full-range loudspeaker cabinets. The units may either be mounted on either side of the stage or rostrum, or preferably they may be mounted centrally over it in the form of a cluster. Use is made of the loudspeaker's directional properties to direct sound into the audience and away from reflecting wall surfaces thereby reducing the degree of reverberant excitation produced by the system and aiding intelligibility and gain before feedback.

Constant directivity (CD) horns are becoming increasingly popular in high quality sound systems applications, or where high sound pressure levels are required. They have many advantages over re-entrant or other horn types which may be described as follows:

(a) They are extremely efficient, typically exhibiting a 1 W/1 m sensitivity of around 113 dB for a 60 × 40° horn.
(b) They can provide extremely uniform coverage as they exhibit a reasonably constant directivity with frequency, e.g. within ±10° over their operational range of 500 Hz to 16 kHz plus, for a well designed large-format device.
(c) They exhibit considerably lower distortion characteristics than a re-entrant horn.
(d) They provide a very much smoother and uniform frequency response of high fidelity (hi-fi) rather than PA quality. CD horns may therefore be used for high quality sound systems in theatres and concert halls as well as large stadia or reverberant exhibition halls. A CD horn usually operates over the range 500 Hz to over 15 kHz (or from 800 Hz for the smaller types). An associated low-frequency loudspeaker is therefore required to form a complete system operating from well below 100 Hz (e.g. 50 Hz) to over 15 kHz. A dedicated cross-over unit is therefore employed to block low frequencies being fed to the horn and high frequencies to the bass driver. The cross-over may either be active or passive, depending on the particular application and system configuration desired.
(e) The controlled dispersion of a CD horn ensures that high frequency beaming and 'hot spots' do not occur, but instead a uniform distribution of sound is created at all frequencies within the working range of the horn. The controlled radiation of the CD horn also enables it to work well in acoustically difficult and reverberant spaces, allowing the sound to be directed onto the absorbing audience or congregation and away from the reflective room surfaces.

Full-range high-power integrated (cabinet) loudspeakers are a compact high quality product. Although a CD horn and bass bin combination can be used for most applications, this can result in a fairly bulky package. A number of proprietary high-quality, compact, high-power loudspeakers are available which include both high and low frequency drivers (some using CD horn principles) within a single cabinet. The speakers are intended for both music and speech applications and generally exhibit smooth frequency response extending from below 100 Hz to over 15 kHz. Both 2-way and 3-way designs are available in a range of sizes, power handling capabilities, frequency ranges and dispersion patterns. Power handling characteristics range from around 100 to 400 W r.m.s. Some models employ dedicated control units, which may incorporate equalization filters, cross-overs and signal limiters to avoid potential loudspeaker overloads.

These loudspeakers may either be portable, transportable or permanently installed. Weather-resistant models are also available for outdoor use. The loudspeakers may be used in almost any high-quality sound system for both speech and music purposes, where highly uniform or directional control across a wide frequency range (e.g. from 500 Hz upwards) is not required. They therefore find application in clubs, discotheques, theatres, auditoria, assembly halls and leisure centres. Care needs to be exercised, however, when employing them in reverberant spaces, e.g. concert halls. Their directional characteristics generally make them less suitable for use in highly reverberant spaces, unless located close to the listener or arrayed to provide additional directional control.

Any of the above types in used need to direct sound away from microphone positions. They also need to provide an even coverage of a room by employing different parts of the cluster to cover different areas, for example the front and rear of the room. Where a column loudspeaker is used, this is generally aimed approximately two-thirds of the way down the room or coverage area. Delayed repeater loudspeakers may also be required, either to fill in areas which cannot be reached from the central cluster, for

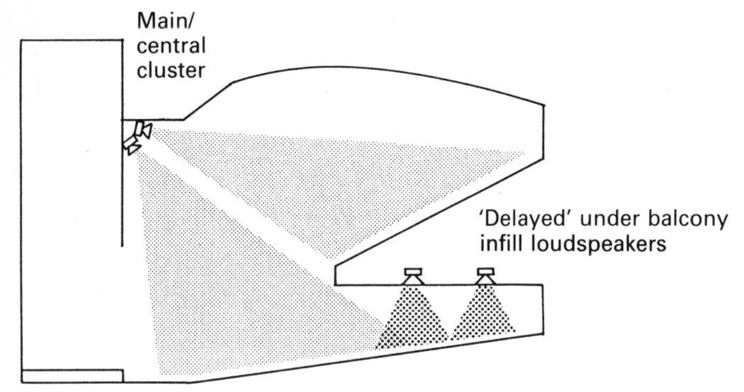

Figure 4.5(a) *Typical auditorium with proscenium loudspeakers*

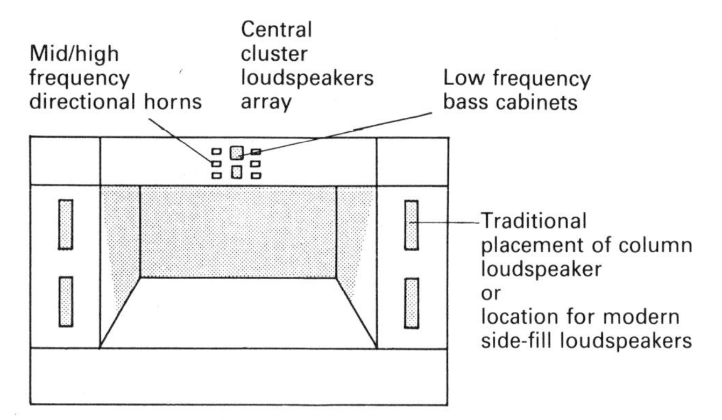

Figure 4.5(b) *Typical stage and loudspeaker arrangements*

example under balconies, or to compensate for inverse square law losses causing too large a variation in direct SPL (see Figures 4.5a and 4.5b).

Low-level distributed sound systems are the alternative to the centralized cluster, consisting of a relatively high density of ceiling-mounted loudspeakers spaced to provide a good overlap between adjacent units. This method can provide extremely uniform coverage of a room and will normally provide better coverage than column or similar loudspeakers, in wide spaces/rooms with low ceilings.

All sound reinforcement systems should be properly equalized in order to obtain a smooth response, natural-sounding reproduction, and optimal gain before feedback. A one-third octave band equalizer should be used in preference to the coarser octave band unit. A temper-proof cover should be fitted over the equalizer controls once the system has been commissioned.

Loudspeaker signal distribution

As was gathered from the previous section, public address and sound reinforcement systems may be divided into two groups associated with their sound distribution and coverage patterns. Firstly, there are high level distribution systems where either a single loudspeaker cluster sound source, or a few such sources, are used to cover an area, with each source radiating a relatively high SPL to do so (for example, a theatre auditorium sound system with a main LS cluster located above the proscenium and covering most of the auditorium). Secondly, there are low level distribution systems, where a large number of loudspeaker sources is used, each operating at a relatively low level of sound output. These systems are often used in areas with relatively low ceiling heights and flat floors, such as conference rooms, exhibition suites or shopping malls. Low level distribution systems are also widely used in large churches and ceremonial halls. Here the coverage is achieved either from loudspeakers distributed along the structural columns on either side of the congregation, or from a localized pew-back arrangement.

Many sound systems in fact make use of both types of distribution. There is often no clear-cut reason for using one type, and so other considerations such as architectural constraints, accessibility and installation costs are often the deciding factors. If design is left to specialist suppliers, they will be influenced by the equipment that they may have in stock, or by the manufacturers of particular equipment, for whom they may act as agents, or by delivery availability.

Distribution of the signal to the loudspeakers can be carried out in one of two ways:

- by standard low impedance connection, i.e. 2–8 Ω, or
- by high impedance (nominally 100 or 70 V line) constant voltage distribution using step-up and step-down transformers.

Each method has its advantages and disadvantages. A low impedance signal distribution generally offers a wide frequency and dynamic range capability, but cable lengths must be kept short to minimize power losses due to cable resistance. Multiple connection of loudspeakers onto one common output can also become unwieldy, often requiring quite complex combinations of series-parallel connections to provide a reasonable load impedance for the amplifier (see Figure 4.6). Installations with widely distributed loudspeakers can present a considerable wiring problem, particularly if one of the units should fail in use.

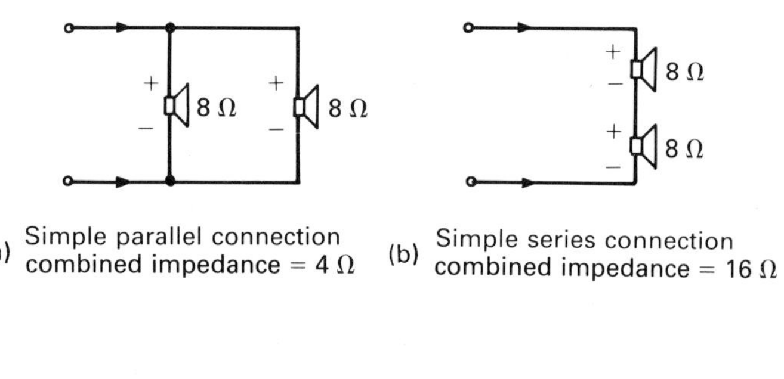

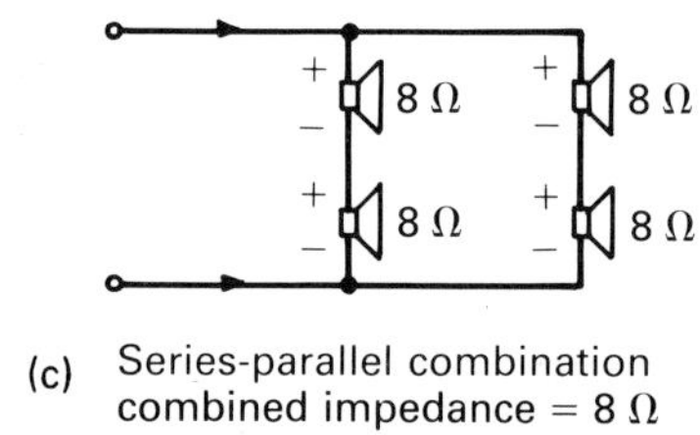

Figure 4.6 *Low impedance loudspeaker connections (networks)*

Constant-voltage, high-impedance (100 or 70 V line) distribution is ideal for large PA installations associated with either long cable runs or large numbers of loudspeakers. It is in essence very similar to electrical mains power distribution. In practice, a step-up transformer is fitted to the power amplifier, rated to take the maximum output capability of the amplifier, e.g. 100 W. Distribution to each loudspeaker is then a simple matter of making numerous parallel connections via matching transformers (see Figure 4.7). Typically, the secondary winding of the matching transformers will be fitted with a number of power tappings, e.g. 1, 2, 5, 10, 20 W, so that the signal level fed to each loudspeaker can be individually adjusted. This allows the LS coverage SPL to be accurately set, taking into account any local acoustic or background noise level conditions.

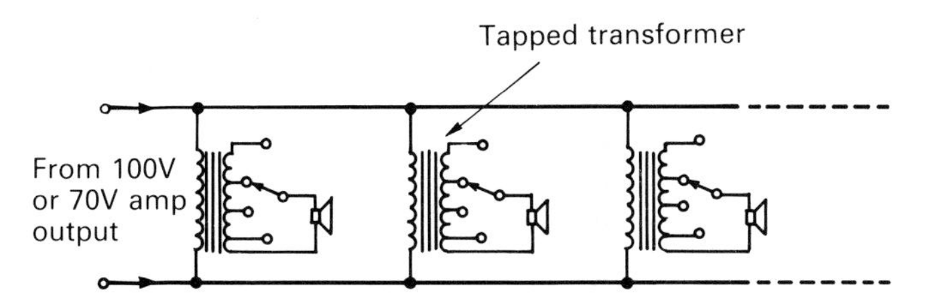

Figure 4.7 *High impedance-constant voltage distribution system*

Theoretically, one can continue to add loudspeakers until the maximum output capability of the amplifier is reached. However, in practice, some contingency/spare power reserve must be left (e.g. 20%). The resistance losses of the distribution cables themselves must also be taken into account, for although transforming the audio signal to a higher voltage for transmission decreases the voltage drop seen at the end of the line, this will still occur to some extent.

The most common nominal line distribution voltages are 70 V (USA and Japan) and 100 V (UK and Europe), but other distribution voltages are to be found, such as 50 V, 30 V and 25 V, to meet certain building and safety codes.

Wiring and installation

All mains equipment and each equipment rack should be earthed. Screened cables interconnecting equipment should be earthed at one point only to prevent the generation of earth-loop currents which may give rise to 'hum' or other electrical interference noise on the system.

Precautions should be taken to minimize electrical interference between audio circuits or between other circuits and other audio circuits, and to minimize the risk of dangerous voltages occurring on audio circuits. Microphone and low signal level cables should be screened and must be physically separated from loudspeaker and power lines.

Audio cables should not be run with cables at extra low, low, or medium voltage, as per IEE Wiring Regulations. Microphone cables should run at least 150 mm and preferably 300 mm from loudspeaker lines and power wiring. Loudspeaker lines should run at least 150 mm and preferably 300 mm from telecommunications wiring. A minimum separation of 500 mm should be observed and may need to be increased to 1000 mm for thyristor-controlled circuits or fluorescent luminaires.

Mains supplies to sound system equipment should not be shared with inductive circuits or thyristor-controlled, motor-driven equipment.

Where a sound system installation is used for emergency communications, consideration must be given to appropriately protecting the cables, e.g. by use of MICS cables.

The installation of equipment in any potentially explosive atmosphere should be avoided.

When telephone lines form part of a sound system, line-isolating transformers to the approval of the telephones statutory undertaker (British Telecom in UK) should be used.

Where a sound system installation is used for emergency communications, cables must be of a fire- and damage-resistant type, e.g. MICS or FP200; see BS 5839 [1] and BS 7443 [2].

Loudspeaker distribution cables should be sized for a maximum of 10%, preferably 5%, voltage drop between power amplifier and any associated loudspeaker. This calculation should be based on the amplifier output of 100 V (nominal), the loudspeaker tappings in use, and the voltage drop due to conductor resistance. Consideration should be given to the use of satellite amplifiers where long cable runs are required to deliver high power loads (see Table 4.2).

'Deaf aid loop' systems

Systems for the hard of hearing may either be based on an audio frequency induction loop system (AFILS), which enables any standard hearing aid with a 'tele pick-up coil' (usually identified by the aid having an additional switch position marked 'T') or via an infra-red (IR) transmission system. The latter system requires a special receiver but has

Table 4.2 *Loudspeaker line losses: maximum permissible line lengths, in feet, for 0.5 dB voltage drop in 100 V/70 V networks*

			Low impedance			High impedance					
			4Ω	8Ω	16Ω	50Ω	100Ω	200Ω	1 000Ω	5 000Ω	10 000Ω
Equivalent conductor size area	AWG	Resistance ohms/ 1000 feet (300 m)				200 W/100 V 100 W/70 V	100 W/100 V 50 W/70 V	50 W/100 V 25 W/70 V	10 W/100 V 5 W/70 V	2 W/100 V 1 W/70 V	1 W/100 V ½ W/70 V
5.2 mm²	10	1.00	120	240	480	1 500	3 000	6 000	30 000	150 000	300 000
3.3 mm²	12	1.59	75	150	300	940	1 800	3 800	18 000	94 000	180 000
2.08 mm²	14	2.50	48	96	190	600	1 200	2 400	12 000	60 000	120 000
1.3 mm²	16	4.02	30	60	90	370	740	1 500	7 400	37 000	74 000
0.87 mm²	18	6.39	19	38	76	230	460	920	4 600	23 000	46 000
0.52 mm²	20	10.1	12	24	48	150	300	600	3 000	15 000	30 000

the advantage of a potentially greater bandwidth and ability to simultaneously transmit up to 12 separate channels, making it very suitable for simultaneous interpretation systems or 'museum tour' systems.

AFILS and IR systems may also be used to transmit speech or warning signals to personnel working in very noisy conditions where hearing protectors are required. Special protectors incorporating a receiver and in-built miniature loudspeaker/earphone are available, enabling 'wireless' communication to be maintained.

Speech intelligibility

Assessment

With the ever-increasing use of public address (PA) systems for fire alarm and emergency warning announcements, more emphasis than ever before has been placed on the intelligibility and clarity of such systems. Whilst it is obvious that an emergency, PA or sound reinforcement system must be intelligible, many specifications currently include such phrases as 'the PA system must be clearly audible' or 'the PA system must be capable of producing clear and intelligible speech'; but what do 'audible' and 'intelligible' mean, for they are far from being the same thing, and under what circumstances has the system to be clearly intelligible and to whom? The answers to these questions are less obvious than at first sight. For example, it is quite possible to design a perfectly intelligible sound system but end up hearing almost totally unintelligible announcements. This may be due to either defective hearing on the part of the listener or poor articulation on the part of the announcer. Equally it is possible that some environmental factors not taken into account could affect the audibility of the system, e.g. intrusive noise from outside or high ventilation noise levels.

The problem is that apart from dealing with the electronic and electroacoustics aspects of the system which are reasonably controllable, a PA system has to interface with people, both at the source and at the receiving end, and people are highly variable. For this reason, when testing a system subjectively it is very important to:

- use a large enough and representative enough sample of listeners, and
- use several announcers.

The effect an announcer can have on the perceived intelligibility from a sound system is profound. Systems with 'good' intelligibility can be transformed into total unintelligibility by different users of the same microphone. One way round the announcer problem would be to use standard recorded messages of known good articulation. However, such messages would need to have an information content and style of delivery similar to that of normal system announcements.

The practicality of setting up statistically valid tests complete with an appropriate listening panel to test out a sound system so that an absolute intelligibility score can be obtained, is generally an impossible goal. Instead, in practice a small group of people, maybe as few as 2 or 3 typically, wander round the designated area and listen to a few test messages. In most cases a consensus view can be reached, but often listeners may well be confusing sound quality with intelligibility, and an impasse over the intelligibility rating poses difficulties.

Ideally, what is required is some totally objective method which can be implemented and produces an easy-to-understand rating. Such methods do now exist and a significant number of sound system specifications are beginning to require specific levels of intelligibility to be met.

Techniques of measurement

At present four generic techniques are available for assessing the potential intelligibility of a sound system. The methods and implementation vary considerably, but only one truly takes both background noise and reverberation into account.

A number of subjective aspects must also be taken into account, particularly when deciding on an appropriate target value. These include:

- the level of difficulty of information that needs to be understood, e.g. the requirements for a supermarket staff announcement system are somewhat less onerous than a system required to reproduce Shakespeare or perhaps relay specific instructions inside a nuclear power station;
- the general hearing ability of the listeners and their environment;
- the articulation or intelligibility of the original signal;
- the rate at which the speech signal is to be delivered;
- the message language and familiarity of the listener to that language.

It is useful to consider also and at greater length the main sound system parameters which have an effect on speech intelligibility. There are a number of parameters, as follows:

- frequency response,
- noise or more precisely signal-to-noise ratio,
- reverberation or more precisely direct-to-reverberant ratio,
- freedom from echo.

Distortion is also an influencing factor but it is generally of secondary importance unless gross overload or similar distortion occurs.

Frequency response

Frequency response was discussed earlier. Figure 4.8 shows the intelligibility range of concern. Although the primary speech information is contained at higher consonant frequencies, the main power of the voice comes from the low and mid frequency vowel sounds with most sound energy centred at around 200–600 Hz. Typically the higher frequency information is 15–20 dB below these levels and can be easily masked or lost if the sound system places too much emphasis on the lower frequency region as often happens when systems are either incorrectly equalized or worse still not equalized at all. Figure 4.9 shows an averaged frequency spectrum for normal speech.

Signal-to-noise ratio

Background noise, such as that from machinery, traffic or other people, for example spectator crowd noise, can adversely affect speech intelligibility by 'masking' the necessary higher frequency components. A number of rules

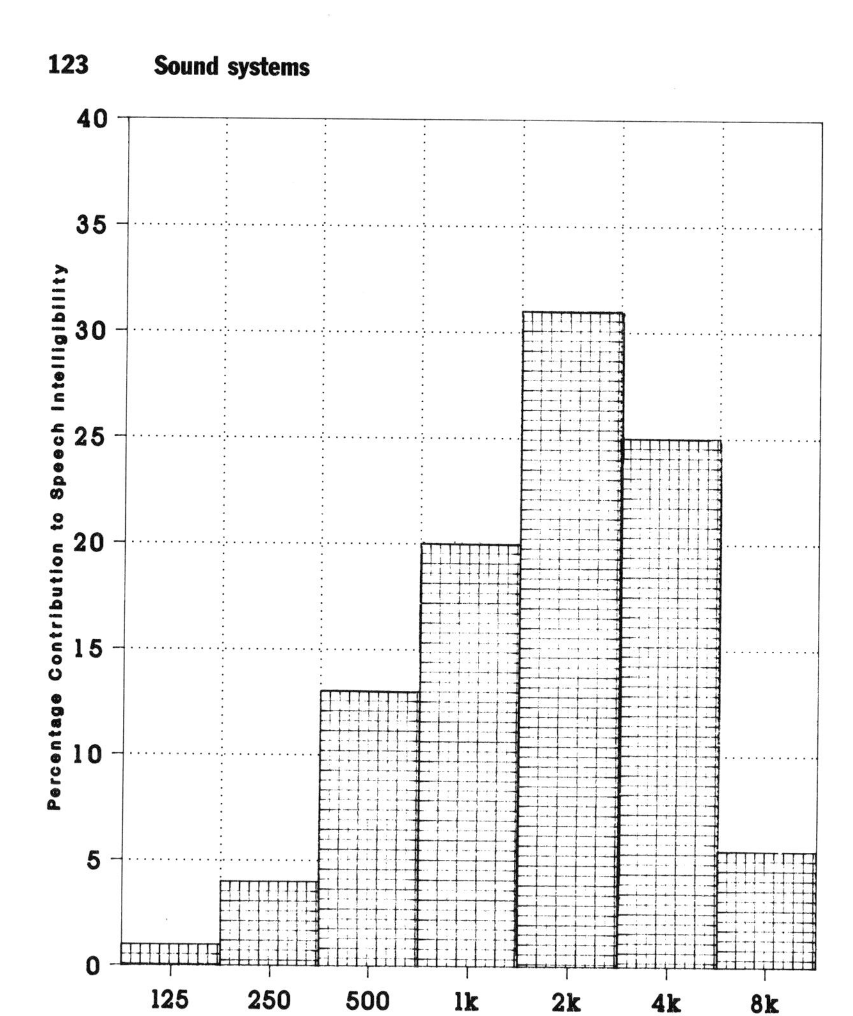

Figure 4.8 *Speech intelligibility versus frequency*

of thumb have been established for producing intelligible speech over background noise and typically range from around +3 dBA to +10 or 12 dBA. The signal-to-noise ratio required depends heavily on the spectral content or frequency make-up of the noise. The effect has been extremely well researched and a standardized method of assessing the degree of intelligibility from a given noise spectrum and level can be used, this being a standard method referred to earlier. Other characteristics of the noise such as its variance with time or the presence of any pure tone components, for example, must also be taken into account.

Apart from the signal-to-noise ratio, the absolute level of the speech signal must also be considered. At levels much over 90 dB, for example, the intelligibility in fact begins to decrease.

Reverberation

Echoes and reverberation are discussed above. For speech it is generally regarded that some reverberation is beneficial with an optimum reverberation time range around 0.8 s minimum to 1.2 s maximum; the volume of the space or room and the way the reverberation time changes with frequency are also important. At around 1.5 s a watershed would appear to occur, with intelligibility decreasing rapidly. Whilst the reverberation time is important, the primary factor is in fact the ratio of direct-to-reverberant signal, i.e. the ratio of that signal heard directly from the loudspeaker to that heard from the reflections off nearby surfaces, and multiple reflections occurring within the body of the room. There is no clear cut-off between 'useful' and deterimental reflections as it very much depends on the circumstances of each case. However, the generally accepted range is from 20 to 50 ms, with 35 ms often being taken.

The required ratio of direct-to-reverberant signal for intelligible speech is highly dependent on the overall reverberation time of the space. For example, a much higher ratio is required in a cathedral with a reverberation time of say 7 s, than a similar-sized building with an RT of only 2 or 3 s. It is interesting to note, as will be shown later, that good intelligibility can be obtained even with negative D/R ratios.

Having now established the basic factors affecting speech intelligibility, measurement techniques can be studied.

Articulation index

The articulation index is a longstanding method for rating the effect of background noise on speech intelligibility with its procedure formally adopted in an American Standard (ANSI 53.5: 1969).

Basically the signal-to-noise ratio between the speech

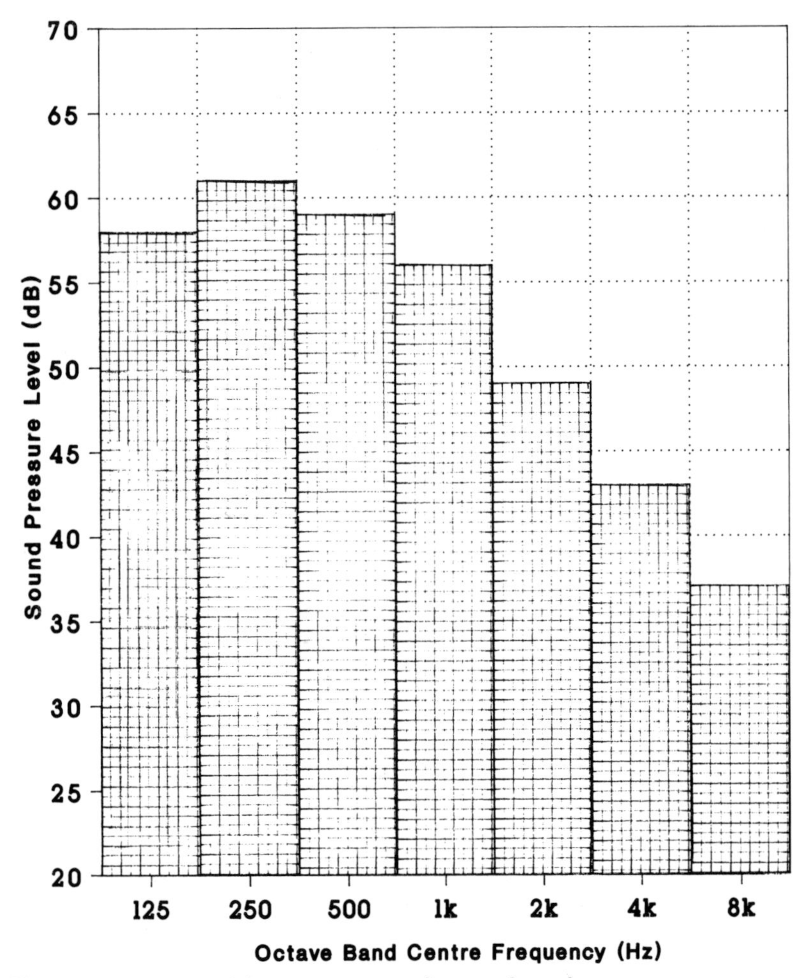

Figure 4.9 *Averaged frequency spectrum for normal speech*

Table 4.3 *AI method/weightings*

Articulation index calculation table

⅓ OBCF (Hz)	Speech peaks minus noise (dB)	Weight	Contribution to articulation index
200		0.000 4	
250		0.001 0	
315		0.001 0	
400		0.001 4	
500		0.001 4	
630		0.002 0	
800		0.002 0	
1 K		0.002 4	
1.25 K		0.003 0	
1.6 K		0.003 7	
2 K		0.003 7	
2.5 K		0.003 4	
3.15 K		0.003 4	
4 K		0.002 4	
5 K		0.002 0	

Relationship between articulation index,
Communication and speech privacy

EXCELLENT
GOOD
FAIR
POOR
VERY POOR
NIL
COMMUNICATION
SPEECH PRIVACY
0.2
0.4
0.6
0.8
ARTICULATION INDEX

signal and the background noise in each of 20 one-third octave bands is measured. The 20 S/N ratios are then individually weighted according to the speech information content contained within a given band. The weighted values are then combined to give a single overall value referred to as the articulation index (AI). The method in fact operates using the rms levels of speech peaks, a +12 dB correction factor being adopted to convert the long-term rms speech level to the equivalent short-term peak level. Table 4.3 illustrates the basic method and weightings. Other corrections are also made for very high background noise levels (above 80 dB) where the effect of masking is more marked. The masking effect of one frequency band on another is also corrected for. It is interesting to note that the method assumes that at S/N ratios of 30 dB or more, no masking of the wanted signal by the noise occurs. This may be compared with the RASTI/STI and % Alcons methods discussed later.

Although a further correction for reverberation time can be applied, for example reducing the overall AI by 0.1 for a 1 s RT or by 0.24 for a 2 s RT, it is generally agreed that these RT corrections are not particularly effective.

The AI uses a scale from 0 to 1.0 where 0 is total unintelligibility and 1.0 is equivalent to 100% intelligibility. Although a generalized subjective response scale can be formulated for example:

0.2 or less	unacceptable
0.2–0.3	marginal
0.3–0.4	acceptable/fair
0.4–0.5	good
0.5–0.6	very good
>0.7	excellent

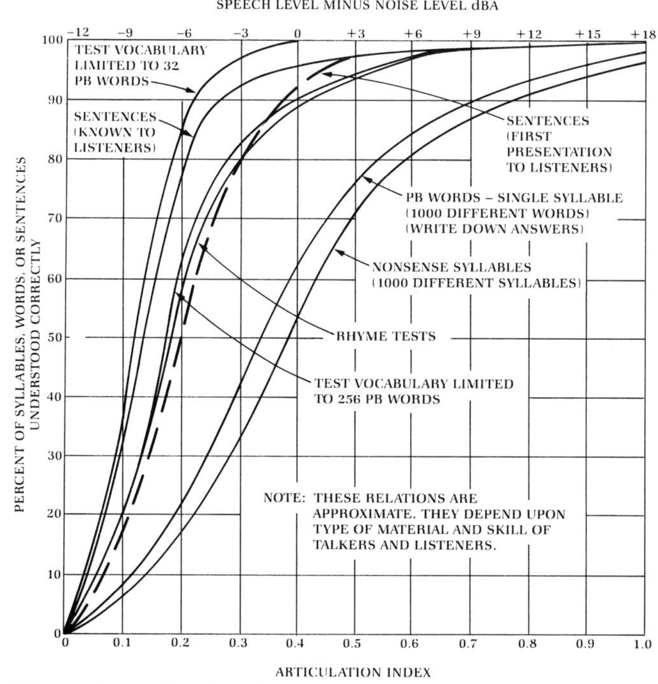

Figure 4.10 *Intelligibility curves. Relationship between articulation index and various measures of speech intelligibility and signal to noise ratio ('A' weighting)*

As previously described, the subjective rating heavily depends upon the type of speech information being imparted. Figure 4.10 compares the AI to various types of speech signal, for example from a known limited vocabulary of words to complete sentences or single words which do not have any contextual clues to help identify them.

Figure 4.10 provides an extremely useful set of intelligibility curves, clearly showing that when the subject is required only to understand a limited number of words or familiar sentences, a much lower AI can be tolerated than when listening for the first time. Equally, the effect of removing the context from around a given word, e.g. when it is used in a sentence, can be seen clearly with much higher AI scores (and hence S/N ratios) being required.

A simplified version of the AI procedure using just 5 octave bands is also available, but its accuracy is not up to the full 20-band test. In the absence of measured speech level data, the AI method provides an idealized speech spectrum (male speech) which has an overall equivalent level of 65 dBC at 1 m (dBC is used rather than dBA as this latter weighting too severely attenuates the lower frequencies).

Subjective testing

A number of procedures are available for objectively testing the intelligibility of a sound system using live listeners. Generaly speaking, specially-prepared phonetically-balanced word lists are used, read out either implanting the test word in a non-contextual carrier sentence so as to excite room reverberation or as individual and separate words with an appropriate spacing between them. The words are then either ticked off on a multiple answer sheet or written down as they are heard. The percentage intelligibility is then calculated by determining the number of correct answers and expressing this as a percentage. By using the curves shown in Figure 4.10, the test scores can also be expressed in terms of other indices such as the articulation index. In order to be statistically valid, lists containing at least 50 words need to be employed, with several sessions, locations and listeners. The quality of the reproduction of the words on the list is of crucial importance.

Percentage alcons: direct/reverberant ratio measurements

Direct percentage Alcons (percentage loss of consonants) testing, although based on concepts dating back to the 1960s and early 1970s, is in fact a relatively new measurement technique. Essentially the method measures the ratio of the direct to reverberant sound components received from a sound system or test loudspeaker at a typical listening position. The measurement is carried out in the 2 kHz band only, although measurements at other frequencies are often carried out to give a more detailed picture of a particular system.

The measurement requires the use of a highly sophisticated and computerized instrument. The % Alcons is obtained by measuring the RT and the direct-to-reverberant (reflected energy) ratio. From these two parameters the equivalent % Alcons can be computed.

One of the most useful features of the % Alcons method is that one can correlate measurement with prediction, and the % Alcons is still effectively the only method we have of predicting the potential intelligibility of a sound system *before* it is installed.

Using the established simple formula:

$$\% \text{ Alcons} = \frac{200\, D^2\, T^2\, (n+1) + K}{QVm}$$

Alternatively:

$$\% \text{ Alcons} = 100\, (10^{-2[(A+BC)-ABC]}) + 0.015$$

where

$$A = -0.32 \log\left(\frac{L_R + L_N}{10L_D + L_R + L_N}\right) \text{ for } A \geqslant 1,\ A = 1$$

$$B = -0.32 \log\left(\frac{L_N}{10L_R + L_N}\right) \text{ for } B \geqslant 1,\ B = 1$$

$$C = -0.5 \log\left(\frac{R_{T60}}{12}\right)$$

where:

D = listener distance to source (in m)
T = Reverberation time (in s)
$n + 1$ = number of like sources (groups of LS) contributing to the reverberant field
Q = loudspeaker directivity (axial factor)
V = volume of the space
m = critical distance modifier where $m = \dfrac{1+\alpha}{1-c}$
α = aberage absorption coefficient of the room
c = average absorption coefficient of surface covered by the loudspeakers
K = listener correction factor, e.g. 1.5% for good listener
L_R = reverberant sound level (dB power ratio)
L_D = direct sound level (dB power ratio)
L_N = ambient noise level (dB power ratio)
R_{T60} = reverberation time: that taken for sound pressure level to fall by 60 dB

The % Alcons scale is not quite as sophisticated or well defined as the AI or the RASTI/STI scales which will be covered next. The concept works more as a series of bands:

>15%	not acceptable except for very simple or well known messages
15–10%	acceptable for general messages of low complexity
5–10%	good
<5%	excellent

The effect of background noise can also be taken into account using a second, more complex, formula.

While very good correlations can be obtained between the D/R ratio at 2 kHz and %Alcons/intelligibility, using this technique only tests the system over one frequency band. The method effectively assumes that the system is well behaved and controlled at other frequencies. It is therefore essential to carry out supplementary measurements such as overall frequency response and impulse response (D/R ratios) at other frequencies, for example 500 Hz, 1 kHz and 4 kHz.

Speech transmission tests (STI and RASTI)

The speech transmission index (STI) and its shorter derivative, RASTI, are the newest and most complex intelligibility measurement techniques currently available. It is only the advent of modern microprocessor and desk top computer technology that have enabled the technique to be implemented on a practical basis.

The STI is not as such based on an impulse/direct reverberant ratio type of technique, but instead measures the modulation transfer function (MTF) between source and receiver. The test signal can be thought of as a complex modulated carrier whereby the modulation depth of the received signal is compared to the perfect (100% modulation) originally transmitted signal. The signal path, both electronic and acoustic, between the origination source and the receiver modifies and degrades the modulation of the signal by adding noise and reverberation components. The reduction in the modulation transfer function is measured over a range of 14 modulating frequencies at the 7 normal octave band 'carrier' frequencies of 125 Hz to 12.5 kHz. The result is that a 98-point measurement matrix is produced, i.e. 7 × 14. Each set of 14 MTFs is reduced down to a single transmission index value. The 7 individual transmission index values are then weighted and further combined to produce a single-figure value, the speech transmission index (STI). A direct and strong correlation between STI and perceived intelligibility has been shown to occur and a simple rating scale has been devised. The scale operates from 0 to 1.0 and is divided into 5 categorized bands from 'bad' to 'excellent' as shown below:

Bad	Poor	Fair	Good	Excellent
0–0.3	0.3–0.45	0.45–0.6	0.6–0.75	0.75–1.0

The values which correspond to 5, 10 and 15% Alcons are 0.65, 0.52 and 0.45, respectively. In use the scales have been found to be extremely sensitive; for example, the author's living room sound system only measuring 0.85. The STI intelligibility measurement technique automatically takes both background noise and reverberation into account. Furthermore, because the measurement is under automatic computer control there are no operator decisions to make, such as where to place integration and divisional cursors. However, setting up a measurement system to give valid results does require some experience and skill, particularly setting up relative levels through a sound system. Setting the correct operating level is obviously essential if the equivalent speech level to background noise is to be accurately accounted for.

As can be imagined from the 98-point measurement matrix, the measurement and calculation of the STI takes considerable computing power. In order to reduce the complexity of the measurement and the measurement time and hence the required computer power, a simpler derivative of the full STI method has been developed. The rapid speech transmission index (RASTI) carries out system measurements at only 500 Hz and 2kHz and at just 9 (shared) modulation frequencies. This has enabled a dedicated hard-wired system to be developed. The advantage of the Bruel & Kjaer RASTI system is that the transmitter and receiver are totally independent units and so require no loop-back interconnection cabling. The B&K RASTI equipment can carry out the foreshortened STI in just 16 to 32 s.

Figure 4.11 *Sound system cluster, The Olympia, East Kilbride*

The major disadvantage of the RASTI method is that no information is gathered over the whole operating range of a sound system, e.g. at 125 and 250 Hz at low to mid frequencies and at 4 and 8 kHz at the upper end. The system inherently assumes that both the system and space it is in are reasonably linear. This, however, is frequently far from being the case and an overly optimistic (or sometimes pessimistic) view of the situation is presented. At present RASTI is the only method that has been formalized and recognized within the International Standards (IEC 268) and so is rapidly gaining authority as a reference. It is subject to misinterpretation as it is often taken as a measure of the overall system quality. This is not what is intended for, now how it should be used. Measurements at only 500 Hz and 2 kHz cannot possibly define the complete performance of a sound system.

A major advantage of STI and RASTI is the ability to readily carry out 'what if' speculations and predictions by post-processing the noise component data; for example, if measurements were taken during a quiet period, it is possible to manually input new noise data and get a recalculated STI. In some implementations of STI and RASTI it is possible to completely isolate out the noise component and so see the effect due to reverberation alone. An increment of 3 dB in the signal-to-noise ratio increases the STI by 0.1 whereas a doubling of early decay time (see Chapter 2 for definition) decreases the STI by 0.15. RASTI/STI assumes that S/N ratios >15 dB do not influence the potential intelligibility any further (cf. 25 dB % Alcons and 30 dB for AI).

In a similar manner to % Alcons, STI can be predicted but it takes a relatively complex computer program to do this. Although RASTI and STI measurements are reasonably straightforward to make, setting up the system particularly with regard to signal-to-noise ratio is highly critical. Under certain conditions it is quite feasible to obtain false readings.

References

1. BS 5839 *Fire detection and alarm systems for buildings*, British Standards Institution, Milton Keynes
2. BS 7443: 1991 *Specification for sound systems for emergency purposes*, British Standards Institution, Milton Keynes

Chapter 5 Technical information

D. J. Saunders

Definitions

Basic concepts

Analogue and digital signals: Analogue signals are signals that vary continuously with time. For example, the sound pressure from a noise source. A digital signal is a signal that only has values at discrete time intervals. It is obtained by sampling an analogue signal at fixed time periods. Much present-day analysis equipment works with digital signals.

Angular frequency: The angular frequency, ω, is related to the frequency, f, by

$$\omega = 2 \times \pi \times f \text{ rad/s.}$$

Anharmonic: A frequency of a system which bears no simple relationship to the fundamental frequency.

Frequency: The frequency of a sound is synonymous with its tone or pitch. The higher the frequency, the higher the pitch. A particular frequency is produced by the regular vibration of the source.

In practice, sources do not generally produce single frequencies although, for example, fans do tend to produce a characteristic tone which is associated with a particular frequency. However, most sources will tend to produce sound that is composed of many different frequencies covering a wide range.

The frequency of a sound is a very important characteristic as many acoustic phenomena are frequency dependent.

Fundamental frequency: The lowest frequency of oscillation of a system.

Harmonic: A frequency of a system which is integrally related to the fundamental frequency.

Hertz (Hz): Hertz is the unit of frequency; it is the same as cycles per second.

Noise: Noise is the term often used to describe unwanted sound, i.e. sound that annoys, interferes with activities or damages hearing. It is also used to describe a combination of sounds which vary randomly with time and which cover a wide frequency range such as the sound produced by a jet engine or many industrial processes.

Pascal (Pa): Pascal is the unit of pressure and so sound pressures are measured in pascals.

The smallest sound pressure that the average human ear can detect is around 2×10^{-5} pascal (Pa). Humans begin to sense sound painful when the sound pressure is around 20 Pa.

Atmospheric pressure has a value of 10^5 Pa.

$$\begin{aligned} 1 \text{ pascal} &= 10^6 \text{ micropascal } (\mu\text{Pa}) \\ &= 1 \text{ newton/metre}^2 \ (\text{N/m}^2) \\ &= 10 \text{ microbar } (\mu\text{bar}) \end{aligned}$$

Range of human hearing: Most healthy young people will be able to hear sounds with frequencies that range from around 20 Hz to 18 000 kHz (K is used to denote a factor of 1000). As people get older the high frequency acuity of the ear deteriorates and this deterioration will be accentuated by prolonged exposure to high noise levels (85 dBA and above).

Prevention of hearing damage: The Noise at Work Regulations 1989 define three action levels for employees at work

First Action Level: 85 dB $L_{\text{Aeq}, 8\text{h}}$
Second Action Level: 90 dB $L_{\text{Aeq}, 8\text{h}}$
Peak Action Level: 140 dB

Protection must be provided to employees exposed to the second or peak action levels. An employee may request hearing protection if they are exposed to the first action level.

Sound: Sound is one form of energy: it is the energy of mechanical vibration of the molecules of a medium through which the sound is passing. Unlike light and heat, sound must have a medium to propagate through. The most common medium is air; however, sound will travel well through most media.

Sound pressure/acoustic pressure: When a sound wave propagates in air there are local variations in the air pressure. At any point the pressure will oscillate about the ambient pressure. These oscillations are known as the sound pressure or acoustic pressure and it is these changes in pressure which the ear detects and which, to a large extent, determine the loudness of a sound. For normal sounds the sound pressures are extremely small compared with the ambient atmospheric pressure. The smallest changes that the human ear can detect are about ten thousand million times less than atmospheric pressure while sounds which cause pain are about one thousand times less than atmospheric pressure (Table 5.1).

Sound sources: Any solid object that vibrates mechanically can communicate the vibrations with the surrounding medium and generate sound. Sound is also commonly generated by objects moving through a medium (e.g. blades of a fan moving through air) and by turbulence in gases and liquids (e.g. jet effluxes and turbulent flow of air and liquids in ducts and pipes).

Sound wave: When sound energy travels through a medium, a sound wave is said to be propagating. The speed of the wave or the speed at which energy travels from one point to another depends on the medium: in general, sound travels faster in solids than in liquids and faster in liquids than in gases. The speed of sound in air is around 340 m/s although it will depend on the air temperature.

$$\text{Velocity (m/s)} = 331 + 0.6 \times \text{temperature (°C)}$$

Units of measurement

Addition of attenuations: The way attenuations are added will depend on whether the attenuations are all in the same acoustic path or not. This can be illustrated by some simple examples.

(a) If a number of silencing units are placed one after another in a length of duct then the total attenuation

Table 5.1 *Sound pressure levels of common sounds*

Source	Typical level (dBA)	Reference distance (m)
Home		
Rural interior, watch ticking, calm breathing	15	1
Refrigerator	45	0.5
Normal voice	60	1
Washing machine, wash cycle	63	0.5
Food blender	84	0.5
Baby screaming	100	0.15
Office		
Computer disc drive	59	1
Open office, general activity	60	
Laser printer	64	1
Dot matrix printer	82	1
Industry		
Metal fabrication workshop	92	
Bottling hall	95	
Gas turbine generator hall	92	
Transport		
Inside stationary car, engine on	57	
Inside medium-size car, 35 mph	68	
Inside medium-size car, 60 mph	76	
Powerboat	63	100
Inside Inter-City Pullman	65	
Inside aircraft	78	
Diesel freight train, passing at 70 mph	80 (peak)	50
Inter-City train at 125 mph	83	50
Motorway (40 000 V/18 h)	80 ($L_{10,\,18\,h}$)	10
Below Boeing 757 taking off	82	650
Below Boeing 757 landing	84	260
Beside racing track	95 (peaks)	10
Leisure		
In a swimming pool	75–85	
Arcade games	82–92	
Firing range: .22 rifle	138 (peak)	0.3
Firing range: .38 pistol	157 (peak)	0.3
Music		
Orchestral music in concert hall, excluding quiet passages	78–95	
Discotheque music	101	
Live pop music venue	107	
Outside		
Suburban street at night	40	
City centre street on pavement	75	

will be the arithmetic sum of the attenuations given by each unit.

(b) If a number of identical ducts all open into the same room and each contains one silencing unit then the total attenuation will only be the attenuation produced by one unit. If the silencing units each give different attenuations R_1, R_2 . . . R_n dB then the total attenuation is

$$\text{attenuation} = 10 \log n - 10 \log (10^{-R1/10} + 10^{-R2/10} + \ldots 10^{-Rn/10}) \text{ dB}$$

This assumes the ducts each carry the same acoustic energy. If this is not so, a total attenuation cannot be obtained and the sound level from each path should be determined separately and the total level obtained by summing the individual levels.

(c) If the attenuations over the top and around the ends of a finite barrier are R_1, R_2 and R_3 dB then the total attenuation provided by the barrier is

$$\text{attenuation} = -10 \log (10^{-R1/10} + 10^{-R2/10} + 10^{-R3/10}) \text{ dB}$$

Attenuations can be added two at a time using the chart given in Figure 5.1.

Attenuation: The reduction in a sound signal produced by some process or device is referred to as attenuation. Attenuation is commonly expressed in decibels in which case

$$\text{attenuation} = 10 \times \log_{10} (I_u/I_a) \text{ dB}$$

where I_u is the intensity at the receiver without attenuation, and I_a is the intensity at the receiver with attenuation.

Averaging noise levels: The way in which a number of noise levels are averaged depends on the circumstances. If, for example, a number of measures are taken of an environmental noise on different occasions, then it would be appropriate to take the arithmetic mean of the levels to find the representative average of the levels, i.e.

$$L_{av} = (L_{p1} + L_{p2} + L_{p3} + \ldots L_n)/n \text{ dB}$$

However, if a number of sound levels are taken in a reverberant field then the logarithmic average is the appropriate average to take as this represents the mean energy value of the field

$$L_{av} = 10 \times \log_{10}((10^{L_{p1/10}} + 10^{L_{p2/10}} + \ldots 10^{L_{pn/10}})/n) \text{ dB}$$

or

$$L_{av} = 10 \times \log_{10}(10^{L_{p1/10}} + 10^{L_{p2/10}} + \ldots 10^{L_{pn/10}}) - 10 - \log n \text{ dB}$$

Combination of sound from several sources: When a number of noise sources operate simultaneously the total sound pressure level is

$$L = 10 \times \log_{10}(10^{L_{p1/10}} + 10^{L_{p2/10}} + 10^{L_{pn/10}}) \text{ dB}$$

L_{p1}, L_{p2}, L_{pn} are the sound pressure levels of the individual sources.

Alternatively, the total level can be found using the chart shown in Figure 5.2. Using this chart, the individual noise levels must be added two at a time. For example, to add 80, 82, 84 and 86 dB, the procedure is:

(i) Difference between 82 and 80 is 2 hence, the correction is 2.1 dB and the total level 84.1 dB.

(ii) Difference between 86 and 84 dB is 2, the correction is 2.1 dB and the total level 88.1 dB.

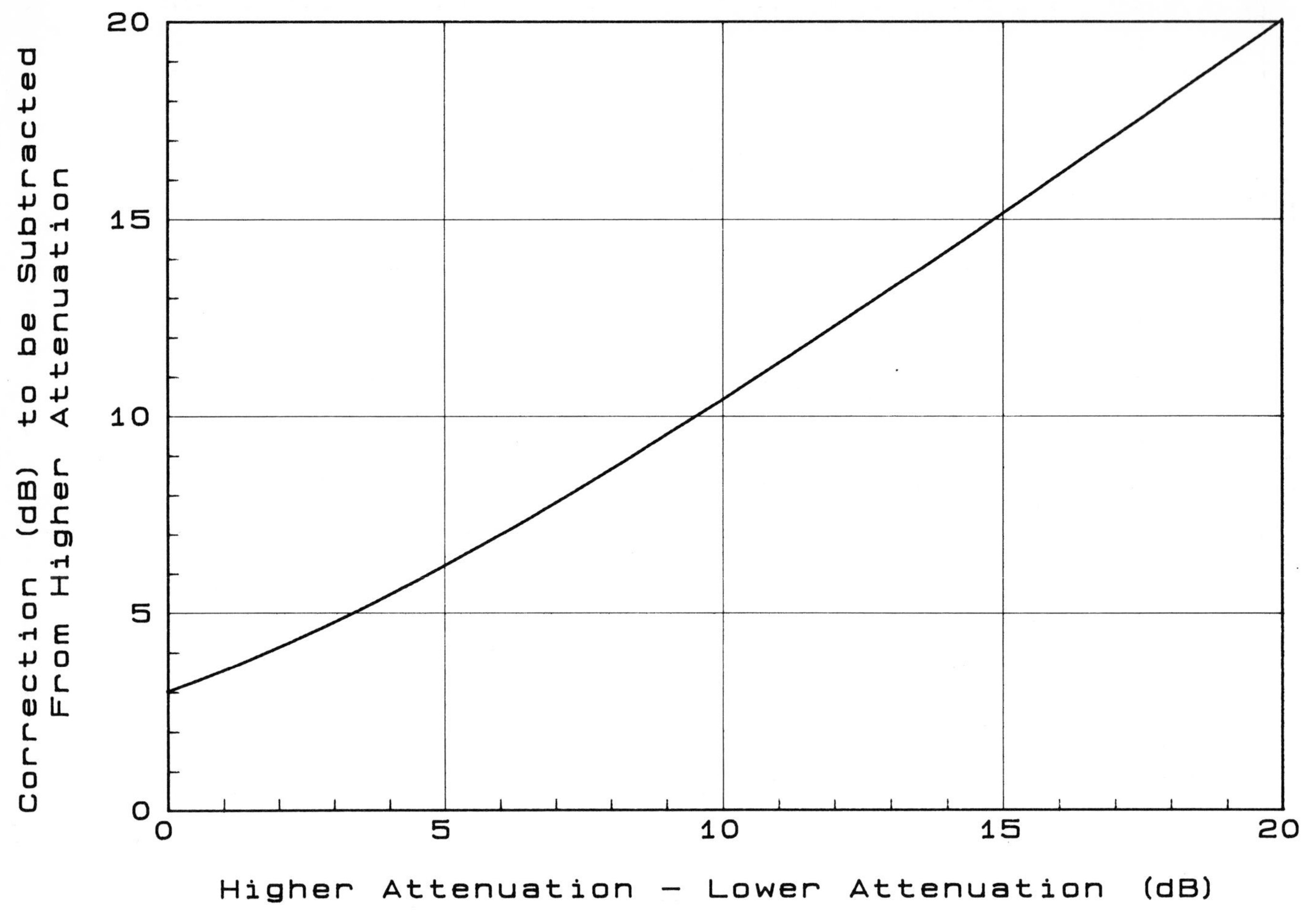

Figure 5.1 *Adding attenuations*

(iii) Difference between 88.1 and 84.1 is 4 dB, the correction is 1.5 dB and the total level 89.6 dB.

An approximate way to add sources is

Difference (dB)	Correction (dB)
0	3
1	3
2	2
3	2
4	2
5	1
6	1
7	1
8	1
9	0

Decibel (dB): The decibel is the unit that is used as a measure of a number of acoustical quantities. The decibel is equal to 10 bels and the bel is a logarithmic unit, strictly the logarithm to the base ten of the ratio of two power related quantities. In the field of acoustics the decibel is commonly used as the unit for sound pressure level, sound intensity level and sound power level. Other assessment units which inherently depend on these quantities will also have the decibel as their unit.

Directivity factor (Q): The directivity factor is the ratio of the intensity in a given direction from a source to the intensity in the same direction had the source radiated uniformly. For a particular source Q would normally be determined by measurement in an anechoic room.

Directivity index (DI): The directivity index is defined as

$$\mathrm{DI} = 10 \times \log_{10} Q \text{ dB}$$

where Q is the directivity factor of the source in a given direction.

Intensity level: Intensity level is often used instead of sound intensity level.

Relation between sound power level and sound pressure level: For a source that radiates uniformly, the sound pressure level, L_p, at a distance r metres from the source is given by:

$$L_p = L_P - 20 \times \log_{10} r - 11 \text{ dB}$$

or

$$L_p = 10 \times \log_{10} P - 20 \times \log_{10} r + 109 \text{ dB}$$

L_p will be the A-weighted sound level if L_p or P is A-weighted.

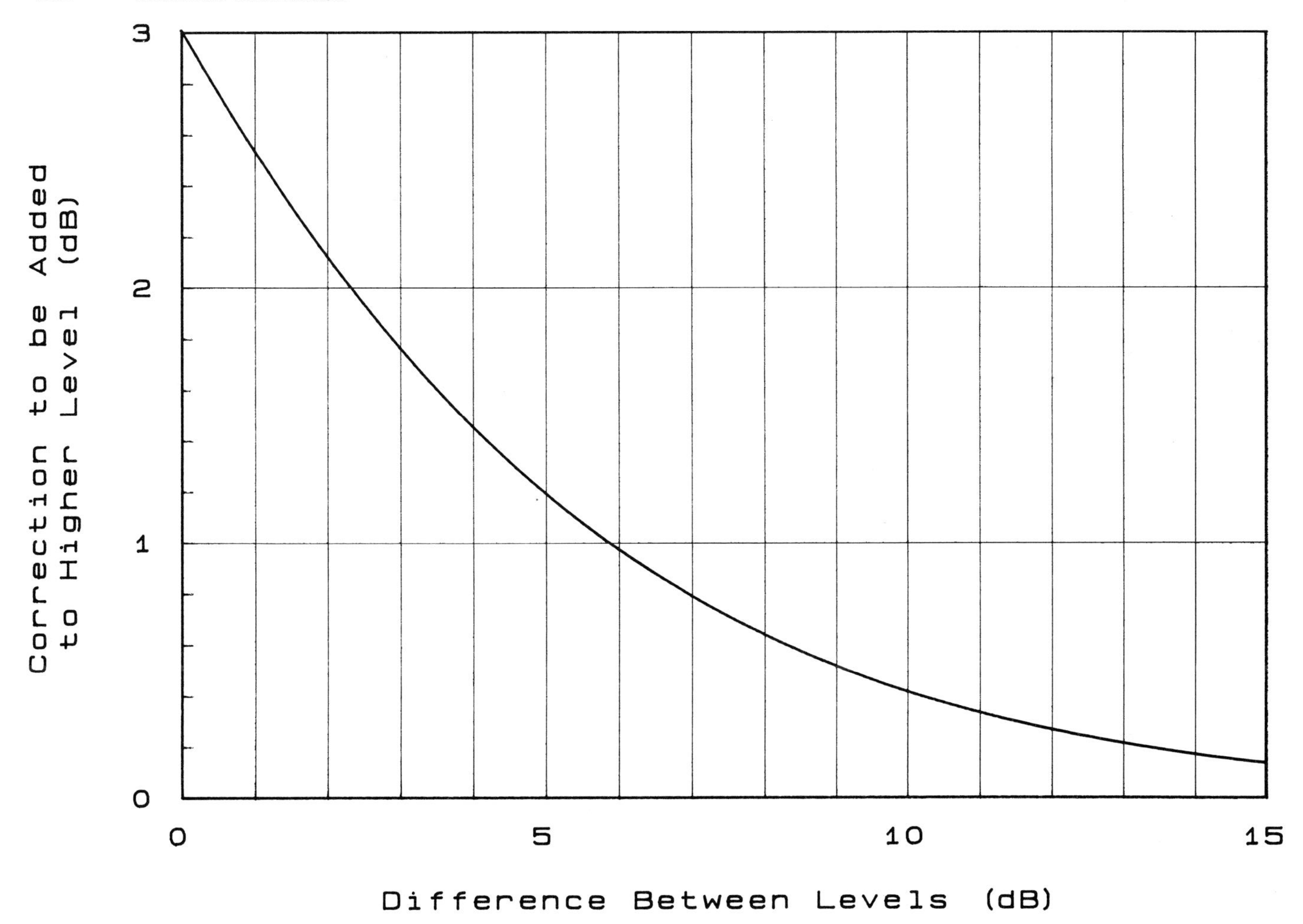

Figure 5.2 *Adding noise sources*

If the source radiates non-uniformly then the equations must be modified to

$$L_p = L_P - 20 \times \log_{10} r - 11 + \text{DI dB}$$

or

$$L_p = 10 \times \log_{10} P - 20 \times \log_{10} r + 109 + \text{DI dB}$$

where DI is the directivity index of the source in the direction of *r*.

Sound level/noise level: Sound level and noise level are often used instead of sound pressure level.

Sound intensity level: The sound intensity level, IL, is defined as

$$L_I = 10 \times \log_{10} (I/I_o) \text{ dB}$$

where *I* is the acoustic intensity which is defined as the power passing through a unit area perpendicular to the direction of travel of the power. Intensity, therefore, has units of watts per square metre (W/m^2).

I_o is a reference intensity which for propagation in air is chosen as 10^{-12} W/m^2.

In many cases the sound pressure level and the sound intensity level have the same numerical value for a given sound and they can be used synonymously. However, circumstances do exist where this equivalence does not hold and thus it is better if the sound intensity level is used exclusively for describing the ratio of two sound intensities as, for example, when measuring sound intensity directly with an intensity probe.

The sound intensity level should always be quoted with a reference quantity, e.g. 120 dB re 10^{-12} W/m^2.

Sound pressure level: The sound pressure level, L_p, is defined as

$$L_p = 20 \times \log_{10} (p/p_o) \text{ dB}$$

where p is the sound pressure and p_o is a reference pressure which for propagation in air has the value 2×10^{-5} Pa.

Hence, as $20 \times \log (2 \times 10^{-5}/2 \times 10^{-5}) = 0$, a sound pressure level of zero dB is equivalent to a sound pressure of 2×10^{-5} Pa.

Also, as $20 \times \log_{10} (20/2 \times 10^{-5}) = 120$, a sound pressure level of 120 dB is equivalent to a sound pressure of 20 Pa. The normal range of human hearing thus covers the range 0 to 120 dB.

When a sound pressure level is given it should always have an associated reference quantity, e.g. 120 dB re 2×10^{-5} Pa.

An increase of 3 dB in the sound pressure level of a noise is thought to be the smallest change that is subjectively definitely noticeable under normal testing conditions. An

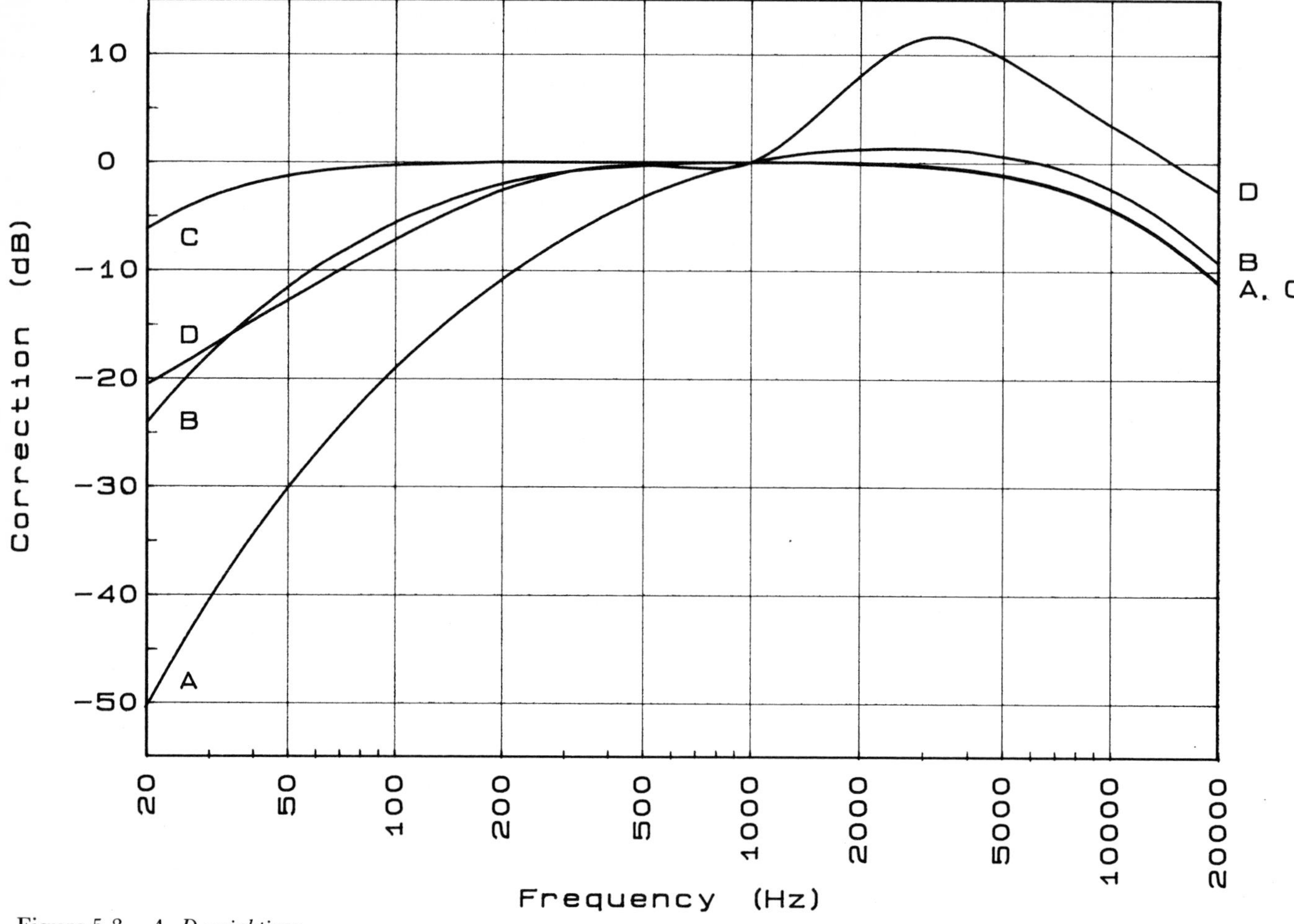

Figure 5.3 *A–D weightings*

increase of 10 dB on average represents a doubling in loudness of the noise.

Sound power level: The sound power level, L_P, is defined as

$$L_P = 10 \times \log_{10} (P/P_o) \text{ dB}$$

where P is the acoustic power of the source in watts (W), and P_o is a reference sound power chosen in air to be 10^{-12} W.

As $10 \times \log_{10} (1/10^{-12}) = 120$, 1 acoustic watt is equivalent to a sound power level of 120 dB re 10^{-12} W.

$$L_p = 10 \times \log_{10} (P) + 120 \text{ dB}$$

Source on a reflective plane: When a source, assumed to radiate uniformly is placed on a reflective plane, e.g. on a concrete surface, the energy radiated above the plane is effectively doubled. The source directivity factor is 2 and the directivity index is 3 dB. Hence,

$$L_p = L_P - 20 \times \log_{10} r - 8 \text{ dB}$$

or

$$L_p = 10 \times \log_{10} P - 20 \times \log_{10} r + 112 \text{ dB}$$

For a source at the junction between two reflecting planes, e.g. a door in a wall on a hard ground, DI is 6 dB.

Weighting networks and frequency bands

A-weighting: Human hearing is not equally sensitive at all frequencies. In addition, the variation with frequency is a function of the sound pressure level. To try and account for this variation when measuring sound, electronic weighting networks are incorporated in the measuring instrument between the microphone and the display. The A-weighting attenuates low and high frequencies relative to 1000 Hz. The standard A-weighting curve is shown in Figure 5.3 and detailed in Table 5.2. The A-weighting is the most frequently-used weighting network, especially for rating environmental noise.

B-weighting: The B-weighting is similar to the A-weighting except that there is less attenuation at low frequencies as shown in Figure 5.3 and Table 5.2. The B-weighting is little used.

C-weighting: The C-weighting is essentially flat except below 50 Hz and above 5000 Hz as shown in Figure 5.3 and Table 5.2. It is not often used although it has been suggested that it should be used to describe low-frequency short-duration events.

Constant bandwidth filters: These are filters that have a bandwidth which is constant independent of the band centre frequency.

Table 5.2 *Specification of weighting networks*

Frequency (Hz)	Curve A (dB)	Curve B (dB)	Curve C (dB)	Curve D (dB)
10	−70.4	−38.2	−14.3	−27.6
12.5	−63.4	−33.2	−11.2	−25.6
16	−56.7	−28.5	−8.5	−23.5
20	−50.5	−24.2	−6.2	−21.6
25	−44.7	−20.4	−4.4	−19.6
31.5	−39.4	−17.1	−3.0	−17.6
40	−34.6	−14.2	−2.0	−15.6
50	−30.2	−11.6	−1.3	−13.6
63	−26.2	−9.3	−0.8	−11.6
80	−22.5	−7.4	−0.5	−9.6
100	−19.1	−5.6	−0.3	−7.8
125	−16.1	−4.2	−0.2	−6.0
160	−13.4	−3.0	−0.1	−4.4
200	−10.9	−2.0	0.0	−3.1
250	−8.6	−1.3	0.0	−1.9
315	−6.6	−0.8	0.0	−1.0
400	−4.8	−0.5	0.0	−0.3
500	−3.2	−0.3	0.0	0.0
630	−1.9	−0.1	0.0	−0.1
800	−0.8	0.0	0.0	−0.4
1 000	0.0	0.0	0.0	0.0
1 250	0.6	0.0	0.0	1.9
1 600	1.0	0.0	−0.1	5.4
2 000	1.2	−0.1	−0.2	8.0
2 500	1.3	−0.2	−0.3	10.0
3 150	1.2	−0.4	−0.5	11.0
4 000	1.0	−0.7	−0.8	10.9
5 000	0.5	−1.2	−1.3	10.0
6 300	−0.1	−1.9	−2.0	8.5
8 000	−1.1	−2.9	−3.0	6.0
10 000	−2.5	−4.3	−4.4	3.0
12 500	−4.3	−6.1	−6.2	−0.4
16 000	−6.6	−8.4	−8.5	−4.4
20 000	−9.3	−11.1	−11.2	−8.1

Digital filters: Digital filters are computer algorithms which filter digital signals in the same way that electrical networks filter analogue signals.

Fast Fourier transform: The Fourier transform is a method for transposing information in the time domain to information in the frequency domain. The fast Fourier transform (FFT) is a computational algorithm which performs the transformation at much greater speeds. The FFT will produce a frequency spectrum where the frequency information is at fixed frequency intervals. The frequency spacing depends upon the time intervals between the original data samples and the total number of time domain samples.

Linear-weighting: There is no standard linear weighting. In general it should have a flat unattenuated response between 2 Hz and 20 kHz.

Octave band: Two frequencies are said to be an octave apart if the frequency of one is twice, or more precisely $10^{0.3}$, the frequency of the other.

Contiguous octave bands have centre frequencies which are also related by a factor of two ($10^{0.3}$). The centre frequencies and bandwidths of standard octave bands are shown in Table 5.3.

An octave bandwidth increases as the centre frequency of the band increases. Each bandwidth is 70% of the band centre frequency.

Octave band filter: An octave band filter is an electrical network which allows frequencies within the octave to pass unattenuated but reduces frequencies outside the octave to an insignificant level. Frequencies just beyond the limits of the band are not always reduced to a level at which they do not contribute to the band. The performance of a filter in these regions will depend much on its design. There are a number of different classes of filter with differing attenuation rates at the filter limits.

The octave band sound pressure level, L_p, can be obtained from the one-third octave level L_{p1}, L_{p2} and L_{p3} by summing the levels as incoherent sources, i.e.

$$L_p = 10 \log [10^{L_{p1}/10} + 10^{L_{p2}/10} + 10^{L_{p3}/10}] \text{ dB}$$

or by adding the levels two at a time using Figure 5.2.

Octave band sound pressure level: The sound pressure level measured when only frequencies within an octave are passed is known as the octave band sound pressure level. Analysis of noise into octave bands is frequently used in the assessment of a noise climate.

Octave band spectrum: When the sound pressure levels in adjacent octaves are plotted against the centre frequencies of the octave bands this is known as an octave band spectrum. It is accepted practice to plot the centre frequencies at equally-spaced intervals, i.e. on a logarithmic scale.

One-third octave band: Two frequencies are said to be one-third octave apart if the frequency of one is 1.26, or more precisely $10^{0.15}$, times the other. There are three one-third octaves in each octave band. The standard centre frequencies and bandwidths of one-third octaves are shown in Table 5.3. The width of a one-third octave band is 23% of the band centre frequency.

Environmental noise measures

A-weighted sound pressure level: This is the sound pressure level measured using an A-weighting network to filter the sound. The sound pressure level has units of dBA so the sound level would be given, for example, as 80 dBA or more correctly as 80 dBA re 2×10^{-5} Pa.

The A-weighted sound pressure level is the basic measure used in most environmental noise assessment indices and schemes.

Calculation of A-weighted sound pressure level from octave band sound pressure levels: The A-weighting attenuation for the centre frequency of each octave is added arithmetically to the octave band sound pressure level and the resulting levels are added together considering them to be separate noise sources (see combination of sound from several sources), e.g.

O.B.C.F (Hz)	125	250	500	1 k	2 k	4 k	8 k
O.B. L_p (dB)	80	80	80	80	80	80	80
A-weighting (dB)	−16.1	−8.6	−3.2	0	+1.2	+1.0	−1.1
A-weighted O.B. L_p (dB)	63.9	71.4	76.8	80	81.2	81	78.9

Table 5.3 *Octave and one-third octave centre frequencies and band limit frequencies*

Band No.	Preferred centre frequency (Hz)	Standard frequencies (Hz)			
		Octave		Third-octave	
		Band limits	Centre	Centre	Band limits
		22.39			22.39
14	25			25.12	
					28.18
15	31.5		31.62	31.62	
					35.48
16	40			39.81	
		44.67			44.67
17	50			50.12	
					56.23
18	63		63.10	63.10	
					70.79
19	80			79.43	
		89.13			89.13
20	100			100.00	
					112.20
21	125		125.89	125.89	
					141.25
22	160			158.49	
		177.83			177.83
23	200			199.53	
					223.87
24	250		251.19	251.19	
					281.84
25	315			316.23	
		354.81			354.81
26	400			398.11	
					446.68
27	500		501.19	501.19	
					562.34
28	630			630.96	
		707.95			707.95
29	800			794.33	
					891.25
30	1 000		1 000.00	1 000.00	
					1 122.02
31	1 250			1 258.93	
		1 412.54			1 412.54
32	1 600			1 584.89	
					1 778.28
33	2 000		1 995.26	1 995.26	
					2 238.72
34	2 500			2 511.89	
		2 818.38			2 818.38
35	3 150			3 162.28	
					3 548.13
36	4 000		3 981.07	3 981.07	
					4 466.84
37	5 000			5 011.87	
		5 623.41			5 623.41
38	6 300			6 309.57	
					7 079.46
39	8 000		7 943.28	7 943.28	
					8 912.51
40	10 000			10 000.00	
		11 220.18			11 220.18

A-weighted level $= 10 \times \log (10^{6.39} + 10^{7.14} + 10^{7.68} + 10^{8} + 10^{8.12} + 10^{8.1} + 10^{7.89})$

A-weighted $L_p = 87$ dBA

Average sound level, $L_{av,T}$: This is the same as the equivalent continuous sound level L_{eq}.

Corrected noise level (CNL): This is the A-weighted sound pressure level which has been modified to take account of any distinguishable characteristics of the noise. It is defined in BS 4142: 1990.

If the noise has a noticeable tonal component, e.g. a whine or hiss, is impulsive or in any way is such that it draws attention to itself, then 5 dBA is added to the measured sound level. If one or many distinguishing features exist, 5 dBA is only added once.

Day/night equivalent sound level, DNL (L_{DN}): This is a rating, based on the equivalent continuous sound level L_{eq}, which has its origins in the USA. The acoustic energy is averaged over a 24-h period but the noise level during the night-time period (22:00–07:00 hours) is penalized by the addition of 10 dBA.

$$L_{DN} = 10 \log \left[\frac{1}{24} \int_{7}^{22} 10^{L_A/10}\, dt + \int_{22}^{7} 10^{(L_A+10)/10}\, dt \right] \text{dBA}$$

where L_A is the instantaneous A-weighted sound pressure level.

For effectively constant noise levels an estimation of L_{DN} can be made in the same way as the normal L_{eq} is estimated (see above). L_{DN} has found widespread acceptance in the USA for environmental noise assessment.

Equivalent continuous sound level, $L_{eq,T}$ (dB): The continuous equivalent sound level, $L_{eq,T}$ is a notional sound level. It is the sound level which if maintained for a given length of time would produce the same acoustic energy as a fluctuating noise over the same time period.

It is defined mathematically as

$$L_{eq,T} = 10 \times \log \left[\frac{1}{T} \int \frac{p^2(t)\, dt}{p_o^2} \right] \text{dB}$$

where $p(t)$ is the acoustic pressure which varies with time; T is the total time over which the $L_{eq,T}$ is calculated; p_o is 2×10^{-5} Pa.

If $p(t)$ is A-weighted before the $L_{eq,T}$ is calculated then the $L_{eq,T}$ will have units of dBA.

The above formula is implemented electronically in all good sound level meters and it is customary to measure rather than calculate the equivalent continuous sound level. It should be remembered that any value of $L_{eq,T}$ should be accompanied by the time over which it was measured. $L_{eq,T}$ is widely used to measure any environmental noise which varies considerably with time.

Estimation of the equivalent continuous sound level: It is possible to estimate the equivalent continuous sound level for a source or number of sources if they have effectively constant noise levels over known periods of time.

If a source of noise level, L_p, is on for a period of time t then the L_{eq} value, over a period T where T is greater than or equal to t is

$$L_{eq,T} = L_p + 10 \log \frac{t}{T} \text{ dB}$$

For example, if a source of noise level 100 dB is on for 0.5 h in 8 h its effective 8-h L_{eq} will be

$$L_{eq,8\,h} = 100 + 10 \log \frac{0.5}{8}$$

$$L_{eq,8\,h} = 88 \text{ dB}$$

Remember t and T must have the same units, whether it is seconds, hours or days.

The corrections to obtain the L_{eq} value can be obtained from the chart shown in Figure 5.4.

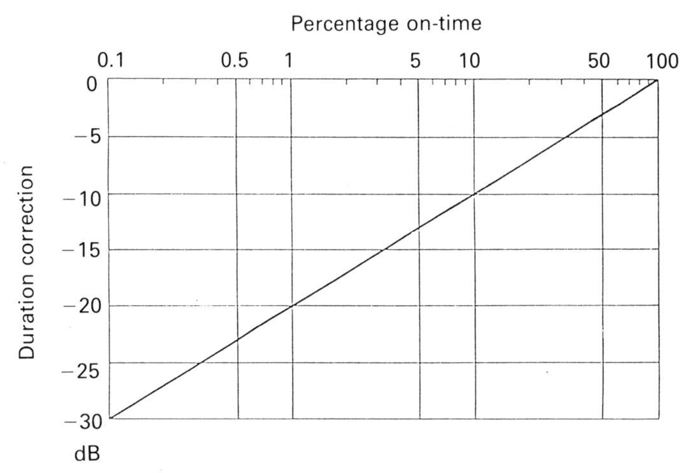

Figure 5.4 *Correction to measured noise level for percentage on-time*

If there are a number of sources the corrections can be applied to each source in turn and then the total $L_{eq,T}$ obtained by decibel addition, e.g.

Source	Level (dBA)	On time (h)	Correction (dBA)	Corrected level
1	80	4	−3	77
2	85	2	−6	79
3	77	8	0	77

8-hour $L_{eq} = 10 \log (10^{7.7} + 10^{7.9} + 10^{7.7}) = 82.5$ dB

Alternatively, one may use

$$L_{eq,8\,h} = 10^{L_{p1/10}} \times t_1 + \ldots 10^{L_{pn/10}} - 10 \log T$$

L_{p1} is sound pressure level which is on for time t_1, etc.

So, for the above example, the 8-hour L_{eq}

$= 10 \log (10^8 \times 4 + 10^{8.5} \times 2 + 10^{7.7} \times 8) - 10 \log 8$
$= 82.5$ dB.

The individual noises can be all on together, on separately, or overlap. It will make no difference to the L_{eq} level.

Maximum sound pressure level, L_{max}: This is the maximum value of the sound pressure level that occurs during any given period. Its value will depend upon the frequency weighting and meter time characteristic. The maximum slow A-weighted sound level during an aircraft flyover is used in the assessment of aircraft noise (see 'perceived noise level').

Minimum sound pressure level, L_{min}: This is the minimum sound level that occurs during any given period. It is little used in assessment procedures.

Noise and number index, NNI: This is an index which was used in the UK for rating aircraft noise until 1990. It is defined as

$$\text{NNI} = \text{average peak perceived noise level} + 15 \log N - 80$$

where N is the number of aircraft having a PNL greater than 80 PNdB in the specified period. The average PNL is given by

$$10 \log \left[\frac{1}{N} \sum_{i=1}^{N} 10^{L_i/N} \right] \text{dB}$$

where L_i is the peak value of PNL during the passage of the ith aircraft.

NNI has been replaced by $L_{eq,T}$.

An approximate relationship between NNI and the 16-h A-weighted L_{eq} is

NNI	16-h L_{eq} (dBA)
35	57
45	63
55	69

Peak sound pressure level: This is the sound pressure level corresponding to the peak sound pressure that occurs in any given period. Its value will depend upon the frequency weighting and the time characteristics of the measuring system.

It is often used to quantify short duration impulses, e.g. gunfire, explosions and high level impact noise. To measure the true peak sound pressure level of an event, great care must be taken in the selection and use of microphone and measuring system.

Perceived noise level, PNL: The perceived noise level is a rating for single aircraft flyovers based originally on jury judgements of perceived noisiness, but it is now commonly derived by an extensive calculation procedure.

For most general purposes the perceived noise level can be obtained from measurements of the maximum slow A-weighted sound level that occurs during a flyover.

$$\text{PNL} = \text{dBA} + 13\,\text{PNdB}.$$

Single event noise exposure level, SENEL or L_{AX}: This is equivalent to the sound exposure level. In the original definition it was assumed that the sound pressure was A-weighted and that the integration was over the time that the signal was within 10 dB of the maximum value.

Sound exposure level, SEL or LAE: The sound exposure level is a notional level. It is the sound level that if maintained constant for 1 s contains the same acoustic energy as a varying noise level.

It is normally used to quantify short duration noise events such as aircraft flyover, single vehicle bypasses, impact or impulsive noise.

It is defined mathematically as

$$\text{SEL} = 10 \times \log \int_0^T \frac{p^2(t)}{p_o^2} \, dt \ \text{dB}$$

where $p(t)$ is the sound pressure which varies with time over the period T. The sound pressure is often A-weighted before the SEL is calculated.

The sound exposure level is related to continuous equivalent level by

$$L_{eq,T} = \text{SEL} - 10 \log T \ \text{dB}$$

where T is the time in *seconds* over which the L_{eq} is required.

If identical SEL values occur in the time T, the L_{eq} is given by

$$L_{eq,T} = \text{SEL} - 10 \log T + 10 \log N \ \text{dB}$$

If the events are not identical then the L_{eq} equivalent of each event must be found and then the total L_{eq} calculated by adding the individual L_{eq} values as decibels are added.

Statistical level, L_1: This is the sound pressure level that is exceeded for 1% of the measurement time. It gives an indication of the maximum sound levels that occur.

Statistical level, L_{10}: This is the sound pressure level that is exceeded for 10% of the measurement time. Consequently it is indicative of the higher levels that occur in the measurement period.

In the UK the A-weighted L_{10} value is used to measure and assess traffic noise. For fairly noisy traffic

$$L_{10} \approx L_{eq} + 3\,\text{dB}$$

Statistical level, L_{90}: This is the sound pressure level that is exceeded for 90% of the measurement time. Consequently it is indicative of the general ambient noise level in the absence of any higher level short-duration events that occur during the period.

The A-weighted L_{90} value is often used as a measurement of background noise in environmental assessment.

Sound insulation

Sound insulation: Airborne sound insulation refers to the process of separating, by a physical barrier, a space to be protected from a space containing a noise source. With noise insulation the sound is effectively prevented from travelling in a specific direction by an impervious barrier. The greater the surface mass of the barrier, the greater the insulation will generally be. Unlike sound absorption, sound insulation does not remove energy from the sound field; it merely redirects it.

Sound reduction index, SRI: The sound reduction index is the measurement generally used to express the insulation properties of a partition in decibels. It is defined as

$$\text{SRI} = 10 \log_{10} (1/\text{transmission coefficient}) \ \text{dB}$$

The sound reduction index is frequency dependent and is usually measured in octave or one-third octave bands.

If the transmission coefficient $\tau = 0.01$, i.e. 1% of the incident sound energy is transmitted by the partition, then the sound reduction index is 20 dB, while if $\tau = 0.001$ the SRI is 30 dB, etc. Hence for 50 dB insulation, which, for example, may represent a reasonable reduction between flats, the incident intensity must be reduced by a factor of 0.000 01.

The octave band sound reduction index, R, is obtained from the one-third octave band indices R_1, R_2 and R_3 as follows (see addition of attenuations, part (b))

$$R = 10 \log 3 - 10 \log [10^{-R_1/10} + 10^{-R_2/10} + 10^{-R_3/10}] \text{ dB}$$
$$R = 4.77 - 10 \log [10^{-R_1/10} + 10^{R_2/10} + 10^{-R_3/10}] \text{ dB}$$
$$R = 4.77 + C$$

Figure 5.1 can be used to find C.

This addition assumes equal energy in all three octave bands.

Sound transmission coefficient, τ: The sound transmission coefficient is a measure of the incident sound energy that 'passes through' a wall, partition or any barrier. The sound does not actually pass through the barrier. Incident sound energy causes the barrier to vibrate and the vibrating barrier then radiates sound into the receiving space. The sound transmission coefficient is defined as

$$\tau = \frac{\text{Intensity incident upon a partition}}{\text{Intensity transmitted by partition}}$$

Transmission loss

Coincident effect critical frequency: When sound waves strike a partition, bending waves are excited in it, the velocity of which depends upon the frequency. As the frequency of the bending wave increases, the bending wave velocity increases and at some frequency, known as the critical frequency, it is equal to the velocity of sound in air. At this frequency the wavelength of the bending wave is equal to the wavelength of the sound in air and if the sound wave impinging on the partition is of the same frequency, resonant excitation can occur and the sound transmission of the panel is increased and the sound reduction index decreased.

Matching of the wavelengths at the critical frequency occurs at grazing incidence and little energy is actually transferred to the partition. However, as the frequency increases above the critical frequency, matching occurs at increasing angles of incidence and energy transference to the partition is significantly increased.

The reduction in performance of the panel is known as the coincidence effect and the performance of the partition can be significantly reduced at frequencies in the region of the critical frequency although at an octave or so beyond the critical frequency the panel will again approach its expected performance.

Field sound transmission class, FSTC: This the STC obtained from values of the sound reduction index measured under field conditions.

Flanking transmission: The transmission of sound energy via paths which bypass a partition is known as flanking transmission.

Noise isolation class, NIC: This is the STC obtained from values of the sound level differences measured between two spaces under laboratory conditions.

Normalized noise isolation class, NNIC: This is the STC obtained from values of the sound level differences measured between two spaces under field conditions. The sound levels in the receiving room are corrected to a standard reverberation time of 0.5 s.

Sound level difference between two spaces: The sound level difference between two spaces separated by a partition depends upon the value of the sound reduction index, the area of the partition and the acoustic properties of the two spaces.

(a) room-to-room

$$L_{p2} = L_{p1} - R + 10 \log_{10} (S/A)$$

(b) inside-to-outside

$$L_{p2} = L_{p1} - R + 10 \log_{10} S - 20 \log_{10} r - 17 + \text{DI dB}$$

(c) outside-to-inside

$$L_{p2} = L_{p1} - R + 10 \log_{10} (S/A) - K + 6 \text{ dB}$$

L_{p1} is the sound pressure level on the source side (dB),
L_{p2} is the sound pressure level on the receiver side (dB),
S is the area of the partition (m^2),
R is the sound reduction index of the partition (dB),
A is the absorption in the receiving room (m^2)
DI is the directivity index of the facade,
r is the distance of the receiver from the partition.

K is a constant, the value of which depends on where, external to the partition, the sound pressure level was measured:

K = 6 dB if measured very close to the partition,
K = 2.5 dB if measurement position is about 1 m,
K = 0 dB if measurement position is far from facade.

Sound transmission class, STC: This is a single figure descriptor used for rating the sound transmission of a partition obtained by laboratory measurements. The measured sound reduction indices in one-third octave bands are compared with a set of standard curves and at each frequency the difference between the measured values and the standard curve values is obtained. The STC value corresponds to that curve for which the sum of the deficiencies is less than or equal to 32 dB and the maximum deficiency at any one frequency is less than 8 dB.

The reference curve for an STC value of 52 is

Frequency (Hz)	Level (dB)
125	36
160	39
200	42
250	45
315	48
400	51
500	52
630	53
800	54
1 000	55
1 250	56
1 600	56
2 000	56
2 500	56
3 150	56
4 000	56

The value of the STC is equal to the dB value of the curve at 500 Hz. Other standard curves are obtained by moving this curve up or down in increments of 1 dB.

SRI of composite construction: A composite construction is one having areas with different sound reduction indices, e.g. a wall with windows.

For a facade consisting of areas $S_1, S_2 \ldots S_n$ with sound reduction indices $R_1, R_2 \ldots R_n$, the value of the sound reduction index for the facade is given by

$$R = 10\log_{10}\left[\frac{S_1 + S_2 + \ldots S_n}{10^{-R_1/10} \times S_1 + 10^{-R_2/10} \times S_2 + \ldots 10^{-R_n/10} \times S_n}\right] \text{dB}$$

The composite sound reduction index may also be obtained using Figure 5.5, taking two areas at a time.

Small areas of very low insulation can drastically reduce the overall performance of a facade, e.g. an opening of area 0.1 m^2 and SRI 0 dB in a facade of area 25 m^2 and SRI 50 dB reduces the overall SRI value to 50 − 26 = 24 dB.

Standardized sound level difference, D_{nT}: The standardized sound level difference is used to assess airborne sound insulation between rooms in buildings. As the sound level difference across a partition will depend upon the absorption in the receiving room it is recommended (BS 5821: 1984) that the measured level difference is corrected to a standard receiving room reverberation time of 0.5 s. Hence

D_{nT} = measured level difference − 10 log $T/0.5$ dB
T = reverberation time in receiving room.

Structureborne sound: Sound which travels from one space to another not through the air but through the fabric of the building is known as structureborne sound. It is one form of flanking transmission. Structureborne sound can travel long distances with little attenuation and be re-radiated, causing problems far from the original source of noise.

Transmission loss, TL: Transmission loss is an alternative name for the sound reduction index.

Weighted apparent sound reduction index, R'_w: This is similar to R_w but is used if it is thought that the value of R_w was obtained with flanking transmission.

Weighted sound level difference, D_w: The weighted level difference is obtained from level differences, measured in one-third octave bands, in exactly the same way at the airborne sound insulation index rating, R_w, is obtained.

Weighted sound reduction index, R_w: This is a weighted single figure descriptor of the sound insulation performance of a partition measured under laboratory conditions. The sound reduction index in each of the one-third octave bands from 100 Hz to 3150 Hz is compared with a standard set of curves. The value of R_w for a given partition is obtained from the standard curve which when compared with the measured SRI values produces an adverse deviation as

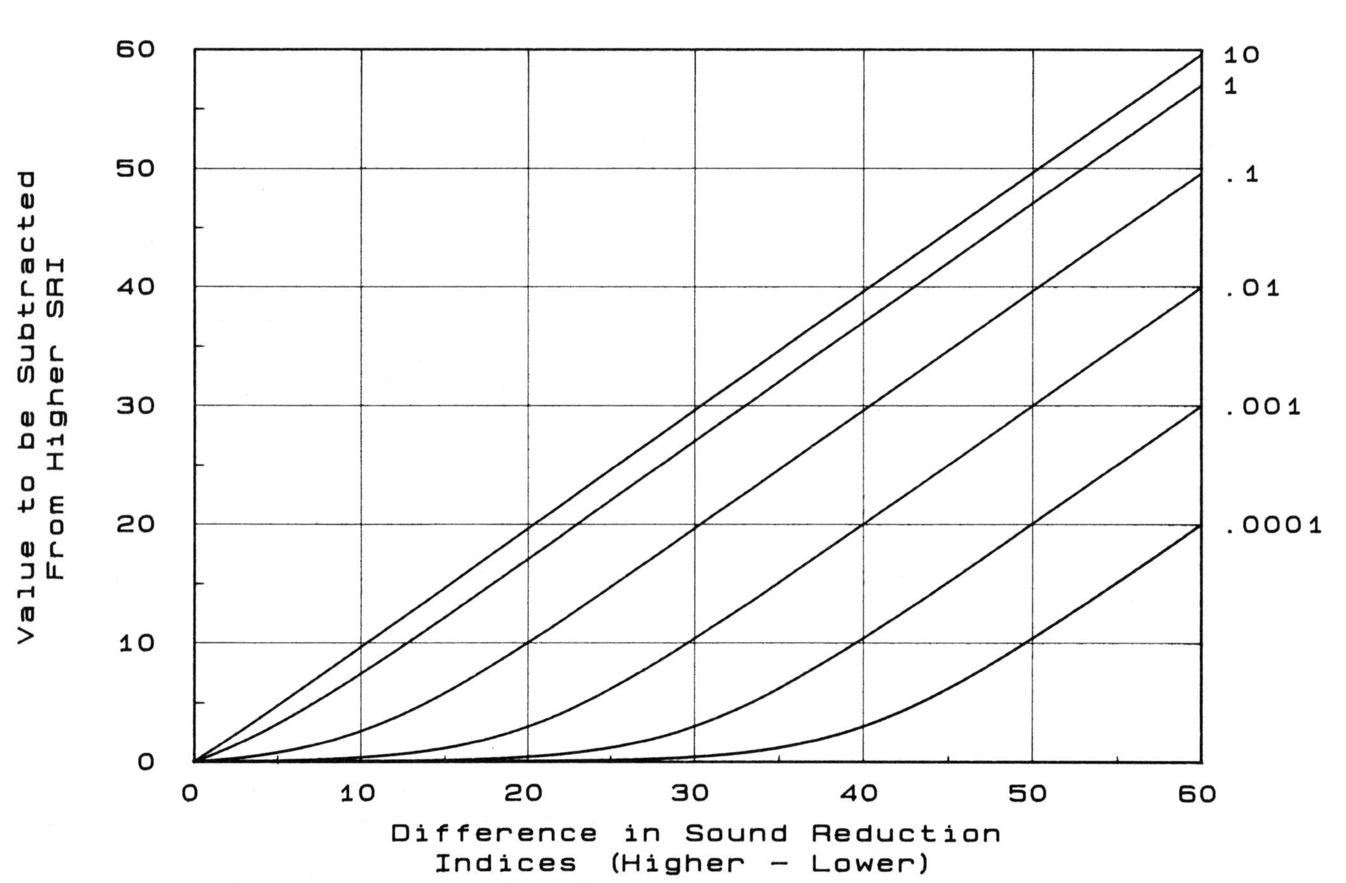

Figure 5.5 *Sound reduction indices loss*

close to -32 dB as possible. Only the SRI values which fall below a particular standard curve are considered in the sum. Positive deviations from the standard curve are not taken into account. The standard values for the curve corresponding to an R_w of 52 are

Frequency (Hz)	Reference value (dB)
100	33
125	36
150	39
200	42
250	45
315	48
400	51
500	52
630	53
800	54
1 000	55
1 250	56
1 600	56
2 000	56
2 500	56
3 150	56

The R_w value is the value in decibels of the reference curve at 500 Hz.

To obtain other reference curves the one-third octave band values are changed in 1 dB steps up or down.

Weighted standardized level difference, D_{NTw}: To obtain a single figure rating value from field measured values of the standardized level difference, D_{nT}, the one-third octave values are weighted using the same method used to obtain the airborne sound insulation index rating, R_w.

Sound in enclosed spaces

Absorption: Absorption is the term applied to the process by which energy is removed from a sound field. Most materials will, to a greater or lesser extent, absorb sound, i.e. convert the acoustic energy into heat. However, to be a good absorber a material should generally have an open surface structure which allows sound to enter and internally it should provide many interconnecting pathways through which the sound may pass to dissipate its energy. Good fibrous absorbents are glass fibre and mineral wool.

In acoustics absorption, A, has a more specific meaning. It is the product of the area, S, of an absorbing material and its absorption coefficient α. So

$$A = S \times \alpha \text{ m}^2 \text{ or Sabines.}$$

Air absorption: The absorption of sound by air is significant for propagation over long distances and in large enclosures. High frequencies are absorbed the most and the absorption is dependent upon both temperature and humidity.

Diffuse sound field: When the sound energy in an enclosure is uniform throughout the space, the sound field is said to be diffuse. This is normally the case for enclosures with conventional aspect ratios and small absorption which is uniformly distributed throughout the enclosure.

Direct sound field: The direct sound field refers to the acoustic energy that arrives at a listener directly from the source without any reflections from nearby surfaces or objects.

Early decay time, EDT: This is the time in seconds, multiplied by six, which the sound in an enclosure takes to decay by 10 dB from its equilibrium value. It is thought to be important in determining the quality of auditoria, especially for music. The EDT is sometimes referred to as the subjective reverberation time.

Noise reduction coefficient, NRC: This is the average, to the nearest multiple of 0.05, of the absorption coefficients measured in the octave bands centred on 250, 500, 1000 and 2000 Hz.

Normal room modes: Sound waves in an enclosure travel around in all directions being reflected obliquely off the walls. Some paths repeat themselves continuously, forming what are known as normal modes.

The normal modes occur at specific frequencies related to the dimensions of the room. For a rectangular room the frequencies are given by

$$fn = \frac{c}{2}\sqrt{\left(\frac{n_x}{\ell_x}\right)^2 + \left(\frac{n_y}{\ell_y}\right)^2 + \left(\frac{n_z}{\ell_z}\right)^2}$$

where c = velocity of sound (m/s), ℓ_x, ℓ_y and ℓ_z are the room dimensions (m), and n_x, n_y and n_z can independently have values 0,1,2 . . .

If the source contains frequencies equal to the normal mode frequencies, resonances occur and large variations in sound level throughout the room can result especially at low frequencies. This is generally to be avoided.

Norris–Eyring equation: This equation is a modified form of the Sabine equation and is suitable for use when the average room absorption coefficient is greater than 0.1.

$$T = \frac{0.161V}{-2.3 \times S \log_{10}(1 - \bar{\alpha})} \text{ s}$$

Reverberance: When a sound source, in an enclosure, is turned off, the sound does not immediately stop but persists for a short time due to the reflection of energy from the walls of the enclosure. Similarly, it takes a finite time for sound to reach its equilibrium value after the source is turned on. This behaviour is known as reverberance and it is a major factor in determining the level and quality of sound in any enclosed space.

Reverberant sound field: This is the sound field within an enclosure due to the continual reflections of the sound energy from the walls of the enclosure.

Reverberant sound pressure level: The steady value of the sound pressure level through the body of an enclosure, and away from the sound source, is referred to as the reverberant sound pressure level, L_{pr}. Its value will depend upon the sound power, P, of the source and the total absorption, A, within the enclosure

$$L_{pr} = 10 \log P - 10 \log A + 126 \text{ dB}$$

or

$$L_{pr} = L_P - 10 \log A + 6 \text{ dB}$$

Reverberation time: The time which is taken for the reverberant sound energy in an enclosure to decay to one millionth of its equilibrium value, i.e. by 60 dB, after the source is turned off, is known as the reverberation time. The reverberation time is frequency dependent and it is customary to measure its value in octave or one-third octave bands. There are a number of simple equations for predicting reverberation times.

Room radius: For a source operating in an enclosure there are two sound fields, the reverberant and the direct, and the value of the sound pressure level at any point is the sum, that is the decibel sum, of the direct and reverberant sound pressure levels. Far from the source the reverberant field will dominate while close to the source the direct field will be greatest.

The distance, r, from the source where the direct and reverberant sound pressure levels are equal is known as the room radius and it can be found from

$$r = \sqrt{\frac{QA}{16\pi}}$$

where Q = source directivity factor, and A = room absorption.

Sabine equation: The Sabine equation gives the reverberation time in terms of the room volume and total room absorption as

$$T = \frac{0.161\,V}{A}\ \mathrm{s}$$

if V is in cubic metres and A in square metres.

The equation is valid for diffuse sound fields only and gives the best results when the average absorption coeffficient is less than 0.1. However, it is often used when this condition is not met. For large enclosures air absorption is included so that

$$T = \frac{0.161\,V}{A + 4mV}$$

where m is a sound attenuation. Coefficient values for $4m$ as given in Table 5.4.

Table 5.4 *Air absorption (values of $4mV$, in m^2 units for a volume of $100\,m^3$ at 20°C)*

Frequency (Hz)	Relative humidity (%)						
	20	30	40	50	60	70	80
125	0.06	0.05	0.04	0.04	0.03	0.03	0.02
250	0.14	0.13	0.12	0.11	0.10	0.09	0.08
500	0.25	0.25	0.26	0.26	0.26	0.25	0.25
1 000	0.57	0.47	0.46	0.46	0.48	0.50	0.51
2 000	1.78	1.21	1.00	0.90	0.88	0.88	0.88
4 000	6.21	4.09	3.10	2.60	2.27	2.08	1.95
8 000	19.00	14.29	11.00	8.95	7.61	6.69	6.04

Sound absorption coefficient: The sound absorption coefficient is the quantity used to describe how well a particular material absorbs sound. It is denoted by α and is defined as

$$\alpha = \frac{\text{Sound energy not reflected from material}}{\text{Sound energy incident upon material}}$$

For a perfect absorber α would have a value of 1 while for a perfect reflector α would equal zero.

The absorption coefficient varies with frequency and also with the angle at which the sound strikes the material. Because of the angular dependence it is usual to measure the absorption coefficient of materials in diffuse sound fields so that sound effectively strikes the material at all angles of incidence. The absorption coefficient measured under these conditions is known as the random incidence sound absorption coefficient and is denoted by $\bar{\alpha}$. It is usually measured in one-third octave or octave bands. There is no accepted way of obtaining the octave band value from the one-third octave band values. The average of the three values or the highest value are both used.

Assessing internal spaces

Articulation index, AI: The articulation index is a weighted fraction representing, for a given speech channel and noise condition, the effective proportion of the normal speech signal that is available to a listener for conveying speech intelligibility. It is obtained from measurements or estimates of the speech spectrum and of the effective masking spectrum of any noise which may be present along with the speech at the ear of the listener.

If the AI is zero then there will be no understanding while if the AI is one there will be complete intelligibility.

Details of the calculation method can be found in ANSI S3.5–1969.

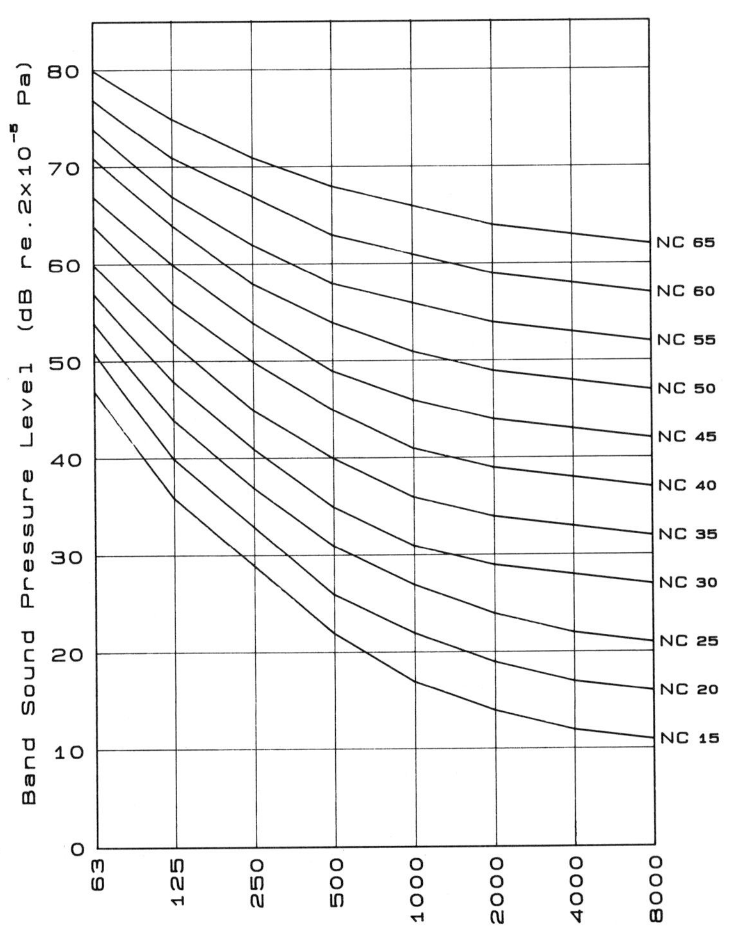

Figure 5.6 *Noise criterion curves*

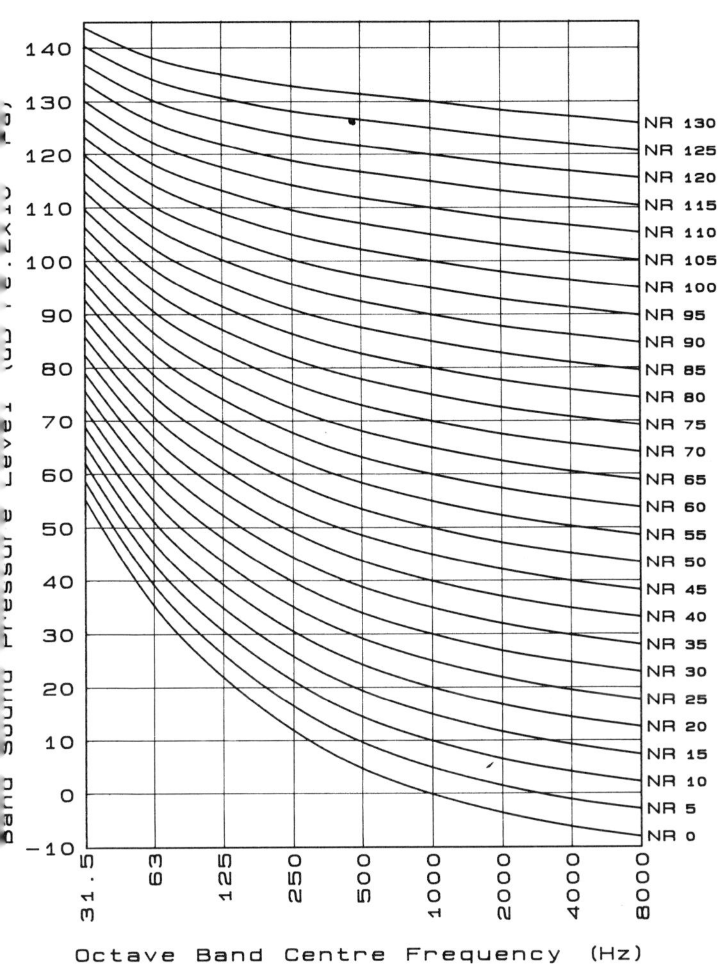

Figure 5.7 *Noise rating curves (octave band)*

Balanced noise criterion curves, NCB: These are a set of curves which have been proposed as an update of the NC curves. They are not as yet generally accepted. Details can be found in a paper by Beranek, *Balanced Noise Criterion Curves*, *J. Acoust. Soc. Am.*, **86**(2), August 1989, pp. 650–664.

Noise criterion, NC: The concept of the noise criterion was originally developed in the United States specifically for application in commercial buildings. Its calculation is based on an octave band analysis of a noise and reference is made to a set of curves which are shown in Figure 5.6.

The noise criterion is obtained by plotting the octave band sound pressure levels onto the reference curves and determining the lowest curve which is nowhere exceeded by the plotted octave band levels.

Noise rating number, NR: The noise rating number is a single figure index obtained from an octave band analysis of a noise. To obtain the NR number the octave band sound pressure levels are plotted onto a set of reference curves which are shown in Figures 5.7 and 5.8. The highest NR curve that is intersected by the curve forming the plotted sound pressure levels gives the noise rating number. The octave band sound pressure levels are normally joined by straight lines.

Preferred noise criterion, PNC: The preferred noise criterion is very similar to the noise criterion in concept. However, the set of curves from which it is obtained extend to a lower frequency than that of the noise criterion curves and more emphasis is placed on low frequency noise. The curves for PNC are shown in Figure 5.9. PNC is not widely used.

RASTI: Rapid speech transmission index (RASTI) is a method for the objective measurement of the speech transmission index.

By restricting the number of noise bands to two and modulation frequencies to five in the calculation, rapid assessments of room speech intelligibility can be made.

Room-noise criterion, RC: This is a single figure rating, used for assessing heating, ventilating and air-conditioning systems. The RC value is obtained by comparing noise levels, made in unoccupied rooms with all systems operating, with a set of curves. In addition the RC number is classified into four categories depending on the overall shape of the spectrum. It is used as an alternative to NC. Details of how to calculate RC values can be found in the ASHRAE Handbook.

Speech interference level: Speech interference level is a simple-to-estimate measure of the masking of speech by noise. It is derived by taking the arithmetic average of the noise levels in the four octave bands centred on 500, 1000, 2000 and 4000 Hz.

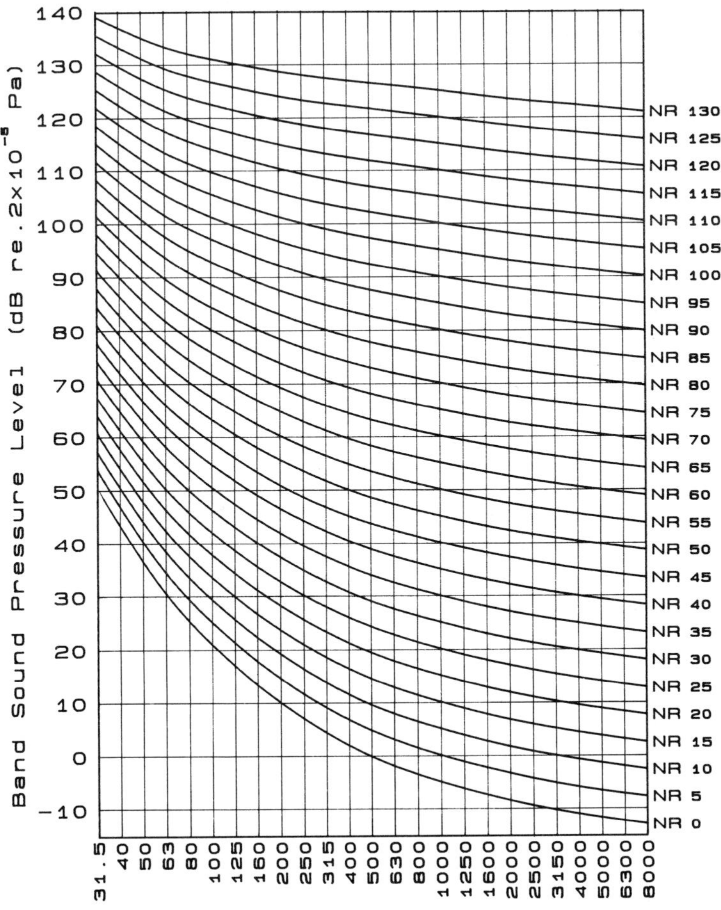

Figure 5.8 *Noise rating curves (one-third octave band)*

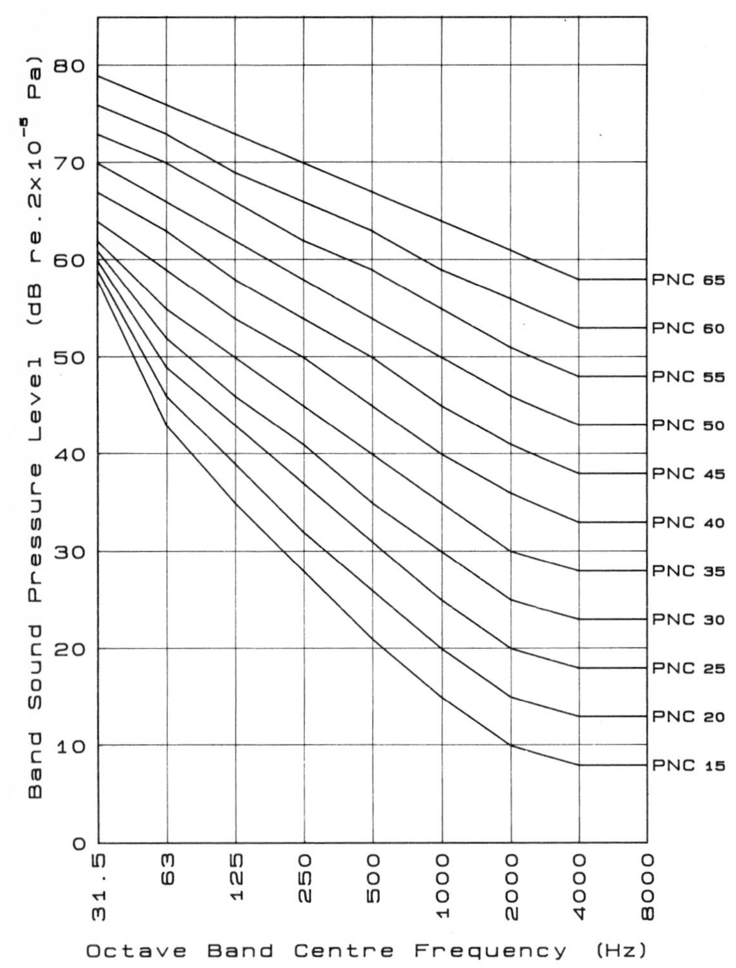

Figure 5.9 *Preferred noise criterion curves*

Speech transmission index, STI: The speech transmission index quantifies the effect of a sound transmission system on speech intelligibility. It is based upon an analysis of the reduction in intensity modulation of a signal which occurs along the transmission path from source to receiver. The analysis is carried out for 8 octave bands of noise, typically 125 Hz to 8 kHz, and 14 modulation frequencies. The results of this analysis are then combined and normalized to give the speech transmission index which has a value of 1 for perfect transmission and a value of 0 for no signal recognition.

Relationships exist between the STI values, the signal-to-noise ratio and the reverberation time of an enclosure, allowing theoretical calculations to be made of the STI value. This is most useful at the design stage of a project.

Details may be found in Bruel and Kjaer Technical Review 3:1985.

Impact sound and vibration

Impact insulation class, IIC: This is a single number rating, used in the United States, which permits easy comparison of the impact isolation performance of floor/ceiling assemblies. Impact sound levels normalized to a room absorption of $10\,m^2$ are compared with a set of standard curves to produce the impact insulation class.

The calculation is exactly the same as that described to determine the weighted normalized impact sound pressure level, L_{nw}, except that in addition to the total adverse deviation being less than or equal to 32 dB, no one deviation must exceed 8 dB.

The standard curves are identical to the L_{nw} curves and the IIC is equal to 110 minus the sound level at 500 Hz on the selected standard curve.

$$L_{nw} = 110 - \text{IIC}$$

See ASTME E492-86 and E989-84.

Impact noise level, L: This is the sound pressure level, measured in a one-third octave band, when a standard tapping machine is operating on the floor above the room.

Impact sound: Impact sound refers to sound produced when a short duration impulse, such as a footfall, acts directly on a structure.

The frequency content of the sound will depend upon the duration of the impact; a short sharp event giving a broadband frequency content while a longer duration event caused, for example, by having a resilient layer over the structure, will contain mainly low frequency sound and will be subjectively less disturbing.

Normalized impact noise level, L_n: The impact noise level will depend upon the acoustic characteristics of the receiving room so to normalize results the measured noise levels are corrected to a constant $10\,m^2$ of absorption.

Hence $L_n = L - 10 \log A/A_o$

A = actual sound absorption in the receiving room in the one-third octave band under consideration.

$A_o = 10\,m^2$

L'_n is used if flanking transmission cannot be eliminated.

See BS 5821, Part 2: 1984 and BS 2750, Parts 6 and 7: 1980.

Resonant frequency, f_r: The resonant frequency of an isolator of stiffness K (Nm^{-1}) which supports a mass M (kg) is

$$f_r = \frac{1}{2\pi} \sqrt{\frac{K}{M}} \text{ . Hz}$$

This is often re-written as

$$f_r = \frac{15.8}{\sqrt{d}}$$

d being the static deflection (in mm) of the isolator when the mass M is placed on it.

For rubber isolators the constant becomes 19.5.

For an isolator to be effective its resonant frequency must be at least three times lower than the lowest frequency to be isolated.

Standardized impact sound level, L'_{nT}: This is the impact sound level measured between two rooms under field conditions and standardized to a reverberation time of 0.5 s,

$$\text{i.e. } L'_{nT} = L' - 10 \log \frac{T}{0.5}$$

L' = measured impact noise level.

See BS 5821, Part 2: 1986 and BS 2750, Parts 6 and 7: 1980.

Standard tapping machine: A standard tapping machine is used to rate the impact noise isolation of floors. The machine has five hammers, each of mass 0.5 kg, equally spaced along a line. The hammers are dropped, from a height of 4 cm, successively to give 10 impacts per second.

Transmissibility: The effectiveness of a vibration isolator is measured in terms of its transmissibility. Two types of transmissibility are generally defined: (1) Force transmissibility which is the ratio of the force transmitted by the isolator to the force applied to the structure on top of it; (2) displacement transmissibility which is the ratio of the displacement transmitted by the isolator to the displacement applied at its base. In both cases the transmissibility, T, is, for lightly damped systems, given by

$$T = \left| \frac{1}{1 - (f/f_r)^2} \right|$$

where f is the frequency of vibrating motion and f_r is the resonant frequency of the isolator together with the structure mounted on it. The variation of the transmissibility with frequency is shown in Figure 5.10.

Vibration acceleration, *a*: The acceleration of a vibrating surface is related to the displacement and velocity by

$$a = d\omega^2 = v\omega$$

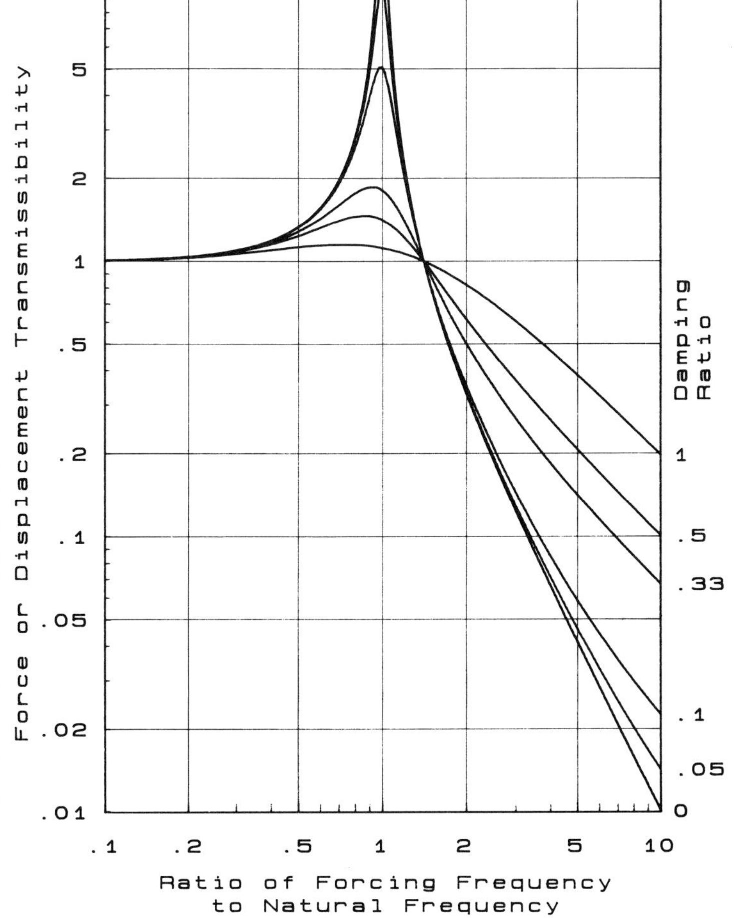

Figure 5.10 *Transmissibility*

Acceleration is what is commonly measured to quantify the vibration of a surface.

Vibration displacement, *d*: When an object vibrates, its surface will oscillate about its stationary position. These changes represent the vibration displacement.

Vibration isolation: Vibration isolation is a means of reducing the transmission of vibrating motions or forces from one structure to another. It is usually achieved by separating the two structures by an elastic element, known as a vibration isolator.

Vibration velocity, *v*: The velocity of a vibrating surface is related to the displacement by

$$v = d\omega$$

$$v(\text{m/s}) = d(\text{m}) \times \omega(\text{rad/s})$$

where ω = angular frequency.

Weighted normalized impact sound pressure, L_{nw}, L'_{nw}: This is a single figure descriptor obtained from one-third octave values of the normalized impact sound levels L_n or L'_n. The normalized levels are compared with a set of weighting curves and the curve found for which the total adverse difference between the normalized levels and the curves is less than but as close to 32 dB as possible. Adverse differences occur when the normalized levels fall above the rating curve.

The weighted normalized impact sound pressure level is the sound pressure level at 500 Hz on the standard curve, which meets the above criterion. For an L_{Nw} or L'_{Nw} of 60 the curve is defined by

Frequency	Level (dB)
100	62
125	62
160	62
200	62
250	62
315	62
400	61
500	60
630	59
800	58
1 000	57
1 250	54
1 600	51
2 000	48
2 500	45
3 150	42

Other curves are obtained by moving the one-third octave bands up or down in 1 dB steps.

See BS 5821, Part 2: 1985 and BS 2750, Parts 6 and 7: 1980.

Weighted standardized impact sound level, $L'_{nT,w}$: This is a single figure descriptor obtained from one-third octave band values of the standardized impact sound levels, $L'_{nT,w}$.

It is obtained in exactly the same way as the weighted normalized impact sound level, L_{nw}.

See BS 5821, Part 2: 1986 and BS 2750, Parts 6 and 7: 1980.

Equivalent standards

Assessing environmental noise

BS 4142: 1990	ASTM E1014-84	ISO 1966: 1986	ANSI S3.23: 1980
BS 5228: 1984		ISO 2204: 1979	ANSI S12.4: 1986
		ISO/DP 7196:	ANSI S12.9: 1988
DIN 18005: 1987			
DIN 45641: 1976/1987		NF S30-010: 1974	
DIN 45642: 1974		NF S30-008: 1984	
DIN 45643: 1984		NF S31-010: 1987	
DIN VDI 2714: 1988		NF S31-050: 1987	
DIN VDI 3723: 1982		NF S31-110: 1985	
DIN VDI 2718: 1975			

Rating of performance of building

BS 5821: 1984	ASTM C1071-86	ISO 717: 1982
	ASTM E413-73	
	ASTM E989-84	
DIN 52210: 1984		
	NF P05-321: 1096	
	NF S31-057: 1982	

Measurement of impact insulation

BS 2750: 1080	ASTM E492-77	ISO 140: 1980
	ASTM E1007-84	
DIN 52210: 1984		
	NF S31-052: 1979	
	NF S31-053: 1979	
	NF S31-056: 1982	

Measurement of sound power

BS 4196: 1986	ASTM E1124-86	ISO 3740: 1980	ANSI S1.23: 1976
		ISO 3741: 1988	ANSI S1.30: 1985
		ISO 3742: 1988	ANSI S1.31: 1986
		ISO 3743: 1988	ANSI S1.32: 1986
		ISO 3744: 1981	ANSI S1.33: 1982
		ISO 3745: 1977	ANSI S1.35: 1985
		ISO 3746: 1979	ANSI S1.36: 1979
		ISO 3747: 1987	ANSI S2.34: 1988
DIN 45635: 1985	NF S31-025: 1977		
	NF S31-026: 1978		
	NF S31-027: 1977		
	NF S31-022: 1989		
	NF S31-024: 1989		
	NF S31-067: 1986		

Noise emission from computers and business machines

ECMA 74: 1981	ISO 7779: 1988	ANSI S12.10: 1985
ECMA 109: 1985	ISO 9296: 1988	
DIN VDI 3729: 1982		

Sound reinforcement systems

BS 6259: 1982	IEC 268: 1985
BS 6840: 1987	
DIN 45589: 1979	

Speech and noise

	ASTM E1041-85	ISO/TR 3352: 1974	ANSI S3.2: 1960
	ASTM E1110-86		ANSI S3.5: 1969
	ASTM E1130-88		ANSI S3.14: 1977
DIN 18041: 1968			
DIN 45621: 1973	NF S31-047: 1975		
	NF S32-001: 1975		
	NF X35-108: 1987	IEC 84(CO)2: 1986	

Vibration – measurement and response

BS 6177: 1982		ISO 2017: 1982	ANSI S3.18: 1979
BS 6472: 1984		ISO 2631: 1985	ANSI S3.29: 1983
BS 6841: 1987		ISO 5805: 1981	ANSI S3-W-39
BS 6611: 1985		ISO 6897: 1984	
		ISO 7849: 1987	
		ISO 4866: 1986	
		ISO 8569: 1989	
DIN 4150: 1986			
DIN 45669: 1981/1989		VDI 2057: 1987	

Mechanical services

BS 848: 1985	ASTM E477-84		ANSI S12.11: 1987
BS 4718: 1971			
DIN 45646: 1988	NF E51-701: 1980		
DIN VDI 2081: 1983	NF P50-402: 1985		
DIN 45635: 1986	NF E51-706: 1988		
DIN VDI 3731: 1982	NF S31-021: 1982		
	NF S31-046: 1988		

Sound insulation in buildings

BS 8233: 1987		ISO/DIS 6242: 1989	
DIN 4109: 1989			
DIN 18165: 1987			
DIN VDI 2569: 1990			
DIN VDI 2571: 1976			
DIN VDI 2711: 1978			
DIN VDI 2719: 1987			
DIN VDI 3728: 1987			

Acoustics of buildings

NF P90-207: 1987

Instrumentation

BS 2475: 1964	IEC 196: 1965	ISO 266: 1975	ANSI S1.4: 1983
BS 3593: 1986	IEC 225: 1966		ANSI S1.4: 1985
BS 5969: 1981	IEC 651: 1979		ANSI S1.6: 1984
BS 6698: 1986	IEC 804: 1985		ANSI S1.11: 1986
			ANSI S1.13: 1971
DIN 45401: 1985	NF S30-002: 1972		ANSI S1.42: 1986
DIN 45651: 1964	NF S31-109: 1983		
DIN 45652: 1964			

Terminology

BS 4727: 1985	ASTM C634-86	ISO 31-7: 1978	ANSI S3.32: 1982
BS 5775: 1979		ISO 131: 1979	
	IEC 50(801): 1984		
DIN 1320: 1990		NF S30-004: 1966	
DIN 45630: 1971		NF S30-101: 1973	
DIN 52217: 1984		NF S30-102: 1973	
		NF S30-103: 1973	
		NF S30-106: 1975	
		NF X02-207: 1985	

Measurement of reverberation time

BS 5363: 1986		ISO 3382: 1975
DIN 52216: 1965	NF S31-012: 1973	

Measurement of absorption coefficients

BS 3638: 1987	ASTM C423-89	ISO 354: 1985
DIN 52212: 1961 DIN 525215: 1963	NF S31-065: 1981	

Measurement of sound insulation

BS 2750: 1980	ASTM E90-83 ASTM E336-84 ASTM E596-78 ASTM E966-84 ASTM E1222-87	ISO 140: 1980
DIN 52210: 1984	NF S31-045: 1989 NF S31-051: 1985 NF S31-049: 1982 NF S34-054: 1982 NF S34-055: 1982	

International standards

International Electrotechnical Commission (IEC)

IEC 50	International electrotechnical vocabulary.
801: 1984	Vocabulary: acoustics and electoacoustics.
IEC 84 (CO) 2: 1986	Sound system equipment; report on the RASTI method for the objective rating of speech intelligibility in auditoria; (Central Office) 2.
IEC 196: 1965	IEC standard frequencies.
IEC 225: 1966	Octave, half-octave and third octave band filters intended for the analysis of sounds and vibrations.
IEC 268	Sound system equipment
Part 1: 1985	General
Part 2: 1971	Explanation of general terms
Part 4: 1972	Microphones
Part 5: 1989	Loudspeakers
Part 7: 1984	Headphones
Part 16 Draft	Report on the RASTI – method for the objective rating of speech intelligibility in auditoria.
IEC 651: 1979	Sound level meters
IEC 804: 1985	Integrating – averaging sound level meters.

ISO Standards

ISO 31-7: 1978	Quantitites and units of acoustics.
ISO 131: 1979	Acoustics – expression of physical and subjective magnitudes of sound or noise in air.
ISO 140	Acoustics – measurement of sound insulation in buildings and building elements.
Part 1: 1978	Requirements for laboratories.
Part 2: 1978	Statement of precision requirements.
Part 3: 1978	Laboratory measurements of airborne sound insulation of building elements.
Part 4: 1978	Field measurements of airborne sound insulation between rooms.
Part 5: 1978	Field measurements of airborne sound insulation of facade elements and facades.
Part 6: 1978	Laboratory measurements of impact sound insulation of floors.
Part 7: 1978	Field measurements of impact sound insulation of floors.
Part 8: 1978	Laboratory measurements of the reduction of transmitted impact noise by floor coverings on a standard floor.
Part 9: 1985	Laboratory measurements of room-to-room airborne sound insulation of suspended ceiling with a plenum above it.

ISO 266: 1975	Acoustics – preferred frequencies for measurements.
ISO 354: 1985	Acoustics – measurement of sound absorption in a reverberation room.
ISO 389: 1991	Acoustics: standard reference zero for the calibration of pure tone air conduction audiometers.
ISO 717	Acoustics – rating of sound insulation in buildings and of building elements.
Part 1: 1982	Airborne sound insulation in buildings and of interior building elements.
Part 2: 1982	Impact sound insulation.
Part 3: 1982	Airborne sound insulation of facade elements and facades.
ISO 1996	Acoustics – description and measurement of environmental noise.
Part 1: 1982	Basic quantities and procedures.
Part 2: 1987	Acquisition of data pertinent to land use.
Part 3: 1987	Application to noise limits.
ISO 2017: 1982	Vibration and Shock – Isolators: procedure for specifying characteristics.
ISO 2204:	Acoustics – guide to International Standards on the Measurement of airborne acoustical noise and evaluation of its effects on human beings.
ISO 2631	Evaluation of human exposure to whole body vibration.
Part 1: 1985	General requirements.
Part 3: 1985	Evaluation of exposure to whole body z-axis vertical vibration in the frequency range 0.01 to 0.63 Hz.
ISO/TR 3352: 1974	Acoustics – assessment of noise with respect to its effect on the intelligibility of speech.
ISO 3382: 1975	Acoustics – measurement of reverberation time in auditoria.
ISO 3740: 1980	Acoustics – determination of sound power levels of noise sources: guidelines for the use of basic standards and for the preparation of noise test codes.
ISO 3741: 1980	Acoustics – determination of sound power levels of noise sources: precision methods for broad band sources in reverberation rooms.
ISO 3742: 1988	Acoustics – determination of sound power levels of noise sources: precision methods of discrete frequency and narrow band sources in reverberation rooms.
ISO 3743: 1988	Acoustics – determination of sound power levels of noise sources: engineering methods for special reverberation test rooms.
ISO 3744: 1981	Acoustics – determination of sound power levels of noise sources: engineering methods for free-field conditions over a reflecting plane.
ISO 3745: 1977	Acoustics – determination of sound power levels of noise sources: precision methods for anechoic and semi-anechoic rooms.
ISO 3746: 1979	Acoustics – determination of sound power levels of noise sources: survey method.
ISO 3747: 1987	Acoustics – determination of sound power levels of noise sources: survey method using a reference sound source.
ISO/DIS 4866: 1986	Mechanical vibration and shock: measurement and evaluation of vibration effects on buildings; guidelines for the use of basic standard methods.
ISO 4871: 1984 (new version in preparation)	Noise labelling of machinery and equipment.
ISO 5805: 1981	Mechanical vibration and shock affecting man: vocabulary.
ISO 6897: 1984	Guidelines for the evaluation of the response of occupants of fixed structures, especially buildings and off-shore structures, to low frequency vibration horizontal motion (0.063 to 11 Hz).
ISO/DIS 6243	Building construction: expression of users' requirements.
Part 3: 1989	Acoustical requirements.
ISO 7235: 1991	Acoustics – measurement procedures for ducted silencers.
ISO 7779: 1988	Acoustics – measurement of airborne noise emitted by computer and business equipment.
ISO/TR 7849: 1987	Acoustics – estimation of airborne noise emitted by machinery using vibration measurement.

ISO 8569: 1989	Mechanical vibration: shock-and-vibration-sensitive electronic equipment; methods of measurement and reporting data of shock and vibration effects in buildings.
ISO 9296: 1988	Acoustics – declared noise emission values of computer and business equipment.

European Computer Manufacturers Association

ECMA 74: 1981	Measurement of airborne noise emitted by computers and business machines.
ECMA 109: 1985	Declared noise emission values of computer and business equipment.

German National Standards

DIN 1320: 1990	Acoustics: terminology.
DIN 1800: 1987	Teil 1. Noise abatement in town planning; calculation methods. Teil 1, Beiblatt 1. Noise abatement in town planning; acoustic orientation values in town planning.
DIN 1804: 1968	Acoustical quality in small to medium size rooms.
DIN 18165: 1987	Teil 2. Fibre insulating building materials: impact sound insulating materials.
DIN 4150: 1986	Teil 3. Structural vibration in buildings: effects on structures.
DIN 4109: 1989	Sound insulation in buildings: requirements and verifications.
Beiblatt 1	Sound insulation in buildings: construction examples and calculation methods.
Beiblatt 2	Sound insulation in buildings: guidelines for planning and execution; proposals for increased sound insulation; recommendations for sound insulation in personal living and working areas.
DIN 45035: 1980	Teil 14. Noise measurement on machines: airborne noise measurement, enveloping surface method, air cooled heat exchangers (air coolers).
DIN 45401: 1985	Acoustic, electroacoustic: standard frequencies for measurements.
DIN 45589: 1979	Requirements for congress microphones.
DIN 45621: 1973	Teil 2. Word lists for intelligibility test. Teil 2. Sentence lists for intelligibility test.
DIN 45630: 1971	Teil 1. Physical and subjective magnitudes of sound.
DIN 45635: 1985	Teil 3. Measurement of airborne noise emitted by machines: engineering method for special reverberation test rooms.
DIN 45635: 1986	Teil 38. Measurement of noise emitted by machines; airborne noise emission; enveloping surface method, reverberation room method and indirect methods; fans.
DIN 45635: 1985	Teil 46. Measurement of noise emitted by machines; airborne noise emission; enveloping surface method; cooling towers.
DIN 45635: 1986	Teil 56. Measurement of noise emitted by machines; airborne noise emission; enveloping surface method and indirect method; fan assisted warm air generators, fan assisted air heaters and fan units of air handling devices.
DIN 45641: 1976	Averaging of time varying sound level; rating levels.
DIN 45641: 1987	Averaging of sound levels; single event level.
DIN 45642: 1974	Measurement of traffic noise.
DIN 45643: 1984	Teil 1. Measurement and assessment of aircraft noise; quantities and parameters. Teil 3. Measurement and assessment of aircraft noise; determination of rating level of aircraft noise exposure.
DIN 45646: 1988	Measurement procedures for ducted silencers; insertion loss, transmission loss, flow noise, total pressure loss.
DIN 45654: 1964	Octave filters for electroacoustical measurements.
DIN 45652: 1964	Third octave filters for electroacoustical measurements.
DIN 45669: 1981	Teil 1. Measurement of vibration emission; requirements on vibration meter.

DIN 45669: 1989	Teil 2. Measurement of vibration emissions; measuring method; amendment 1.
DIN 52210: 1984	Teil 1. Tests in building acoustics; airborne and impact sound insulation; measuring methods. Teil 2. Tests in building acoustics; airborne and impact sound insulation; laboratories for measuring of the sound reduction of building elements.
DIN 52210: 1987	Teil 3. Testing of acoustics in buildings, airborne and impact sound insulation; laboratory measurements of sound insulation of building elements and field measurements between rooms.
DIN 52210: 1984	Teil 4. Tests in building acoustics; airborne and impact sound insulation; determination of single-number quantities.
DIN 52210: 1985	Teil 5. Testing in building acoustics; airborne and impact sound insulation; field measurements of airborne sound insulation of exterior building elements.
DIN 52210: 1989	Teil 6. Tests in building acoustics; airborne and impact sound insulation; determination of the level difference.
DIN 52210: 1989	Teil 7. Tests in building acoustics; airborne and impact sound insulation; determination of the lateral sound reduction index.
DIN 52212: 1961	Testing of architectural acoustics; measurement of sound absorption coefficient in a reverberation room.
DIN 52215: 1963	Testing of architectural acoustics; determination of sound absorption coefficient and impedance in a tube.
DIN 52216: 1965	Testing of architectural acoustics; measurement of reverberation time in auditoria.
DIN 52217: 1984	Test in building acoustics; flanking transmission; terms and definitions.
VDI 2057 Blatt 4.1: 1987	Effect of mechanical vibrations on human beings; measurements and assessment for workshop places in buildings.
VDI 2081: 1983	Noise generation and noise reduction in air-conditioning systems.
VDI 2566: 1988	Noise reduction on lifts.
VDI 2569: 1990	Sound protection and acoustical design in offices.
VDI 2571: 1976	Sound radiation from industrial buildings.
VDI 2711: 1978	Noise reduction by enclosures.
VDI 2714: 1988	Outdoor sound propagation.
VDI 2718: 1975	Noise abatement in town planning.
VDI 2719: 1987	Sound isolation of windows and their auxiliary equipment.
VDI 3720 Blatt 1: 1980	Noise abatement by design; general fundamentals.
VDI 3720 Blatt 2: 1982	Noise abatement by design; compilation of examples.
BDI 2723 Blatt 1: 1982	Application of statistical methods for the description of variating ambient noise levels.
VDI 3728: 1987	Airborne sound isolation of doors and movable walls.
VDI 3729 Blatt 1: 1982	Characteristic noise emission values of technical sound sources; office machines; basic directions.
VDI 3729 Blatt 2: 1982	Characteristic noise emission values of technical sound sources; office machines, typewriters.
VDI 3729 Blatt 3: 1982	Characteristic noise emission values of technical sound sources; office machines, duplicating machines and copiers.
VDI 3729 Blatt 5: 1982	Characteristic noise emission values of technical sound sources; office machines; mail processing (preparation) machines.
VDI 3729 Blatt 6: 1990	Characteristic noise emission values of technical sound sources; computer and business equipment; computer.
VDI 3731 Blatt 1: 1982	Characteristic noise emission values of technical sound sources; compressors.
VDI 3731 Blatt 2: 1988	Characteristic noise emission values of technical sound sources; fans.

VDI 3733: 1983	Noise at pipes.
VDI 3744: 1983	Noise control in hospitals and sanatoriums; instructions for planning.

American National Standards Institute

ANSI S1.4: 1983	Specification for sound level meters.
ANSI S1.4a: 1985	Sound level meters.
ANSI S1.6: 1984	Preferred frequencies and band numbers for acoustical measurements.
ANSI S1.11: 1986	Octave-band and fractional octave-band analog and digital filters, for.
ANSI S1.13: 1971 (R 1986)	Methods for the measurements of sound pressure levels.
ANSI S1.23: 1976 (R 1983)	Method for the designation of sound power emitted by machinery and equipment.
ANSI S1.30: 1979 (R 1985)	Guidelines for the use of sound power standards and the preparation of noise test codes.
ANSI S1.31: 1980 (R 1986)	Broad-band noise sources in reverberation rooms. Precision methods for the determination of sound power levels.
ANSI S1.32: 1980 (R 1986)	Discrete-frequency and narrow-band noise sources in reverberation rooms, precision methods for the determination of sound power levels.
ANSI S1.33: 1982	Engineering methods for the determination of sound power levels of noise sources in a special reverberation test room.
ANSI S1.35: 1979 (R 1985)	Determination of sound power levels of noise sources in anechoic and semi-anechoic rooms.
ANSI S1.35: 1979 (R 1985)	Survey methods for the determination of sound power levels of noise sources.
ANSI S1.42: 1986	Design response of weighting networks for acoustical measurements.
ANSI S2.8: 1972 (R 1986)	Guide for describing the characteristics of resilient mountings.
ANSI S3-W-39	Effects of shock and vibration on man. (A special report – not a standard.)
ANSI S3.2: 1960 (R 1982)	Method for measurement of monosyllabic word intelligibility.
ANSI S3.5: 1969 (R 1986)	Methods for the calculation of the articulation index.
ANSI S3.14: 1977 (R 1986)	Rating noise with respect to speech interference.
ANSI S3.18: 1979 (R 1986)	Guide for the evaluation of human exposure to whole-body vibration.
ANSI S3.23: 1980 (R 1986)	Sound level descriptors for determination of compatible land use.
ANSI S3.29: 1983	Guide to the evaluation of human exposure to vibration in buildings.
ANSI S3.32: 1982	Vibration and shock affecting man, mechanical-vocabulary.
ANSI S12.4: 1986	Method for assessment of high energy impulsive sound with respect to residential communities.
ANSI S12.9: 1988	Quantities and procedure for description and measurement of environmental sound (part 1).
ANSI S12.10: 1985	Methods for the measurement and designation of noise emitted by computer and business equipment.
ANSI S12.11: 1987	Methods for the measurement of noise emitted by small air-moving devices.
ANSI S12.34: 1988	Engineering methods for the determination of sound power levels of noise sources for essentially free field conditions over a reflecting plane.

American Society for Testing and Materials Standards

ASTM C 384-88	Test method for impedance and absorption of acoustical materials by the impedance tube method.
ASTM C 423-89	Test method for sound absorption and sound absorption coefficients by the reverberation room method.
ASTM C 634-86	Definitions of terms relating to environmental acoustics.

ASTM C 1070-86	Standard specification for thermal and acoustic insulation (mineral fibre, duct lining material).
ASTM E 90-87	Method for laboratory measurement of airborne sound transmission loss of building partitions.
ASTM E 336-84	Test method for measurement of airborne sound insulation in buildings.
ASTM E 413-87	Classification for determination of sound transmission class.
ASTM E 477-84	Method of testing duct liner materials and prefabricated silencers for acoustical and air flow performance.
ASTM E 492-86	Method of laboratory measurement of impact sound transmission through floor–ceiling assemblies using the tapping machine.
ASTM E 497-87	Practice for installing sound-isolating gypsum board partitions.
ASTM E 596-88	Method for laboratory measurement of the noise reduction of sound-isolating enclosures.
ASTM E 597-81 (1987)	Practice for determining a single-number rating of airborne sound isolation for use in multi-unit building specifications.
ASTM E 795-83	Practices for mounting test specimens during sound absorption tests.
ASTM E 966-84	Guide for field measurement of airborne sound insulation of building facades and facade elements.
ASTM E 1007-84	Test method for field measurement of tapping machine impact sound transmission through floor–ceiling assemblies and associated support structures.
ASTM E 1014-84	Guide for the measurement of outdoor A-weighted sound levels.
ASTM E 1041-85	Guide for measurement of masking sounds in open offices.
ASTM E 1110-86	Classification for determining of articulation class.
ASTM E 1124-86	Test method for field measurement of sound power level by the two-surface method.
ASTM E 1130-88	Test method for objective measurement of speech privacy in open offices using articulation index.
ASTM E 1222-87	Test method for laboratory measurement of insertion loss of pipe-lagging systems.

French Standards

NF E51-701: 1980	Controlled mechanical ventilation components. Code for aerodynamic and acoustic testing of extract air terminal devices.
NF E51-706: 1988	Controlled mechanical ventilation components. Code for aerodynamic and acoustic testings of extraction units for private houses. Simple flux.
NF P05-321: 1986	Performance standard in building. Presentation of the performances of facades made of components from the same source.
NF P50-402: 1985	Ventilation components. Code for aerodynamic and acoustic testing of facade air inlets.
NF P90-207: 1986	Sport halls. Acoustics.
NF S30-002: 1972	Acoustics. Standard frequencies for acoustic measurement.
NF S30-004: 1966	Acoustics. Expressing the physical and psychophysiological properties of a sound or a noise.
NF S30-008: 1984	Acoustics. Guide to standards on the measurement of airborne acoustical noise and evaluation of its effects on human beings.
NF S30-010: 1974	Acoustics. NR curve for the assessment of noise.
NF S30-101: 1973	Acoustics. Terminology: general definitions.
NF S30-102: 1973	Acoustics. Terminology: transmission systems and transducers for sound and vibrations.
NF S30-103: 1973	Acoustics. Terminology: instruments.

NF S30-106: 1975	Acoustics vocabulary: architectural acoustics.
NF S31-010: 1987	Acoustics. Description and measurement of environmental noise: investigation of complaints against noise in inhabited areas.
NF S31-012: 1973	Acoustics. Measuring the period of reverberation in auditoria.
NF S31-021: 1982	Acoustics. Platform measurement of the noise emitted by ducted fans. Reduced chamber on discharge method.
NF S31-022: 1989	Acoustics. Determination of sound power levels of noise sources. Precision methods for broad-band sources in reverberation rooms.
NF S31-025: 1977	Acoustics. Determination of sound power levels of noise sources. Part 4: Engineering method for free field conditions over a reflecting plane.
NF S31-026: 1978	Acoustics. Determination of the sound power emitted by noise sources. Part V: Laboratory methods in anechoic and semi-anechoic rooms.
NF S31-027: 1977	Acoustics. Determination of the level of acoustic power emitted by nosie source. Part 6: Control method for on-site measurements.
NF S31-045: 1989	Acoustics. Measurement of the acoustic insulation of building and building components. Laboratory measurement of insulation against airborne noise of small sized building components.
NF S31-047; 1975	Acoustics. Assessment of speech intelligibility distances in noisy conditions.
NF S31-049: 1982	Acoustics. Measurement of the acoustic insulation of buildings and building components. Precision specifications.
NF S31-050: 1979	Acoustics. Measurement of the acoustic insulation of building and building components. Specifications relating to laboratories.
NF S31-051: 1985	Acoustics. Measurement of the acoustic insulation of buildings and building components. Laboratory measurement of insulation against airborne noise of building components.
NF S31-052: 1979	Acoustics. Measurement of the acoustic laboratory measurement of transmission of impact noise by floors.
NF S31-053: 1979	Acoustics. Measurement of the acoustic insulation of building and building components. Laboratory measurement of the reduction in transmission of impact noise due to floor covering and floating floors.
NF S31-054: 1982	Acoustics. Measurement of the acoustic insulation of buildings and building components. Investigatory method for the in-situ measurement of airborne sound insulation between rooms.
NF S31-055: 1982	Acoustics. Measurement of the acoustic insulation of buildings and building components. Investigatory method for the in-situ measurement of airborne sound insulation of rooms from road traffic noise.
NF S31-056: 1982	Acoustics. Measurement of the acoustic insulation of buildings and building components. Investigatory method for the in-situ measurement of impact sound transmission.
NF S31-057: 1982	Acoustics. Verification of the acoustic quality of buildings.
NF S31-059: 1983	Acoustics. Test code for the measurement of noise emitted by bar guide tubes (screw cutting industry).
NF S31-065: 1981	Acoustics. Testing of architectural acoustics. Determination of sound absorption coefficient and impedance in a tube.
NF S31-067: 1986	Acoustics. Determination of sound power levels of noise sources. Part 7: Survey method using a reference sound source.
NF S31-109: 1983	Acoustics. Integrating sound level meters.
NF S31-110: 1985	Acoustics. Description and measurement of environmental noise. Basic quantities and general evaluation methods.
NF S32-001: 1975	Acoustics. Sound signal for emergency evacuation.

NF X02-207: 1985	Fundamental standards. Quantities, units and symbols of acoustics.
NF X35-108, NF ISO 7731: 1987	Danger signals for work places. Auditory danger signals.

British Standards

BS 648: 1964	Schedules of weights of building materials.
BS 848	Fans for general purpose.
Part 2: 1985	Methods of noise testing.
BS 1042	
Part 1: Various dates by section	Pressure differential devices.
BS 2475: 1964	Specification for octave and one-third octave band pass filters.
BS 2750:	Measurement of sound insulation in buildings and of building elements.
Part 3: 1980	Laboratory measurements of airborne sound insulation of building elements.
Part 4: 1980	Field measurements of airborne sound insulation between rooms.
Part 5: 1980	Field measurements of airborne sound insulation of facade elements and facades.
Part 6: 1980	Laboratory measurements of impact sound insulation of floors.
Part 7: 1980	Field measurements of impact sound insulation of floors.
Part 8: 1980	Laboratory measurements of the reduction of transmitted impact noise by floor coverings on a standard floor.
Part 9: 1980	Method for laboratory measurement of room-to-room airborne sound insulation of a suspended ceiling with a plenum above it.
BS 3593: 1986	Recommendation on preferred frequencies for acoustical measurements.
BS 3638: 1987	Method for measurement of sound absorption in a reverberant room.
BS 4142: 1990	Method of rating industrial noise affecting mixed residential and industrial areas.
BS 4196	Sound power levels of noise sources.
Part 0: 1986	Guide for the use of basic standards and for the preparation of noise test codes.
Part 1: 1991	Precision methods for determination of sound power levels for broad-band sources in reverberation rooms.
Part 2: 1991	Precision methods for determination of sound power levels for discrete-frequency and narrow-band sources in reverberation rooms.
Part 3: 1991	Engineering methods for determination of sound power levels for sources in special reverberation test rooms.
Part 4: 1981	Engineering methods for determination of sound power levels for sources in free field conditions over a reflecting plane.
Part 5: 1981	Precision methods for determination of sound power level for sources in anechoic and semi-anechoic rooms.
Part 6: 1986	Survey method for determination of sound power levels of noise sources.
Part 7: 1988	Survey method for determination of sound power levels of noise sources using a reference sound source.
Part 8: 1991	Specification for the performance and calibration of reference sound sources.
BS 4718: 1971	Methods of test for silencers for air distribution systems.
BS 4727: Part 3 Group 08: 1985	Acoustics and electroacoustics terminology.
BS 4773: 1989	Methods for testing and rating air terminal devices for air distribution systems.
Part 2: 1989	Acoustic testing.
BS 4856	Methods for testing and rating fan coil units, unit heaters and unit coolers.
Part 4: 1978	Acoustic performance, without additional ducting.
Part 5: 1979	Acoustic performance, with ducting.
BS 4857	Methods for testing and rating terminal reheat units for air distribution systems.
Part 2: 1978 (1983)	Acoustic testing and rating.

BS 5108	
Part 1: 1991	Sound attenuation of hearing protectors.
BS 5228	Noise control on construction and open sites.
Part 1: 1984	Code of practice for basic information and procedures for noise control.
Part 2: 1984	Guide to noise control legislation for construction and demolition including road constructions and maintenance.
Part 3: 1984	Code of practice for noise control applicable to surface coal extraction by opencast methods.
Part 4: 1992	Code of practice for noise control applicable to piling operations.
BS 5363: 1986	Method for measurement of reverberation time in auditoria.
BS 5727: 1979	Method for describing aircraft noise heard on the ground.
BS 5775, Part 7: 1979	Specification for quantities, units and symbols: Acoustics.
BS 5793	Industrial process control valves.
Part 8: 1991	Noise conditions.
BS 5821	Methods for rating the sound insulation in buildings and of building elements.
Part 1: 1984	Method for rating the airborne sound insulation in buildings and of interior building elements.
Part 2: 1984	Method for rating the impact sound insulation.
Part 3: 1984	Method for rating the airborne sound insulation of facade elements and facades.
BS 5969: 1981	Specification for sound level meters.
BS 6083	Hearing aids.
Part 3: 1991	Methods for measurement of electroacoustical characteristics of hearing aid equipment.
BS 6177: 1982	Guide to selection and use of elastomeric bearings for vibration isolation of buildings.
BS 6259: 1982	Code of practice for planning and installation of sound systems.
BS 6472: 1984	Guide to evaluation of human exposure to vibration in buildings (1 Hz to 80 Hz).
BS 6611: 1985	Guide to evaluation of the response of occupants of fixed structures, especially buildings and offshore structures, to low-frequency horizontal motion (0.063 Hz to 1 Hz).
Part 1: 1986 Part 2 Sect 2.1: 1990 Part 2 Sect 2.2: 1990 Part 2 Sect 2.3: 1991	Methods for determination of airborne acoustical noise emitted by household and similar electrical appliances.
BS 6698: 1986 Amd 1: 1991	Specification for integrating-averaging sound level meters.
BS 6840	Sound system equipment.
Part 1: 1987	Methods for specifying and measuring general characteristics used for equipment performance.
Part 2: 1988	Glossary of general terms and calculation methods.
Part 3: 1989	Methods for specifying and measuring the characteristics of sound system amplifiers.
Part 4: 1987	Methods for specifying and measuring the characteristics of microphones.
Part 6: 1987	Methods for specifying and measuring the characteristics of auxiliary passive elements.
Part 8: 1988	Methods for specifying and measuring the characteristics of automatic gain control devices.
Part 9: 1987	Methods for specifying and measuring the characteristics of artificial reverberation, time delay and frequency shift equipment.
Part 11: 1988	Specification for application of connectors for the interconnection of sound system components.
Part 12: 1987	Specification for applications of connectors for broadcast and similar use.
Part 13: 1987	Guide for listening tests on loudspeakers.
Part 14: 1987	Guide for circular and elliptical loudspeakers; outer frame diameters and mounting dimensions.
Part 15: 1988	Specification for matching values for the interconnection of sound system components.
Part 16: 1989	Guide to the 'RASTI' method for the objective rating of speech intelligibility in auditoria.

BS 6841: 1987 (under review)	Guide to measurement and evaluation of human exposure to whole body mechanical vibration and repeated shock.
BS 7385	
Part 1: 1990/ISO 486 1990	Evaluation and measurement for vibration in buildings.
BS 7443: 1991	Specification for sound systems for emergency purposes (IEC 849) (largely updates BS 5839 'Fire detection and alarm systems for buildings').
BS 7445	Description and measurement of environmental noise.
Part 1: 1991	Guide to quantities and procedures.
Part 2: 1991	Guide to the acquisition of data pertinent to land use.
Part 3: 1991	Guide to the application to noise limits.
BS 7458	Test code for the measurement of airborne noise emitted by rotating electrical machinery.
Part 1: 1991	Engineering method for the free-field conditions over a reflecting plane.
Part 2: 1991	Survey method.
BS 8233: 1987	Code of practice for sound insulation and noise reduction for buildings.

Index

A-weighting, 123
Abatement notice, *see* Noise abatement notice
Absorption, 36, 47, 52, 53, 97, 139
Absorption coefficients, 47, 49, 50, 104, 146
Acceleration, 143
Acoustic appraisal, 7, 8, 9, 13
Acoustic curtains, 99
Acoustic louvres, 98, 99
Acoustic shielding, 8, 9
Aircraft noise, 18, 41
Airfields, military, 18
Air-handling plant, 89
Ambient noise, 85
American Society for Testing and Materials Standards (ASTMS), 150
Amplifiers, 117
Analysis, site, 8
Anechoic chambers, 51, 104
Anti-vibration, 101
Articulation index, 76, 123, 124, 140
Association of Noise Consultants (ANC), 7
Atria, 74
Attentuation, 22, 93, 129
Attenuators, 94, 98, 104
Audio Frequency Induction Loop System (AFILS), 121
Auditoria, 30, 36, 53, 62, 65, 98, 107, 108
  modelling, 53
  ventilation, 65

BBC, 86
Background noise, 10, 44, 77, 78, 85, 105, 136
Banners, 63, 79
Barriers, 14, 22, 24, 99
Bass traps, 51
Berlin Philharmonie, 60, 62
Best practical means, 23
Blasting, 10, 26, 29
Boilers, 89
Bowling alleys, 28, 53, 78
British standards, 8
Builder's work, 103, 105, 113
  ducts, 95
  penetrations, 97
Building regulations, 2, 36, 113

Calibration, 11, 12
Ceilings, 42, 43, 50
  suspended, 42, 43
Ceiling voids, 42
Cinemas, 28, 40, 53, 85, 88
Civil Aviation Authority, 21
Clarity, 61
Clarity index, 62
Clay pigeon shooting, 29
Coincidence effect, 41
Commissioning tests, 105, 113
Complaints, 9, 26, 28, 87
Composite construction, 41, 138
Compression ratio, 116
Computer model, 28
Computer rooms, 77
Concert halls, 30, 55, 57, 63, 85, 108, 119
Concertgebauw (Amsterdam), 56
Condenser units, 89, 97, 99
Conference rooms, 30, 71, 72, 74, 108
Consent, 23
Constant directivity horns, 119
Construction noise, 22, 24, 25
Control of Pollution Act, 24, 25
Corrected Noise Level (CNL), 27, 135
Court order, 33
Courts, 62, 67, 85, 108
Cross-talk, 93, 95, 98

Daily personal noise exposure level, 70
Deaf aid loop systems, 121
Decibel, 130
Deutlichkeit, 62
Diffuse sound field, 139
Diffusion, 52, 53
Directivity
  factor, 130
  index, 130
Discontinuity, 41
Discothéques, 29, 53, 67, 108, 109
Dispersion, 114
Distortion, 107, 122
Docklands Light Railway, 16
Doors, 44, 72, 92
Dry construction, 40
Dry lining, 44
Duct,
  shape, 93
  velocities, 94
Dynamic range, 62

Early decay time, 61, 139
Early lateral energy fraction, 62
Echoes, 107, 113, 114
Edge damping, 47
Education building, 68
Electroacoustics systems, 79, 80, 81, 114
Enclosures, 99, 100
Ensemble, 62
Entertainments license, 68
Envelopment, 114
Environmental
  assessments, 30
  Health Department (EHD), 9, 32
  noise, 1, 11, 133, 144
  Protection Act, 25, 26, 27, 33
  statement, 7, 31, 32
Equalizer (graphic), 117, 120
Equipment, 11, 12, 23, 26, 88, 112
Equivalent continuous sound level ($L_{eq,T}$), 135
Escalators, 92
European Broadcasting Union (EBU), 86
Event noise exposure level (L_{Ax}), 17
Existing noise climate, 87
External noise, 86

Fan coil units, 96
Factories, *see* Industrial buildings
Fan noise, 83, 88, 89, 92, 93
Fast fourier transform (FFT), 133
Finishes, 51, 52
Flanking, 34, 35, 47, 137
Floors, 47
  floating, 47, 49, 69
  party, 69
Footfall, 29, 67, 81
Free-field, 9

Frequency, 12, 128
natural, *see* Natural frequency
resonant, *see* Resonant frequency
response, 113, 118, 122

Gas turbines, 90
Generators, 90

Harmonic, 128
Health and Safety Executive, 70
Health buildings, 68
Hearing damage, 67, 128
Hertz, 128
Hospitals, 108
Hotels, 68, 107, 109
Housing, 17, 69
Hydraulic mast, 9

Ice rinks, 85, 107, 110
Impact sound, 47, 67, 69, 104, 142
Industrial buildings, 70, 85, 108
Industrial noise, 26, 70
Inertia bases, 104
Insertion loss, 93, 99
Intensity levels, 130, 131
International Electrotechnical Commission (IEC), 146
Inverse square law, 114
Isolators, 103
Inspection, site, 8
Institute of Acoustics, 7
Intermittent noise, 87

L_{A10}, 12
$L_{Aeq,T}$, 13, 18, 23, 24, 29
L_{Ax}, 17, 21, 136
$L_{eq,T}$, 135
L_{max}, 136
L_{min}, 136
L_{pAmax}, 23, 24
Laboratory tests, 104
Lagging, 98
Lateral efficiency, 61
Lecture rooms, 71, 85, 107, 109
Legislation, 3
Leisure noise, 28
Libraries, 72
Lifts, 92
Lighting, 92
Linköping Concert Hall, 64
Locations (for measurement), 8, 9
Locomotives
diesel, 16, 19
electric, 16, 19
Loudness, 113
Loudspeakers, 112, 114, 115, 117, 118, 119
signal distribution, 120
line losses, 121
Louvres, acoustic, 98, 99

Maintenance, 113
Mass law, 37
Materials handling, 26
Maximum noise level (L_{pAmax}), 23, 24
Maximum sound pressure level (L_{max}), 136
Measurement locations, 8, 9
Measurement
standards for, 144
techniques, 122
units, 11, 23, 26, 128

Metal fabrication, 26
Microphones, 9, 11, 78, 80, 113, 114, 115, 116, 119
Mineral extraction, 26, 28, 31
see also Quarries
Minimum sound pressure level (L_{min}), 136
Multiuse, 62
Museums, 72
Music practice rooms, 72, 73, 74, 75, 76

National measurement accreditation service (NAMAS), 11
NNI, *see* Noise and Number Index
Natural frequency, 88
Night clubs, 28
Noise,
aircraft, 18, 21, 41
construction, 22, 24, 25
door, 92
external, 86
fan, 83, 88, 89
industrial, 26, 70
intermittent, 87
leisure, 28
railway, 16, 17, 18, 19
regenerated, *see* Regenerated noise
road traffic, 12, 15, 70
transportation, 12, 22
ventilation, 53, 68, 85, 86
Noise abatement notice, 27, 32
Noise Advisory Council, 18, 29
Noise and Number Index (NNI), 18, 19, 21, 69, 136
Noise at Work Regulations, 2, 70, 128
Noise break-in, 37
Noise break-out, 37, 67, 70, 83, 87, 94
Noise control, 25, 26, 28, 32, 83, 86, 97
Noise Criteria (NC), 85, 140, 141
Noise insulation regulations, 13, 21
Noise Isolation Class (NIC), 137
Noise-limiting devices, 68
Noise Rating (NR), 85, 141
Noise Reduction Coefficient (NRC), 139
Norris Eyring equation, 139

Octave band filter, 133
Odense, 56
Offices, 13, 17, 74, 77, 85, 88, 96, 109, 129
Opera houses, 79
Opera hall, 85
Optical equipment, 30

Partition resonances, 41
Partitions, 42
folding, 42
Pascal, 128
Path difference, 14
Perceived Noise Level (PNL), 136
Percentage alcons, 125
Piling equipment, 29
Plant rooms, 97
Plenum chambers, 95
Prediction, 14, 17, 21, 24, 25, 28, 30
Preferred Noise Criterion (PNG), 85, 141, 142
Privacy, 68, 79
speech, 44, 86

Quadratic residue sequence, 62
Quarry, 10, 26

Radio studios, 78
Railways, 16, 17, 18, 19, 29
Docklands Light, 16

Rain, 10
Rapid Speech Transmission Index (RASTI), 141
Recording studios, 78
Reflection, 52, 53
  facade, 9
Refrigeration units, 89
Regenerated noise, 94
Regulations,
  building, 2, 36, 47, 113
  noise at work, 2, 70, 128
  noise insulation regulations, 13, 21
Resonant frequency, 39, 41, 47, 88, 142
Reverberant field, 36, 52
Reverberant sound pressure level, 52, 139
Reverberation, 114, 123
Reverberation time, 52, 56, 62, 66, 67, 76, 78, 111, 139, 146
Riser ducts, 95
Road traffic, 12, 15, 70
Roofs, 41
  flat, 42
  lightweight, 42
Room criteria, 85, 141
Room radius, 140
Royal Albert Hall (London), 61, 62

Seating, 56
Segestrom Hall (California), 56, 64
Sensitivity, 115, 117
Shielding, acoustic, 8, 9
Signal processing, 116
Signal-to-noise ratio, 81, 107, 113, 122, 123
Silencers, 91
Single Event Noise Exposure Level (SENEL or L_{Ax}), 136
Site analysis, 8
Site inspection, 8
Site noise, *see* Construction noise
Site survey, 8, 111
Sleep disturbance, 17
Sound, 1, 128
  insulation, 22, 34, 39, 136, 145, 146
  level difference, 34, 35, 36
  lobby, 45
  pressure levels, 129
  reduction index, 36, 38, 39, 44, 104, 136, 138
  reinforcement systems, 144
  systems, 53, 68
  transmission class (STC), 137
Speech,
  intelligibility, 67, 80, 107, 111, 117, 122, 123
  interference level, 141
  privacy, 44, 86
  reinforcement, 67, 107, 111
  Transmission Index (STI), 142
  transmission tests, 126
Sports stadia, 107
Standards,
  American, 150
  British, 153
  French, 151
  German, 148
  international, 146
Standby diesels, 102
Static deflection, 101, 103
Stiffness, 41, 47
Studios, 29, 31, 36
  recording, 78
  tv and radio, 78
Subjective testing, 125
Survey Procedure, 9
Survey, site, 8
Swimming pools, 78, 107, 109

Television studios, 78, 85
Temperature inversions, 10
Tests,
  see under type Commissioning, Laboratory, Speech, Transmission, Subjective, Works
Theatres, 68, 79, 80, 85, 107, 109
Trading rooms, 80
Transducer, 9
Transformers, 102
Transmission suites, 104
Transportation noise, 12

Valves, 91
Velocity, 128
Ventilation, 13, 25, 62, 65
  auditorium, 65
  mechanical, 14
  noise, 63, 68, 85, 86
Vibration, 11, 30, 81, 83, 87, 88, 92, 99, 142, 145
  acceleration, 143
  causes, 29
  groundborne, 29
  isolation, 101, 102
  measurements, 9
  monitoring, 10
  velocity, 30
Volume, 56

Weather, 10
Weighting networks, 132, 133
Weighted standardized level difference ($D_{\mathrm{nT,w}}$), 36, 139
Wellington Town Hall (New Zealand), 61
Windows, 13, 45, 68
Winds, 10
Works tests, 104
Wycombe Entertainments Centre (High Wycombe), 64

Young's modulus, 41